THE RENEWABLE ENERGY LANDSCAPE

The Renewable Energy Landscape is a definitive guide to understanding, assessing, avoiding, and minimizing scenic impacts as we transition to a more renewable energy future. It focuses attention, for the first time, on the unique challenges solar, wind, and geothermal energy will create for landscape protection, planning, design, and management.

Topics addressed include:

- Policies aimed at managing scenic impacts from renewable energy development and their social acceptance within North America, Europe and Australia
- Visual characteristics of energy facilities, including the design and planning techniques for avoiding or mitigating impacts and improving visual fit
- Methods for assessing visual impacts of energy projects and the best practices for creating and using visual simulations
- Policy recommendations for political and regulatory bodies.

A comprehensive and practical book, *The Renewable Energy Landscape* is an essential resource for those engaged in planning, designing, or regulating the impacts of these new, critical energy sources, as well as a resource for communities who may be facing the prospect of development in their local landscape.

Dean Apostol is Senior Landscape Architect and Restoration Ecologist for MIG Inc, a consulting firm with offices in California, Oregon and Colorado, USA. He researches, consults, and does environmental analysis on energy projects and aesthetic impacts.

James Palmer has had a distinguished professional career in landscape architecture spanning thirty-five years, focusing on the assessment of landscape character and aesthetic quality. Through publications, peer reviews, court expert testimony, and teaching, he has raised the standards in the field. He frequently consults on wind energy aesthetic impacts in the Northeast.

Martin Pasqualetti is Professor in the School of Geographical Sciences and Urban Planning at Arizona State University, USA, and a Senior Sustainability Scientist in the Julie Ann Wrigley Global Institute of Sustainability, USA. For over forty years, he has been examining the relationships between energy and environment, especially the formation and mitigation of renewable energy landscapes.

Richard Smardon is a SUNY Distinguished Service Professor Emeritus at the SUNY College of Environmental Science and Forestry in Syracuse, New York, USA, and has over thirty-five years of experience with visual impact assessment methodology development, twenty years of project management, and has testified in over fifteen cases with visual impact assessment issues.

Robert Sullivan is an environmental scientist in Argonne National Laboratory's Environmental Science Division, USA. He conducts research on the visual impacts of fossil fuel and renewable energy systems, and develops guidance documents for federal agencies on visual resource inventory, management, and protection.

THE RENEWABLE ENERGY LANDSCAPE

Preserving scenic values in our sustainable future

Edited by Dean Apostol, James Palmer, Martin Pasqualetti, Richard Smardon and Robert Sullivan

Routledge
Taylor & Francis Group
LONDON AND NEW YORK

First published 2017
by Routledge
2 Park Square, Milton Park, Abingdon, Oxon OX14 4RN

and by Routledge
711 Third Avenue, New York, NY 10017

Routledge is an imprint of the Taylor & Francis Group, an informa business

British Library Cataloguing in Publication Data
A catalogue record for this book is available from the British Library

Library of Congress Cataloging-in-Publication Data
Names: Apostal, Dean, editor.
Title: The renewable energy landscape : preserving scenic values in our sustainable future / edited by Dean Apostal, James Palmer, Martin Pasqualetti, Richard Smardon and Robert Sullivan.
Description: Abingdon, Oxon ; New York, NY : Routledge is an imprint of the Taylor & Francis Group, an informa business, [2016] | Includes bibliographical references and index.
Identifiers: LCCN 2016007800 | ISBN 9781138808980 (hardback : alk. paper) | ISBN 9781315618463 (ebook)
Subjects: LCSH: Renewable energy sources.
Classification: LCC TJ808 .R4135 2016 | DDC 333.7/16—dc23
LC record available at http://lccn.loc.gov/2016007800

ISBN: 978-1-138-80898-0 (hbk)
ISBN: 978-1-315-61846-3 (ebk)

Typeset in Bembo
by Apex CoVantage, LLC

CONTENTS

ILLUSTRATIONS

PLATES

FIGURES

TABLES

FOREWORD

Dean Apostol and Martin Pasqualetti

This book results from a collaboration that began in 2012 during planning for a special symposium on visual impact assessments at the meeting of the National Association of Environmental Professionals. Several of us discussed the possibility that, if we could assemble the right group, we could create a timely book that would look toward a future that was fast approaching, a future when renewable energy would be appreciably replacing conventional energy.

As we were assembling our team and getting started, we took note that the United Nations was thinking along complementary lines. After years of negotiations among 193 countries, the climate conference scheduled in Paris at the end of 2015 (COP21, as we went to press) would likely be a turning point in the emerging global emphasis on renewable energy. The UN planned to showcase a "renewable energy track" that would demonstrate that the accelerated development of renewable energy would be the fastest, cleanest, most reliable and most economically beneficial way to meet climate goals.[1] The book that you hold in your hands, while the result of an idea hatched in 2012, supports the prediction of the United Nations in 2015. It reflects our collective determination to preserve scenic quality in valued landscapes while the world makes this necessary transition.

The field of visual and scenic resources, as noted in Chapter 2, has developed over many years, yet reached something of a stalled point in the 1990s, at least in the US, when federal agencies that had been crucial in nurturing the field began laying off or reassigning landscape architects. This was partly a consequence of budget cutting, but it also resulted from a decrease in the types of projects that threatened scenic values, especially on Forest Service lands when the annual timber harvest quotas were significantly reduced to protect a wide variety of wildlife.

Well before the turn of the twenty-first century the renewable energy movement had taken hold, and very large projects were proposed. This was particularly true of wind and solar development on lands managed by the US Bureau of Land Management, but it was also increasingly common on non-federal lands across much of the United States and elsewhere. A spate of policy shifts – including tax credits to energy developers, plus the rising concern over climate change helped drive this development forward.

A number of us with long experience working on scenic resource conservation were involved in this effort, either on the side of developers trying to address impacts on scenery, or on the side of opponents who resisted these changes, or in some cases as consultants to the regulators who had to develop policies that addressed both camps.

We found that prior ideas about scenic resources were difficult to apply to renewable energy. The facilities were so large, and their appearance so inherently contrasting, that the normal rules and mitigation techniques seemed ill suited. We witnessed proposals that claimed there would be no visual impact because turbines were to be five miles or more distant, previously considered "background" in visual resource impact assessment. Yet the scale, bright color, motion and lighting often made turbines high contrast even at these distances. Entire landscapes were transformed from undeveloped and open into new energy landscapes with dozens or even hundreds of gleaming white turbines set on ridges or covering plains.

The sheer scale and the wide geographic range of these facilities made them unlike anything in our prior experience. We realized it was more on the order of the interstate highway system; that is, a transformation that would change the very character of much of the country and developed world. It was apparent that this shift needed more attention before it had progressed too far.

We were fortunate that some of the pioneers of visual resources were delivering papers at the symposium. One by one we discussed our ideas, and a number eagerly signed up to make our initial idea a reality.

Our sincere hope is that this book will find its way into the hands of practitioners, regulators, and developers who together are reshaping landscapes in countries all around the world. We also see it as a good tool for communities faced with the prospect of their surroundings being transformed by energy projects. A lot is at stake if we are going to retain landscape quality, and we hope we can help meet this challenge. Renewable energy, even with all its promise of helping us create a more sustainable future, should not simply be developed wherever the wind blows or the sun shines. We believe this book will help us avoid placing renewable energy where it should not be, while also helping it "grace" the landscape where it should be.

Note

1 From a press release issued by IRENA. http://irenanewsroom.org/2015/07/15/retrack-at-cop21/

PREFACE

Ian Bishop

I am writing in a train – on a four-hour journey. I make this trip quite often – sometimes by car and often by train. The landscape is familiar, sometimes too familiar, and one looks for changes. I've been thinking about, and doing research on, wind turbines in the landscape for over ten years, but it is only in the last few years that I began to think about the cumulative effect that will be the consequence of a rapid expansion of renewable energy capacity.

Journeys have been a part of our life and culture for millennia – at least since Odysseus. Travel has been epitomized in the USA by the road movie and for many years families have enjoyed a summer pilgrimage to the west or to the Keys. Chapter 1 of this volume paints a vivid picture of a road trip through American landscapes doing a new job. For millennia we have tamed our landscapes for habitation and food production. In isolated areas we have dug great holes or built huge dams to support energy and mineral needs. Now we need to use the landscape for energy in a new way. The new infrastructure will be more visible – but more revocable, less isolated – but less damaging, than the old. On my four-hour journey I may, before too many years have passed, have wind turbines in my line of sight for 30 percent of the time. If I travelled in a different direction, to windier places, it might be 60 percent (Bishop, 2015).

Such radical change is going to meet opposition. In my country the Prime Minister called wind turbines 'visually awful', and our Treasurer said they are 'utterly offensive', fuelling the arguments of those who may support renewable energy in principle – as long as its somewhere else. It is therefore extremely timely for a book to emerge that encapsulates our knowledge and current thinking about the visual impacts of renewable energy infrastructure – especially wind energy as the public encapsulation of the alternative to fossil fuels.

As Chapter 1 also makes clear, we have to consider the visual impact of the alternatives. Carry on along our pathway of the last 150 years and we are also very likely to see radical changes in our landscape, not only visually but also in terms of both ecological and agricultural productivity. A comprehensive assessment must consider all options and all kinds of effects on the economic, social and environmental fabric. There is scope for several books, but the visual effects of the renewable options is a necessary component and one for which the authors and editors of this volume are very well equipped.

Planning for a future that involves radical changes to our view of the world requires skills and processes from multiple disciplines. A combination of natural and social sciences, engineering, planning and design, supported by effective visualization and communication are essential in support of reasoned public debate and sound decision-making within an appropriate regulatory framework.

Central to all this is the recognition that visual impacts are a tractable issue, made much easier by the similarity of view that people have about what makes a landscape beautiful, and (to a lesser extent) the ways in which infrastructure can detract from this beauty. This makes the type of analysis reviewed in Chapter 2, and described in more detail in Chapters 6, 7 and 8, both possible and meaningful. Over the years the developed theory and analytical techniques have been applied to landscapes at a wide range of scales. The difference now is that the size of area to be considered may be much more than can be encompassed in a single view. We need to be more aware of cumulative visual effects experienced through movement.

While people move they also typically have somewhere to call home. No more is it just one's home that is one's castle, it is the greater environs, the 'backyard', that people seek to project and keep in some idealized state. There is debate in the literature about whether NIMBY movements are really narrow-minded or tend also to take in the bigger picture, whether their complaints are against the process or the outcomes. Whatever the case, we need to be very aware of the potential trade-off between local loss and wider national or global benefit. There is a rich body of relevant research and experience from parts of Europe, where renewable energy infrastructure is already ubiquitous, which is the basis of Chapter 5.

What constitutes best practice in the planning for, and design of, windfarms, large-scale solar projects, and other renewable energy development is still an open question. What is clear is that an adversarial paradigm in which one party proposes and another opposes is not in anyone's interest. In Chapter 10 a number of case studies are reviewed in the expectation that lessons can be learned, by all parties, about developers and communities working together. These discussions need, of course, to be about more than just visual effects – although that is often the starting point. Just as research, and acceptance of research, on the visual component is central to this volume, other factors need their own rigorous review and establishment of best practice.

Returning to the visual, there is a solid base of empirical information about people and their landscape response, and a more specialized body of research on the visual impacts of wind energy, but there remain unanswered questions that require more research. As that body of knowledge develops we should continue to apply best information and best practice to the emerging energy landscape. We should be wise in recognizing that these decisions cannot be considered in isolation of a bigger picture, and we should not fear this future but should be prepared for it and recognize that it may require us all to shift our perspectives on landscape qualities. This book can only help with that process.

Reference

Bishop, I.D. (2015) Sequential experiences in energy producing landscapes, in C. Nunes Silva (ed.) *Emerging Issues, Challenges and Opportunities in Urban E-Planning*, IGI Global, Hershey, PA.

ACKNOWLEDGEMENTS

This book has resulted from a number of years of development, research and application. The authors and contributors cannot claim it entirely as our project, since we learned our craft from others, and have had many collaborators along the course of our careers that led us to this project. We acknowledge many of the landscape practitioners and educators that have contributed to the foundation of knowledge regarding landscape assessment, visual resource management, and impact assessment for the past thirty-five years. This list includes Professor Donald Appleyard, R. Burton Litton Jr. (former US Forest Service LA), Professor Kenneth Craik, Professor Julius Fabos, Professor John Felleman, Grant Jones (Jones & Jones), Bob Leopold (former BLM head LA), Professor Merlyn Paulson, Professor Angel Ramos (Spain), Robert Ross (former BLM head LA), Professor Sally Schauman (former NRCS LA), Professor Stephen R.J. Sheppard, Professor Carl Steinitz, Ed Stone (former US Forest Service head LA), Professor Joanne Vining, William Yeomans (former British Columbia Forest Service LA) and Professor Ervin Zube. Hopefully we have built well upon this foundation and so honored their past contributions. We also acknowledge the members of the Scenic Resources Working Group established after the 2012 NEAP Visual Impact Assessment (VIA) short course who strive to increase the calibre of VIA practice. Our individual and collective debt of gratitude is substantial.

Dean Apostol offers thanks to MIG Inc., which provided him a platform for professional practice and initial support in allowing him to use precious time to develop ideas for this book and get it started. He thanks his clients who provided him opportunities to apply his expertise to visual resource projects, including Friends of the Columbia River Gorge, Portland General Electric, Oregon Department of Transportation, the Bureau of Land Management, Peter Goldman of the Washington Forest Law Center, and Oregon Energy Department. Dean thanks his former colleagues at the US Forest Service who taught him the trade of scenic resource management. In particular Warren Bacon, a pioneer of scenic resources provided early training and encouragement. Robert Ribe, University of Oregon has been a close colleague of Dean's who has been a frequent collaborator and has demonstrated the value of research on the field of landscape aesthetics. Stephen Sheppard of University of British Columbia is also engaged in important research and has been an inspiration and occasional colleague. Many thanks of course to Dean's life partner Marcia Sinclair, who has put up with an idiosyncratic partner who wanders off on poorly paid projects and fails to finish household projects in a timely way.

Bob Sullivan would first like to thank his mother, Elizabeth J. Sullivan for telling him the names of all the flowers in the garden, and for letting him skip school sometimes to spend the day at science and art museums. Thanks also to Professor Brian Orland for first introducing Bob to the scientific study of landscape aesthetics, and to John McCarty, Tom and Sherry Lahti, and Ray Brady of the BLM, for opening doors, offering encouragement, and reviewing numerous lengthy visual resource reports. Bob would also like to thank Carol McCoy of the National Park Service, whose vision and leadership has allowed Bob to pursue the most rewarding work of his career, and for the best of causes. Bob is also grateful for the opportunity to work with Mark Meyer of the National Park Service, with whom Bob has visited some of the most spectacular landscapes in our beautiful country. Bob would also like to express his gratitude to his co-editors, Dean, Rick, Mike and Jim, for their patience with the 'newbie' editor, and for all that Bob has learned from them about landscapes. Finally, and most importantly, Bob would like to thank his wife Susan who has patiently endured thirty years of Bob complaining either about not having enough work, or more often, having too much.

1

INTRODUCTION TO THE CHANGING LANDSCAPES OF RENEWABLE ENERGY

Dean Apostol, James Palmer, Martin Pasqualetti, Richard Smardon, and Robert Sullivan

Highlights

- Provides an overview of scenic challenges presented by renewable energy.
- Illustrates that different renewable energy types have different impacts.
- Summarizes themes presented in following chapters.

Lester Brown's new book (2015) *The Great Transition: Shifting from Fossil Fuels to Solar and Wind Energy* promotes the 'solar revolution' in residential and commercial land use in the US, Canada, Italy, Germany, China and Australia. It also presents the "age of wind" as it is occurring in China, UK, Italy, France, Canada, Demark, Germany, Portugal and Massachusetts. And it discusses the potential for tapping the Earth's heat (geothermal energy) in Chile, Peru, Columbia, Japan, China, Philippines, Indonesia, the Great Rift Valley in Africa, the Eastern Mediterranean, and of course, Iceland, where geothermal energy is already highly developed.

Brown makes a strong case for how the world is now making a significant shift to renewable energy and why this is important, but he avoids the significant and growing controversy over locating utility scale renewable energy facilities in visually sensitive landscapes. These may be highly valued nationally for their scenic qualities (i.e. a national park), or may have local importance that generates resistance to significant changes in landscape and community, such as the everyday pastoral or historic settings near our homes.

While the authors support the trend away from fossil fuels and to renewable energy, our work in landscape and scenic conservation over many years has directed us to the subject of how our world can develop renewable energy while at the same time conserving what we value in our landscapes. This book represents our attempt at addressing this issue as it is being experienced around much of the world, with special attention to North America, Europe and Oceania. We offer both analysis of the challenge, and proposed approaches and solutions that should be useful to energy developers, regulators, and communities where renewable energy is being contemplated.

1.1 Driving across America in the year 2030

Picture a road trip across the United States in the year 2030. We begin at the coast of Maine and journey across the Appalachians through the mid-section of the country to the West Coast. Off in the distance near Mohegan Island a cluster of floating wind turbines punctures the horizon. Closer to shore other groupings appear on some of the small islands in Penobscot Bay. They dwarf the quaint fishermen homes and villages. If you continued along the coastal highway and side trip to the beaches, you would see wind turbines offshore all the way south past the Carolinas. If instead you turned inland and head west or southwest, your encounters would be fewer and more widely spaced, with turbines appearing intermittently on ridgelines of Maine, New Hampshire, Vermont, Pennsylvania, and West Virginia. In addition, many small rural community fields are covered by arrays of photovoltaic panels, perhaps 10 to 30 acres each in size. These may coexist with sheep grazing in or near them.

Somewhere not too far from Cleveland, fields of turbines would be more frequently in view. There will be many on or along Lake Erie, and near Sandusky or Lorain, a distance of 10 to 15 miles north of the Interstate 90. If you continue through Indiana, near South Bend and Elkhart there will be evenly spaced grids of turbines right alongside the freeway. South of Joliet, Illinois, these will grow more frequent and densely spaced, and may extend to the horizon in any direction. By the time you reach Marshalltown, Iowa, wind turbines will have been near-constant roadside companions. It's likely that you will now have giant wind turbines in view for the rest of your journey across the prairies and plains, through Wyoming if you stay north, or if you turn Southwest, through eastern Colorado and north Texas into New Mexico and especially in California at either Tehachapi Pass or San Gorgonio Pass near Palm Springs.

FIGURE 1.1 Wind turbines integrated into ranchland, near Rio Vista, Solano County, California. (Photo: M. Pasqualetti.)

FIGURE 1.2 Wind turbines adjacent to pre-existing residence, San Gorgonio Pass, California.
(Photo: M. Pasqualetti.)

You may have a visual respite from turbines if you wind across the Rockies and out into the Western deserts and Ranges, but instead will encounter the eerily glowing towers of concentrated solar power plants, easily visible at 20 or more miles distant. If your route takes you northwest on I-84, you will see wind turbines in Idaho intermittently along the Snake River Plain. And if you stay on I-84 through Oregon, you will encounter wind turbines for much of the 250-mile (417-kilometer) ride from the eastern boundary of the state to the edge of the Columbia Gorge National Scenic Area. Traveling through the Gorge, if you look up at surrounding ridges only 2 or 3 miles from the Interstate you may see yet more turbines crowding the boundaries of the designated scenic area. You will get another turbine break as you pass through Portland, which is where I-84 ends (we hope you stopped for micro brew or some Northwest cuisine, or whatever gluten-free food people are eating and drinking in 2030). If you continue across the northern Willamette Valley west to the Oregon coast, you will see turbines again once you are most of the way through the Coast Range. The first line of mountains in from the sea will have had the ridgeline partly shorn of forest to make way for more turbines. And of course you will likely see turbines out in the ocean, along with large yellow wave buoys, lights blinking to the watery horizon. The wave buoys in particular may be very close to the beaches that Governor Oswald West declared to be public in the 1890s.

If, instead of taking I-84, you had followed I-80 through Utah to California, you would encounter neat rows of turbines, possibly with solar panels under them, all through the Great Salt Lake Desert. In Nevada the wind resource is intermittent, although geothermal development is prominent. You will likely encounter massive solar projects on the valley floors, partially powering the Tesla battery factory

FIGURE 1.3 Cluttered energy facilities at San Gorgonio Pass – transmission lines, wind turbines, solar panels, and a gas-fired generating station, with Desert Hot Springs, California in the background.
(Photo: M. Pasqualetti.)

FIGURE 1.4 A landscape in transition. New wind turbines in a legacy oil field. Forty miles north of Dallas-Ft Worth, this is Wolf Ridge, Texas.
(Photo: M. Pasqualetti.)

and the electric cars once they are on the road (we hope the price has come down by then). These large solar developments will accompany you through the Mojave Desert if you head south or southwest. Alternatively, if you traveled north or south to transect the Great Plains, what had once been empty grassland now has tall turbines in view much of the way from Texas to the Canadian border and beyond (along with a few remaining tired oil wells of course).

Connecting all this new, renewable, clean energy will require thousands of miles of new above-ground high-voltage transmission lines, running north-south and east-west across many if not most states, transporting electricity from areas of sparse population to large urban areas. Similar future stories can be imagined for journeys across Europe, Australia, China, and Canada.

How much land will all these renewable energy facilities require? A lot. According to research from the National Renewable Energy Lab, a division of the US Department of Energy, a 50 megawatt (MW) commercial wind project needs a dedicated perimeter of about 1,000 hectares (2,500 acres) of land area (Denholm et al. 2009). As of this writing, larger windfarms approved for development in the US have 1,000 or more turbines occupying more than 20,000 hectares (50,000 acres) of land.

Although these numbers may indeed seem large, there are some important caveats to consider. The wind support structures require only a small amount of the land, and other land uses, such as agriculture, can co-exist if care is taken during construction to minimize ground disturbance. Economically, this can be a win-win since the owners also get rent or a share of the energy value produced while they continue

FIGURE 1.5 Visual impacts can also be significant from the transmission needed to carry the electricity generated by renewable energy.
(Photo: M. Pasqualetti.)

FIGURE 1.6 Dixon, Illinois farm growing maize and soybeans. The wind turbine array has little effect on the agriculture and more than doubles the income from the land.
(Photo: M. Pasqualetti.)

farming. As an example, a farmer in Dixon, Illinois (see Figure 1.6) reported that he made more from the turbines per year than from his soybeans and corn. This is also reportedly the case for wheat farmers in the Columbia Basin of Oregon and Washington.

But a key reason the turbines and commercial solar plants will be so visible in the future, so much a part of the landscape, is that these forms of energy require such large areas of land. To generate enough wind energy to provide 20 percent of US electricity by the year 2030, the amount of land required adds up to around 6 million hectares, equal to the entire state of West Virginia. Adding 16 kilometers around each project, and considering the number of turbines required, increases the potentially visually impacted area to nearly half the size of the continental United States. How much area ultimately is visually impacted depends on a range of factors, including terrain and how turbines are spaced. But there is little question that views of turbines will be a lot more common than they are at this writing, when renewable energy provides only a small fraction of energy in most nations.

One strategy for reducing visual impacts on land is locating wind turbines offshore. Sea-based turbines are generally more expensive to install, but often require less surface area since winds over the water are consistently stronger and steadier, resulting in fewer total turbines needed to generate a given amount of electricity. In most cases, offshore turbines will be placed close enough to shore to be seen by coastal visitors because mounting them on the sea floor in shallow waters is less costly than floating turbines in deeper waters. The visual impact of offshore turbines depends on many factors, including their size, the angle of the sun, and the elevation of the viewer on shore. Turbines can be visible to the unaided eye at distances greater than 40 kilometers (Bishop 2002; Bishop & Miller 2007). At a distance of 16 kilometers they can be a major focus of visual attention, a common threshold distance for judging visual impacts (Sullivan et al. 2013)).

As we go to press, the US Bureau of Ocean Energy Management (BOEM) is leasing hundreds of thousands of hectares of ocean for wind energy development on the east coast of the US, and is attempting to develop pilot projects in deep waters off the west coast. Meanwhile Europe has already developed substantial offshore wind energy, its first project dating from 1991 in Denmark. Worldwide, the Wind Power database lists over 700 offshore wind projects either in planning, production or operation as of this writing (mid 2015), and some are very large, as are the installations in the Thames estuary.[1]

Commercial scale solar energy also requires large land areas, as sunlight, like wind is a low-density source of energy. According to estimates by the Bureau of Land Management (BLM), a commercial solar facility located in the desert southwest needs about 4 hectares (10 acres) for every megawatt of production from photovoltaic panels. Thus, in order to reach a 20 percent electrical generation threshold, the US could need to dedicate 1.2 million hectares (3 million acres) of land to solar energy. (When the entire fuel cycle is considered, this amount of land is comparable to what is required for coal.) The issue from a scenic impact perspective is that, in the case of solar, all the land impact is in one place instead of being widely distributed.

FIGURE 1.7 Nysted Wind Facility, 8–12 miles (13–20 kilometers) offshore, Denmark, the North Sea. (Photo: National Renewable Energy Laboratory.)

FIGURE 1.8 Recreational beach at Carpentaria, California with offshore oil rigs at 15 miles. (Photo: M. Pasqualetti.)

As with wind projects, significant visual impacts can extend 40 or more kilometers for some types of solar energy, such as power towers. However, photovoltaic (PV) installations, even large ones such as the 500 MW facility near Desert Center, California, can have low visual contrast and appear as shadows or bodies of water from a distance. Because PV panels are low to the ground, visibility is highly dependent on the elevation of the viewer and the surrounding terrain (Sullivan et al. 2012; Sullivan & Abplanalp 2012, 2015).

The potential near future of landscapes and seascapes in the US is for renewable energy installations stretching from sea to shining sea and then some. Similar pictures can be painted for much of Canada, Europe, North Africa, China and Australia. The Renewable Energy Age has dawned, and it will reshape the landscape around us. If the world is going to slow, stabilize, and hopefully reverse climate change, extensive development of new energy sources is inevitable and, arguably, both necessary and desirable. But as nations continue to plan for and deploy these clean energy devices, how can they best avoid major regional- and national-scale impacts to landscapes? Can they avoid turning highly valued scenic landscapes into industrial scale energy harvest landscapes?

1.2 The challenge

Many landscapes and seascapes around the world are highly valued for their natural and/or cultural scenic features, and have been granted some form of protected status reflecting this value. Some examples include: UNESCO World Heritage landscapes, National, Provincial, and State Parks, Wilderness Areas,

National Scenic Areas, Wild and Scenic Rivers, National Seashores, and Historic Landmarks. Other landscapes have local or regional recognition as being scenically or culturally important, but are not officially protected. Scenic Byways and designated tourist routes come to mind. Still other areas have no official recognition at all, but have scenic characteristics, openness, interesting topography, mosaics of farms and woodlands, that many or most people would likely find appealing and worthy of conservation at some level, if only local. However, the issue of conservation of these lands often does not arise until a large energy project is proposed that threatens the resource, and by then it is often too late.

Commercial scale renewable energy, particularly wind turbines and large solar arrays, have already encroached on protected scenic and cultural landscapes. Wind turbines press up against the Columbia Gorge National Scenic Area in Oregon and Washington states. The proposed (but as yet not built) Cape Wind in Nantucket Sound has been highly controversial on scenic impact grounds and permits have been held up for years (Rooney et al. 2012; Williams & Whitcomb 2007). The US Bureau for Ocean Energy Management is presently identifying leasing tracts for offshore wind projects from Beaufort to Nags Head, North Carolina, a distance of over 100 nautical miles that parallels the Outer Banks, including the Cape Hatteras National Seashore. They are also identifying leasing tracts off Massachusetts, Virginia, New Jersey and Rhode Island. Texas is planning windfarms directly adjacent to Padre Islands National Seashore. There are proposals for offshore wind in Oregon, near scenic headlands south of Coos Bay.

Large solar arrays are being developed adjacent to Joshua Tree National Park and Mojave National Preserve, as well as various National Historic Trails. Wind and solar facilities encroach on Native American cultural sites in Southern California. In the UK, dozens of offshore windfarms are visible from scenic areas, and wind energy is being pushed ever farther offshore to avoid impacts (Hill et al. 2001). Onshore wind facilities in the UK are routinely subject to fierce opposition based on potential scenic impacts. Germany and the Netherlands have experienced increasing opposition to local wind projects (Sellman 2010). Many additional examples of opposition building can be found.

The authors have encountered a number of challenges that the existing scenic resource conservation toolbox may be insufficiently equipped to handle. These challenges include:

- *The sheer scale of renewable energy harvest systems*. Both solar and wind facilities may include towers 500 feet or more in height, and highly reflective industrial devices spread over thousands of acres of rural or semi-wild land, visible from tens of kilometers distance. Very little in past scenic resource management has prepared regulators and professionals in the scenic resource field for the scale and visibility of renewable energy facilities.
- *Seascapes and ocean views*. Up until now, with the exception of oil platforms in a few areas, there has been little placement of industrial facilities in our near-shore ocean environment. Most scenic assessment experience has thus been with landscapes, not seascapes, where topography, vegetation, and other mitigating visual factors are not found.
- *The dynamic visual qualities of renewable energy*. Blades turn, lights flash on and off, wave buoys bob up and down, and solar glare changes intensity and direction with the course of the sun across the sky. These all challenge the usual approaches to visualizing and measuring impacts, including simulation technology.
- *Lack of meaningful mitigation measures*. Reducing or mitigating the impacts of hundreds of white towers, each over 500 feet tall, or massive arrays of highly reflective mirrors placed in what had been an undeveloped landscape with few or no signs of human impact, is very challenging. There are very few reliable ways to lessen these impacts.
- *Lack of relevant landscape perception studies*. Scenic resource management over the last few decades has focused primarily on landscape changes from forestry, transmission lines, and highways. Research

was able to tease out the degree of change that would be acceptable to most people, largely by using various visual preference rating systems and surveys. But up until recently there have been few perception studies on renewable energy systems. We do not yet know the extent to which people view renewable energy through the filter of their bias on energy policy (i.e. clean = good, global warming = hoax).

- *Lack of an integrated policy framework.* In the US, federal and state policies have focused on facilitating rapid deployment of renewable energy as it becomes technically and financially feasible. But the present regulatory infrastructure is not designed to manage the visual impacts of this development. The federal government can and does limit direct impacts to national parks, forests, and other public lands. States protect some limited areas but tend to leave land use rules and decisions to local authorities. Yet rural counties, where most renewable energy development will take place, especially east of the Rocky Mountains, rarely have any ordinances that deal with scenic impacts.
- *Night sky visual impacts.* Because of the height of wind turbines and solar towers, they must be lit day and night to help pilots avoid crashing planes into them. Valued dark, starlit skies in remote, previously undeveloped deserts and grasslands are at risk from thousands of blinking lights spreading as far as the eye can see. This is a problem that is new to scenic resource management, one that is difficult to evaluate impacts to night skies.

This book includes the thoughts and experience of leading experts on scenic conservation. We have striven to address the main issues renewable energy development presents to our visual environment. Our expertise and focus is primarily on North America, with excursions to Europe, and Australia, but we believe these issues will eventually also apply in Asia, Africa, the Pacific Island states, and Latin America as well. We do not pretend to have all the answers, but believe we have identified the most pressing questions, and have identified a number of technically defensible and widely applicable solutions. These solutions include avoidance, good design, and mitigation, described in Chapters 7 and 8. We have summarized the state of the art of scenic landscape assessment and conservation, and how this knowledge can be applied to renewable energy.

Some caveats are important to note. Given the magnitude of the challenge, we know that every scenic landscape cannot be preserved in its current state. In fact this is true whether or not renewable energy is developed as extensively as it appears it will be. All landscapes change, often too slowly to notice at first. But if nothing, or not enough is done to limit climate change, many highly valued parks, cultural landscapes, and even remote wilderness areas will transform beyond recognition, perhaps within the lifetimes of many readers, if not we aging authors.

According to climate modeling, ecosystem changes resulting from climate change will shift species and change entire plant assemblages (Kliejunas et al. 2008). Many forests will be stressed, plant diseases will spread, wildfires will increase in size and intensity, glaciers will continue to shrink, and rivers may dry up. As we write this, California is in the fourth year of a severe drought that may have initiated irreversible changes to foothill woodland and forest ecosystems (Tansey 2015).[2] Thus preserving valued scenery must not mean fighting against renewable energy altogether. It cannot even mean significantly impeding its development by over-arguing or exaggerating scenic impacts. It has to mean finding ways to identify the most important landscapes and making sure these are protected, not just to their boundaries, but taking account of what we call "the viewshed," or "zone of visual influence." And it should mean limiting impacts in landscapes of lesser importance, but still valued for their aesthetics. In essence, the need is to determine the right and wrong places for renewable energy development, encourage development only in the "right places," and then to use design and mitigation to reduce scenic impacts wherever renewable energy facilities are sited.

While many valued landscapes and seascapes will no doubt be impacted visually, the authors believe there are ways to lessen these by designing energy facilities to better harmonize with their surroundings, by limiting the number of devices in a given area, or in some cases by adjusting our aesthetic expectations to include renewable energy as a new cultural feature (Sellman 2010; Sheppard 2012).

The chief concern that led us to dedicate the time to research and write this book is that renewable energy development is proceeding at such a rapid pace and has an unprecedented potential to change our surroundings. This is happening in the absence of a sufficient regulatory or analysis framework to address impacts related to scenic resources. While we do not mean to suggest that other impacts, most notably to wildlife, are any less of a concern, we do believe that landscape aesthetic impacts need to be addressed on their own merit, as well as within the context of a more comprehensive environmental analysis at both the individual project and cumulative scales.

There are multiple facets to this issue, and this book tries to address all of the most important ones. **Part One** lays the groundwork by offering a broad survey of the field of visual and scenic resource assessment, and how this applies to renewable energy.

Chapter 2 is for readers who may have little knowledge of or experience in the field of scenic conservation, and may falsely assume that scenery is simply "in the eye of the beholder" and lacks measurable objective criteria. This is not the case. The study of landscape scenery and human perceptions of it goes back very far; with roots in garden art thousands of years ago.

Beginning primarily in the 1960s, landscape architects in the UK, US, Canada, and Australia began developing the modern system of scenic resource assessment, which later became folded into environmental impact analysis (Smardon et al. 1986). This has become a largely objective, technical field, albeit with some subjective judgment applied strategically. These systems are rooted in a deeper history of aesthetics, including both cultural (art) and scientific analysis that shed light on how people view landscapes and why. Though imperfect, they provide a good basis for understanding scenery and scenic impacts.

Chapter 3 summarizes the regulatory and legal framework and tools that apply to scenic resource conservation for siting renewable energy facilities for the US, Canada, and Australia. We find that this framework is highly variable, from careful pre-consideration of land suitability for renewable energy at the federal US level, to a near absence of regulations at many local levels. We term this a regulatory mess, and hope regulators can find places to improve by emulating the more successful examples.

Chapter 4 covers the European experience in managing scenic impacts from renewable energy, with an emphasis on the British Isles. Given Europe's more dense population, richness of cultural landscapes, and rapid deployment of renewable energy, especially wind, our contributors provide an overview of approaches, including case studies and how Europeans are adjusting to an emerging renewable energy landscape.

Chapter 5 focuses on social acceptability of commercial scale renewable energy from an aesthetic perspective. It discusses the paradox of general support for renewable energy, but resistance in "my back yard" largely due to aesthetics. Opinion surveys are reviewed as a very important component of scenic management, including information from recent surveys on public perceptions regarding renewable energy development in multiple nations.

Part Two shifts the focus to the toolbox of scenic conservation and how it applies to renewable energy.

Chapter 6 discusses the visual characteristics of renewable energy facilities: their size, unique visual features, visibility from varying distances, and the degree of visual contrast they present. This chapter incorporates recent research findings that are changing the way planners understand and can best address visual impacts from renewable energy.

Chapter 7 discusses design issues and approaches that can, at least to an extent, harmonize large-scale commercial energy facilities with landscapes. It includes good up-front planning to determine where such facilities should not be built, then moves into design issues such as composition, fitting facilities to the land, clustering, glare reduction, shape, and avoidance of skylines.

Chapter 8 reviews methods to measure impacts objectively by combining elements such as visual sensitivity and contrast. It builds largely on the US federal experience in scenic impact analysis as a component of environmental impact statements. It includes a discussion of compensatory mitigation for situations where scenic impacts cannot be reduced to levels below significant.

Chapter 9 provides a primer on the very important issue of visualization. Providing mapped and pictorial representations of renewable energy is a key element in understanding impacts and reducing them. The state of the art of visualization is changing rapidly with new technologies. We utilize recent work done by one of our authors for the US National Park Service that suggests standards for visual simulations and how to use them that could be adapted by local regulators.

Chapter 10 analyzes the social and community issues surrounding scenic impacts from renewable energy projects. We offer recommendations and case studies that can help communities stave off or cope with landscape impacts, and so can help energy developers work more effectively with local communities.

Chapter 11 closes the book with a summary of recommendations for regulators and fellow practitioners to improve and modernize visual resource assessment, and frankly to help raise the regulatory and performance bar. While renewable energy presents challenges, these are solvable using a best practices framework.

Throughout the book, case studies are provided to illustrate issues and point to solutions to typical scenic resource problems.

It is the hope of the authors that this book will help communities, energy developers, environmental practitioners, and regulators all better understand, manage, and limit scenic impacts related to renewable energy projects.

Notes

1 See www.thewindpower.net/windfarms_offshore_en.php
2 See www.fs.fed.us/research/climate-change/

References

Bishop, I.D. 2002. Determination of thresholds of visual impact: the case of wind turbines. *Env. and Planning B: Planning and Design* 29(5): 707–718.

Bishop, I.D. and D.R. Miller. 2007. Visual assessment of offshore wind turbines: the influence of distance, contrast, movement and social variables. *Renewable Energy* 32: 814–831.

Brown, Lester. 2015. *The Great Transition: Shifting from Fossil Fuels to Solar and Wind Energy*. W.W. Norton, New York and London.

Denholm, Paul, Maureen Hand, Maddalena Jackson, and Sean Ong. 2009. Land Use Requirements of Modern Wind Power Plants in the United States. Technical Report NREL/TP-6A2-45834, August 2009.

Hill, M., J. Briggs, P. Minto, D. Bagnall, K. Foley, and A. Williams. 2001. *Guide to Best Practice in Seascape Assessment*. Countryside Commission of Wales, Brady Simpson Marsh and University College Dublin. Available at www.marine.ie/NR/rdonlyres/683C8CDo-3367-4704-8542D3091607C9C2/0/interreg5_1.pdf

Kliejunas, John T., Brian Geils, Jessie Micales Glasser, Ellen Michaels Goheen, Paul Hannon, Mee-Sook Kim, Harry Kope, Jeff Stone, Rona Sturrock, and Susan Frankel. 2008. *Climate and Forest Diseases of Western North America: A Literature Review*. USDA Forest Service. Available as at www.fs.fed.us/wwetac/projects/PDFs/Climate_Forest_Disease_Lit_Review.pdf

National Renewable Energy Lab. 2008. *20% Wind Energy by 2030: Increasing Wind Energy's Contribution to the US Electricity Supply*. US Department of Energy, Oak Ridge, TN.

Rooney, S., P. Dascomb, S. Korjett, and R.C. Smardon. 2012. *Technical Bulletin #12-001 Visual Impact Assessment (VIA) Methodology for Offshore Development*. Cape Cod Commission, Barnstable, MA.

Sellman, P.H. 2010. Learning to love the landscapes of carbon-neutrality. *Landscape Research* 35(2): 157–171.

Sheppard, S.R.J. 2012. *Visualizing Climate Change: A Guide to Visual Communication of Climate Change and Developing Local Solutions*. Routledge, London.

Smardon, R.C., J.F. Palmer, and J.P. Felleman (eds.). 1986. *Foundations for Visual Project Analysis*. Wiley, New York. Available at www.esf.edu/via

Sullivan, R.G., L.B. Kirchler, C. McCoy, J. McCarty, K. Beckman, and P. Richmond. 2012. Visual impacts of utility-scale solar energy facilities on southwestern desert landscapes. Paper presented at the National Association of Environmental Professionals, 37th Annual Conference (May 21–24, 2012, Portland, OR), pp. 897–942.

Sullivan, R.G., L.B. Kirchler, J. Cothren, and S.L. Winters. 2013. Offshore wind turbine visibility and visual impact threshold distances. *Environmental Practice* 15(1): 33–49.

Sullivan, R.G., and J. Abplanalp. 2012. Utility-Scale Solar Energy Facility Visual Impact Characterization and Mitigation Project Final Report. Argonne National Laboratory, Argonne, IL, December.

Sullivan, R.G., and J. Abplanalp. 2015. Visibility and Visual Characteristics of the Ivanpah Solar Electric Generating System Power Tower Facility Project Final Report. Argonne National Laboratory, Argonne, IL, March.

Tansey, Michael. 2015 Personal Communication. (Michael Tansey is a Climate Change Modeler with the Bureau of Reclamation in California.)

Williams, W. and R. Whitcomb. 2007. *Cape Wind: Money, Celebrity, Class Politics and the Battle for Our Energy Future*. Public Affairs, New York.

PART I

2

CONSERVING SCENERY DURING AN ENERGY TRANSITION

Martin Pasqualetti and Richard Smardon

Highlights

- A rise in the sophistication of assessment techniques has accompanied the rising public concerns about landscape changes and intrusions.
- Most improvements have focused on wind energy landscapes.
- Geothermal landscape changes have so far been isolated from public scrutiny and lack assessment methods.
- The wind industry has made several adjustments to reduce landscape impacts.

> The new map consists of the immense constructions of the oil refineries, the nuclear reactors and power stations, and radiating from them, the great latticed towers and overheard lines of the electricity grid, forming a network which covers the whole country and spreads even to the most remote rural areas
>
> Dame Sylvia Crowe, *The Landscape of Power* (1958)

2.1 Introduction

Dame Sylvia Crowe was prescient in referring to our energy infrastructures as "landscapes of power," a phrase now adopted by others in reference to wind turbines (Pasqualetti et al. 2002). Her aim was to illuminate these landscapes so that we could see them more clearly as a burden we were tacitly agreeing to bear to obtain the lifestyles we desired. She included energy-siting, zones of influence, scale, new shapes in the landscape from mining and power plants, coal trains and electric transmission lines. She was sharp in noticing that energy quest had "destroyed . . . the landscape and left us the legacy of thousands of acres of ugly and derelict land." And she worried about the future:

> but this is nothing compared to the havoc we shall leave to our descendants unless we take avoiding path[s] now and find a means of reconciling our need for power with our need for a landscape fit to live in.
>
> (Crowe 1958, p. 10)

Crowe's reflections, over half a century ago, were a warning that was heeded in some ways but overlooked in others. For example, we are more aware nowadays of landscape environmental costs. We have laws that require reclamation, remediation, and public review. While we have become more attentive to the changing nature of energy landscapes, change and conflict continue accumulating.

Two reasons help explain this trend: (1) Simply put, the population of the planet is growing. This means not only that we have a more complex and larger energy infrastructure, but also that people encounter it more regularly. (2) Our recent enthusiasm for renewable energy has introduced new landscapes of power which we did not fully anticipate, and some of which the public has not come to embrace.

Large-scale renewable energy development – especially wind and solar installations – is facing opposition from the public in many regions around the world where developers are trying to diversify power generation, reduce landscape damage, and avoid the emission of greenhouse gases (see Plate 2). The greater goal of such projects is what Stremke and Dobbelsteen describe in their book *Sustainable Energy Landscapes* (2013). They start with a history of energy impacts on the land (Pasqualetti 2013). They then proceed to discuss necessary elements of accommodating these impacts, the educational needs of the next energy transition, and end with a series of case studies of renewable energy development and landscape assessment that would be part of a more sustainable energy future.

2.2 Visual elements of renewable energy landscapes

Achieving a sustainable energy future must include measuring public responses to the imposition of energy facility on the landscape, identifying acceptable renewable energy development goals, and

FIGURE 2.1 Appreciable landscape remolding is required because of the site-specific nature of geothermal energy and the hilly topography at The Geysers, Northern California.

(Photo: M. Pasqualetti.)

managing the landscapes that reaching this goal will create. Of the several resources that may play a role, wind, solar, and to a much lesser degree geothermal energy development each create landscape-siting issues. All three of them have intrinsic characteristics that will strongly influence the appearance of the landscape, public reactions to the associated generation facilities, and how we can assess their visual impacts. Each produces its own unique landscape signature and we address the visual characteristics of the technologies in detail in Chapter 6.

In a nutshell, windfarms tend to produce extensive visual impact because of their large size, turning blades, and visual contrast. Solar development arrays may take up large land areas and produce glare depending on lighting direction and atmospheric conditions.

Geothermal energy is available at a wide range of temperatures in most places of the world. In some places, most often near fault-lines, it is especially hot and near the surface. The gathering fields of pipelines tend to be extensive, producing visual impact, while power plants are scattered and produce plumes of steam. Individual geothermal power plants exceeding 150 megawatts (MW) are rare, but are particularly valuable to utility companies because they provide base load power, unaffected by the limitations of solar and wind energy.

Geothermal resources are also site-specific. For this reason, they must be developed where found, or not at all. This means that successful developers of geothermal resources must integrate their equipment into the natural and human landscape, whether hilly or flat, vacant or populated, wasteland or bucolic. Because of such restrictions, the landscape signatures of geothermal energy development are unmistakable (Figures 2.1 and 2.2).

FIGURE 2.2 Integrating geothermal energy into the flat Imperial Valley presents minimal difficulties, although it can interfere with the irrigated agriculture that dominates the local economy.

(Photo: M. Pasqualetti.)

FIGURE 2.3 Wind turbines installed immediately adjacent to an existing house in North Palm Springs, California can generate opposition from home owners who are impacted by the visual and auditory changes they bring.

(Photo: M. Pasqualetti.)

Wind power offers another site-specific challenge. As with geothermal power plants, wind turbines are positioned where there is an adequate resource; like hot water and steam, wind cannot be packaged and sent somewhere else more convenient or less challenging. Owing to lower energy density and the spacing required by downwind turbulence, moreover, the layout of wind turbines requires substantially more land than conventional energy resources like coal or uranium. Wind facilities are also becoming more noticeable because natural laws and economies of scale favor taller hub heights and larger rotor diameter. More than many other objects on the landscape, wind turbines are noticeable because they rotate, stop and start, spin, glint, and produce some noise. Recent research has shown that size and color of windfarms stand out over a broad range of visible distances and blade movement is more noticeable at shorter distances (Sullivan et al. 2012b). These are the characteristics that attract attention and often generate opposition as power (Pasqualetti 2001, 2011), although they are ironically not part of visualizations used in public hearings (Figures 2.3 and 2.4 and Plates 1, 2 and 3).

Solar energy produces three additional landscape impacts from photovoltaic (PV) arrays, reflecting mirror arrays, and concentrating solar towers, each of them different in many ways but similar in some others. The similarities come from the low energy density of sunlight. This translates into a

FIGURE 2.4 Wind power resources are commonly located at a substantial distance from load centers, requiring additional transmission lines. These are in San Gorgonio Pass, near Palm Springs, California, directly in a power line corridor to Los Angeles.
(Photo: M. Pasqualetti.)

requirement, at least in centralized deployments, for large and relatively flat open spaces for PV facilities. The new Topaz Solar Farm, a 550-MW PV project, currently the largest in the world, requires 9 million solar panels and covers 9.5 square miles (25.6 square kilometers) of California's Carrizo Plain (Figure 2.5 and 2.6). In contrast, the 4000-MW Palo Verde Nuclear Generating Station sits on about 6.5 square miles (16.8 square kilometers), while the 2250-MW Navajo Generating Station, near Page, Arizona, occupies about 3 square miles (7.8 square kilometers).

Such photovoltaic projects, while generally appearing to be black against much lighter desert backgrounds, have some visual advantages in that they are low to the ground, usually do not move and operate in almost complete silence.

Concentrating solar power (CSP) plants all use a thermal cycle to produce electricity. The facility structures are taller, may require some form of cooling (wet or dry), and track the sun. Parabolic concentrators track the sun in a single axis, i.e. from east to west, from sunrise to sunset, with no adjustment for the seasonal variation in the sun's angle.

Power towers (sometimes called central receivers) track the sun in two axes, i.e. both diurnally and seasonally. They require tall structures near the center of each field of heliostats to receive the reflected sunlight, a characteristic that has attracted unexpected public reaction to the 392-MW Ivanpah Solar Electric Generating System, 40 miles southwest of Las Vegas.

FIGURE 2.5 Nine million cadmium telluride solar modules now cover part of Carrizo Plain in southern California. The modules are part of Topaz Solar Farm, one of the largest photovoltaic power plants in the world. At 9.5 square miles (25.6 square kilometers), the facility is about one-third the size of Manhattan island, or the equivalent of 4,600 football fields.

(Photo: Earth Observatory image by Jesse Allen, using EO-1 ALI data provided courtesy of the NASA EO-1 team.)

While the landscape signatures of CSP installations might be more noticeable than PV installations, CSP solar plants have several advantages: they require less land and they are appropriate for energy storage (often using molten salt to store heat) for generation after sunset. Parabolic trough and solar power tower facilities use mirrored surfaces to reflect and concentrate sunlight that can be visible at long distances. They require more personnel for plant operations and maintenance than PV facilities, and so produce more visible personnel traffic.

FIGURE 2.6 Parabolic concentrators near Tucson, Arizona.
(Photo: M. Pasqualetti.)

The question that is now before us is how we can assess the visual impacts of such renewable energy developments and what we will need to do going forward. We begin with the approaches half a century past.

2.3 The early years

The 1960s and 1970s witnessed many funded and/or permitted federal projects come to a standstill because of lack of public involvement and/or incorporation of environmental and social values (Smardon 1982, 1986). The passage of the National Environmental Policy Act (NEPA) in 1970 helped change that pattern by requiring evaluation of a wide range of potential environmental impacts, including those affecting visual aesthetics.

Visual landscape management systems evolved and matured during the 1970s and 1980s in a number of Federal agencies (Elsner & Smardon 1979, Smardon 1982, 1986). The US Forest Service (USFS), US Department of the Interior (USDI) Bureau of Land Management and the Soil Conservation Service (predecessor to the Natural Resources Conservation Service) were the early leaders among US federal agencies in this regard (Bacon 1979, Ross 1979, Schauman & Adams 1979).

The USFS introduced their Visual Management System (VMS) regulations in 1976, and these were gradually introduced across many national forests as part of timber management and land use management planning from 1975 to 1976 and thereafter (Bacon 1979, Smardon 1986). The USFS updated this system to the Scenery Management System (SMS) in 1995 (USDA FS 1995). Both the VMS and SMS systems required landscape classification, assessment of visibility of activities to sensitive users, and the ability of the landscape to *absorb* visual impact. A series of handbooks were developed by the USFS starting in the latter half of the 1970s; the major one of interest here reached print in 1975 (USDA 1975). It addressed the visual impacts of the many electric power transmission lines and pipelines running through national forests.

The Bureau of Land Management (BLM) has purview over oil and gas development on 243 million acres of BLM-administered surface lands as well as 700 million acres of subsurface federal mineral estate that includes both surface lands under private ownership and surface lands administered by other federal agencies. The BLM is a principal agency in the nation's strategy to diversify the US energy portfolio. Since 2009 and after rigorous environmental impact analyses, the BLM has authorized approximately 14,000 MW of utility-scale renewable energy generation facilities for development on as many as 124,000 acres of BLM-administered surface acres.

In the mid-1970s the BLM developed its own systematic process for inventory and management of scenic values. The BLM first issued the Visual Resource Management (VRM) manuals and handbooks in 1975, and then retooled and reissued them in 1978 after the passing of the Federal Land Policy and Management Act 1976. Much like the USFS VMS system, the VRM classified scenic attributes based on scenic quality, public sensitivity for the scenic quality, and visibility distance zones. The VRM system also included a visual impact assessment (VIA) methodology that evaluated how readily a project could be distinguished from the background in which it was to be sited. For example, it screened the degree of contrast of the introduced project to existing landforms, vegetation and land use via shifts in line, form, color, and texture impacting BLM lands. The level of contrast created is then examined for the degree of noticeability from an observer's position based on 10 factors (distance, duration of view, angle of observation, lighting conditions, relative size and scale, spatial relationship, atmospheric conditions, season of use, recovery time, and motion). It has proved particularly useful in the assessment of renewable energy projects such as windfarms, solar arrays, geothermal development and, of course, transmission.

The Federal Highway Administration (FHWA) provides financial support and technical assistance for highway construction and maintenance throughout the US. Naturally, these projects often have major environmental impacts upon the landscape. Based upon the work of Jones and Jones landscape architects out of Seattle, Washington, the FHWA published their visual resource guidance document for highway projects in 1988 (Blair et al. 1979, Jones et al. 2012). This document offered basic landscape and highway project visual inventory guidance as well as a visual impact method that could be used to compare transportation alternatives based on the attributes of *landscape unity*, *vividness*, and *integrity*. Updated guidance was issued by FHWA in 2015 (FHWA 2015). In a related contribution, Jones and Jones also pioneered several visual assessments for larger power line projects associated with large hydroelectric and nuclear projects (Jones et al. 1975, Murray et al. 1979).

The US Army Corps of Engineers (USACOE) also became involved in visual aesthetic assessments, publishing *Visual Resources Assessment Procedure* (VRAP) in 1988 (Henderson et al. 1988, Smardon et al. 1988). The VRAP was based upon methods and training work developed by Smardon and Palmer from 1983 until 1988 (Henderson et al. 1988), and it was partly in response to a National Academy of Sciences panel that questioned why the Corps had not been addressing aesthetic and other environmental values as part of environmental assessment.

This guidance was embraced with the development of better procedures to ensure reliability and validity as part of the VRAP method. As a result, the VRAP approach is fairly time-consuming and costly to implement, which explains why it has not been heavily utilized by the Corps, albeit it has been used by other federal and state agencies for power plants and water related development.

The US energy licensing agencies – the Nuclear Regulatory Agency (NRC) and the Federal Energy Regulatory Commission (FERC) – have no established methods for reviewing power plant or transmission line visual impact on the landscape. However, many state regulatory bodies – such as those in California and New York – require some consideration of landscape impacts. More recently the US Bureau of Ocean and Energy Management (BOEM) permits offshore wind, solar, wave and ocean current energy. (See Chapter 3 for more detail).

Whereas generation units such as dams and power plants are often large, they are localized facilities on the landscape. Moving the power they generate is another matter, and for this reason the early focus of visual impact assessment was on siting of electric transmission power lines and support structure design. Henry Dreyfus Associates published aesthetic guidelines for electric transmission in 1972. These guidelines were enhanced in 1980 when Charles T. Main worked with the Institute on Man and Science to develop a method to assess transmission visual impacts.

2.3.1 Early wind assessment

One of the earliest assessment studies of wind turbine visual impact was the National Science Foundation Study (Survey Research Lab 1977), which carried out a nationwide effort to obtain people's reactions to simulations of various designs of wind turbines, depicted in various settings. The survey revealed that designs favored, in order from 'good' to 'bad', were:

Good	Traditional Dutch windmill
	Horizontal axis, twin blade rotor on an Old Dutch Tower
	Horizontal axis, twin blade on a solid pole
	Vertical axis 'egg beater' (Darrieus type)
	Horizontal axis, twin blade with lattice tower
Bad	Giromill

Only the traditional windmill was judged to improve the landscape by more than 50 percent of the respondents. The Giromill was considered a distraction on the landscape by a majority of respondents (Figure 2.7). The Giromill wind turbine is a type of vertical axis wind turbine which is used to produce power. The turbine consists of three straight blades, which is technically an airfoil, connected to the rotating main shaft. All designs were shown at the same scale, which may have biased the responses. Aside from traditional windmills, there was preference for modern designs without cluttered or lattice structures (Survey Research Lab 1977) (Figure 2.8).

A 1980 Swedish study described the effects of wind turbines on the landscape where critical distance zones that affected the visual dominance of wind turbines were determined (Engstrom & Pershagen 1980) This report describes the effects of 'area dominance' where the observer is surrounded by wind turbines (rather than viewing a group from outside) and all views were intruded upon. The relative effects of four different groupings of large turbines (at that time) were appraised with the aid of artist sketch simulations in a flat, open landscape.

The 'densest' array, equivalent to a maximum build-out of 28 machines at 1,000 meter spacing produced a disordered and cluttered landscape in various views from 0 to 3 miles away. Area dominance was widespread but a sparser array of 12 machines at 2–400 meter spacing tended also to look

FIGURE 2.7 H-type Giromill wind turbine.
(Photo: Stahlkocher, Wikipedia.)

disordered, but was less widely dominating. A double row of 14 machines at 500–1,000 meter spacing appeared more orderly in some views, but dominated views from up to 2 miles away. An array of two clusters of 6 and 7 machines at 500 meter spacing within clusters generally appeared most ordered and dominated the least number of views (Engstrom and Pershagen 1980).

They also compared arrays of 12 machines at 1,500–2,000-meter spacing with an array of 54 similar but smaller turbines (50 meter rotor diameter) at 500–700 meters. The arrays were equally visible; the smaller machines yielded a more cluttered landscape but was less dominating. Engstrom and

FIGURE 2.8 Lattice being replaced by monopoles near Palm Springs, California. Monopoles are less intrusive and preferred by the public. They also avoid nesting opportunity for birds.
(Photo: M. Pasqualetti.)

Pershagen (1980) concluded that fewer large turbines are preferable to more small turbines (arranged in large arrays), and that arrays should be structured to provide visual order and balance with existing landscape features. These early studies were, along with simulation guidance, utilized in the San Gorgonio California Wind Resource Study visual impact assessment, completed by Wagstaff and Brady in 1982.

CASE STUDY 2.1 THE SAN GORGONIO CALIFORNIA WIND STUDY

The San Gorgonio California Wind Study was an early assessment of wind energy impacts on the landscape. For this project Wagstaff and Brady inventoried the project area and divided the landscape into visual units, which were then evaluated in terms of the landscape unit distinctiveness, diversity and unity. Visibility of the landscape units was compiled using mapping with notes, photographs taken from key viewpoints, and computer-generated viewshed maps. Visual vulnerability of each landscape unit was evaluated by combining appraisals of scenic quality, visibility, and visual absorptive capacity. Visual quality objectives to maintain scenic quality were set based on maximum acceptable levels of visual contrast (a la BLM) and visual dominance (Smardon et al. 1984) (Figure 2.9).

FIGURE 2.9 Looking west toward San Gorgonio Pass from Edom Hill just after sunrise. Windfarms are about 15 miles distant (oval) and barely noticeable. 50 mm lens.
(Photo: M. Pasqualetti.)

The rapid expansion of windfarms, particularly in California in the mid-1980s, substantially ramped up the degree of public concern about their visual impacts. The National Wind Coordinating Committee published their *Handbook on Permitting of Wind Energy Facilities* in 1998 with a brief section on visual impact assessment (Therkelsen et al. 1998). It was not until the National Research Council's *Environmental Impacts of Wind-Energy Projects* (NRC 2007) that we had some guidance and recognition of visual impacts of commercial-scale wind development in the US. The second document was *Guidelines for Protecting Scenic Resources* and *Guidelines for Evaluating Cumulative Aesthetic Impacts*. Both documents are in the 2007 NRC study as appendices.

Scenic America's website on *Wind Energy Landscapes* (See Table 2.1 below) includes many examples of scenic impacts. More recently where there were high quality scenic landscapes features, there was potential opposition to siting windfarms. In the wide-open Great Plains or mid-western landscapes this controversy appears to be less common, unless the sites are near national parks or other highly sensitive areas. Table 2.1 below highlights more recent commercial-scale windfarms in the US with varying levels of controversy.

In response to public outcry about the 'intrusion' of wind turbines on their crowded landscape, several European countries, particularly the UK, have aimed to standardize the visual impact assessment process. *Guidelines for Landscape and Visual Impact Assessment* was published by the Landscape Institute and the Institute of Environmental Management and Assessment in 2001, followed by a third edition that came into force in April 2013 (Landscape Institute and Institute of Env. Mgmt. and Assessment

TABLE 2.1 Wind controversy case studies in the US

Windfarm	*Location*	*Scale*	*Developer*	*Controversy*	*Status*
Elk River	Butler County, KS	100 turbines 150 MW	Iberdrola	Moderate	Completed
Forward Wind Ctr.	Found Du Lac, Dodge Co., WI	86 turbines 129 MW	Invenergy	Highly	Completed
Horse Hollow	Taylor & Nolan, Counties, TX	421 turbines 735 MW	FPL Energy	Moderate	Completed
Wild Horse Wind	Kittitas County, WA	127 turbines 230 MW	EDP Renewables	Little	Completed
Cohocton Wind	Steuben County, NY	35 turbines 87 MW	First Wind	Highly	Completed
Highland Wind	Highland County, VA	20 turbines 40 MW	Highland New Wind	Highly	Approved/Under construction
Valley County	Valley County, MT	6 turbines 10 MW	Green Hunter	Moderate	Scaled back/ Completed
Redington/Black Nubble	Franklin County, ME	18 turbines 54 MW	Maine Mtn. Power	Highly	Delayed
Jordanville Wind Farm	Jordanville, NY	49 turbines 98 MW	Community/Iberdrola	Highly	Scaled back/ Delayed
Cascade Wind	Wasco County, OR	40 turbines 60 MW	First Wind	Highly	Cancelled

(*Source:* www.macalester.edu/windenergy/casestudies/casehome.html)

2013). This basic VIA procedural guide has been utilized for large projects throughout the UK – particularly large windfarm projects. Specific chapters include:

- Description of the proposed project,
- Baseline studies,
- Identification and assessment of landscape and visual effects,
- Presentation (visualization) techniques, good practice,
- Consultation and review.

2.4 Evolving visual impact assessment methods

In the last 5 to 10 years, we have seen a resurgence of VIA – particularly for large-scale wind turbine farms – both on land and off shore. Some of these projects such as the proposed Cape Wind project between Cape Cod and Martha's Vineyard have incited prolonged controversy, largely over aesthetic issues (Williams and Whitcomb 2007). Cape Wind and other projects in various countries have prompted a continuing refinement in VIA methods (Bishop 2002, Bishop & Miller 2007, DTI 2005, Hill et al. 2001, Horner & Maclennan/Envision 2006, Palmer 2012, Rooney et al. 2011, Sullivan et al. 2012a, 2012b, 2012c, 2013, Sullivan & Abplanalp 2013, Sullivan & Meyer 2014). We describe three of these methods in more detail below.

Largely in response to the public opposition to land-based windfarms on their small and crowded lands, the UK has been especially active in developing VIA methods for offshore windfarms. We summarize three:

- *Guidance on the Assessment of the Impact of Offshore Wind Farms: Seascape and Visual Impact* (DTI 2005),
- *Guide to Best Practice in Seascape Assessment* (Hill et al. 2001),
- *Visual Representation of Wind Farms: Good Practice Guidance* (Horner & Maclennan/Envision 2006).

DTI (2005) is more of an overall VIA methodology framework, whereas Hill et al. (2001) focused on landscape description and classification, and Horner & Maclennan/Envision (2006) is more along the lines of technical guidance for doing visibility mapping, photography, visual simulation and presenting the visual analysis. Hill et al. (2001) provide a detailed and systematic method on how to inventory and characterize existing seascape and coastal landscape qualities of Scotland and Wales with some guidance on how to evaluate landscape quality and capacity to accommodate change.

Of these approaches, DTI's is the most comprehensive VIA method and includes the following steps:

- Scoping of seascape and visual effects,
- Baseline studies of existing seascape and visual resources,
- Assessing sensitivity,
- Key considerations in siting, layout and design,
- Predicting impacts and assessing their magnitude,
- Evaluation of seascape and visual impacts – assessing significance,
- Cumulative impact assessment, and
- Monitoring.

In addition, it includes quantitative factors of:

- Distance from development, <13 kilometers (8 miles) possible major effects, 13–24 kilometers (8–14 miles) possible moderate effects, >24 kilometers (14 miles) minor effects,

- Number and proportion of turbines visible,
- Duration of effect,
- Angle of view in relation to receptor activity,
- Number and extent of resources affected, and
- Amount of navigational lighting required.

Less quantifiable parameters include:

- Scale of change,
- Arrangement of turbines,
- Degree of contrast,
- Nature of effect – positive/neutral/negative,
- Background – sea, sky or landform,
- Extent/prominence of onshore substation, and
- Extent/prominence of other vertical structures.

A review of six major offshore projects in the UK was done to ascertain their means of assessment. The two major criteria used most often for assessing *degree of visual impact* were *magnitude of impact* and *effect on sensitive viewers and/or activities*. Some projects considered *cumulative* visual impact.

The rapid growth of renewable energy installations around the world has drawn increasing attention to their visual landscape impacts and how to assess them. The approaches to VIA vary widely:

- Torres-Sibille et al. (2009) emphasize the aesthetic aspects of integrating windfarms into the landscape by using photographs and interviews to develop objective indicators. This approach has been extended as well to large-scale PV arrays, utilizing the indicators of visibility, color, fractality and concurrence of fixed and mobile panels for the quantification of visual impact (Torres-Sibille et al. 2009). This is the only PV solar array method that has utilized public perception testing.
- Ladenburg (2009), on the other hand, focuses attention on assessing the observer's prior experience with the renewable technology, relying on public surveys to develop the analysis of impact. Such an approach may help with broader social impact issues but not so much with site-specific issues.
- Bishop and Miller (2007) highlight the importance of parameters such as distance, contrast and motion in the visual impact assessment with use of photographs, computer digital simulations and interviews. Such methods are useful for the assessment of visual impact of a single technology (such as windfarms) on a local scale because design mitigation measures become more apparent.
- Graham et al. (2009) assess the public perception of wind energy concerning social reluctance for siting based on physical context, political and socio-economic factors. Again, such an approach is good for regional identification of social impact issues.
- Higgs et al. (2008) examine the use of visualization and decision support tools such as GIS and Multi-Criteria Decision-Analysis (MCDA) to facilitate public participation in windfarm siting. This is a catalog of specific techniques that could be utilized within a visual impact assessment framework.
- Newer methods use digital imagery and web-based surveys (Berry et al. (n.d.), Chiabrando et al. 2009, 2011, Higgs et al. 2008, Rodrigues et al. 2010, Roth & Gruehn 2014). Such methods may be useful in gaining an understanding of public perception regarding regional to site-specific visual impact assessment.
- Under the assumption that all lands have inherent scenic value, the USDI BLM procedures now require conducting a quantitative assessment of the potential reduction in the landscape's scenic

value, and methods for evaluating opportunities for replacing these reduced values at offsite locations within a defined region. The procedures include an evaluation of the regional condition, trend, and scarcity of the scenic resource; impact to viewsheds and affected stakeholders within these viewsheds; and conformance to projected desired future outcome defined in the BLM land use plans (McCarty et al. 2014).

2.5 Visibility assessment techniques

Much recent VIA method work has concentrated on digital techniques to assess potential visibility of proposed renewable energy development. The visual assessment of whether an installation is visible is typically done with the help of a geographic information system (GIS) that utilizes computer algorithms to project lines of sight from the proposed project to the surrounding landscape and from the project outward to the affected landscape viewers.

Such an operation is usually termed *potential viewshed* or the *zone of visual influence* (ZVI), which is utilized in the international literature and practice. This means that if we know the location of visually sensitive areas (such as near homes, historic sites, or recreational areas), we can project toward the introduced windfarm to produce such a viewshed. This is what was done for the Cape Wind project (Cape Wind & EDR 2006) and such visibility analyses are described in the literature (Bishop 2003, Ervin & Steinitz 2003, Felleman 1986, Hurtado et al. 2004). There are unique problems with projecting viewsheds with intersecting landmasses or islands. There are also atmospheric affects such as fog, haze and wave action, which can affect distance perception and visibility for offshore projects.

Similar GIS-based models for visibility assessment of renewable energy facilities have gained in popularity, both for wind and for solar installations. These methods have been discussed in several publications: Germany (Roth & Gruehn 2014), Hungary (Jombach et al. 2014), Italy (Chiabrando et al. 2011, Fabrizio & Garnero 2012), Lithuania (Depellegrin et al. 2014), Spain (Chias & Abad 2014, Garcia-Garrido et al. 2012, Molina-Ruiz et al. 2011, Rodrigues et al. 2010), and the United Kingdom (Higgs et al. 2008). Refinements continue.

Bishop and Miller (2007) used an online survey to gauge reactions to simulations of an 18-turbine hypothetical windfarm at three distance zones (4, 8 and 12 kilometers). Their analysis indicated that two out of five atmospheric conditions at 4 kilometers were significant, but none at 8 or 14 kilometers in terms of visibility and degree of visual contrast. Cape Wind & EDR (2006) utilized three distance zones 0–6 miles, 6–12 miles and 12–18 miles for visibility analysis for the Cape Wind project.

CASE STUDY 2.2 CAPE WIND, MASSACHUSETTS

To do the seascape horizon visibility analysis for the Cape Wind project, Cape Wind & EDR (2006) performed a viewshed analysis from 12 viewpoints at a 5.1-foot viewer height with a 10-mile radius, utilizing a digital elevation (terrain) model (DTM) with a 100-foot grid resulting in a viewshed with uninterrupted views of seascape horizon. Any area in which the visibility to the seascape was interrupted by land was subtracted. The percentage of the visible seascape and an arc (in degrees) occupied by the proposed project from each viewpoint were calculated. For each viewpoint, the total visible seascape was then calculated in degrees and as a percentage of visible seascape horizon. It should be noted that this is a very straight-forward, but simplified analysis.

It should also be noted that Cape Wind & EDR (2006) obtained base data from the Massachusetts GIS database (www.mass.gov/mgis/) and included only sound-facing shoreline within the limits of Nantucket Sound. Interior views with small bays, coves and tidal creeks with limited or no views to the ocean were excluded. Areas such as large bays enclosed by spits or headlands would show up as shadow zones in visibility and were excluded. Results included three layers: (1) total sound-facing shoreline in linear miles; (2) sound-facing shoreline with potential views of the proposed project in linear miles; and (3) proposed project boundary. The character of the visibility from each of the 12 viewpoints was discussed.

Few audits have been performed (Wood 1999, 2000, Wood et al. 2000) to ascertain the degree of accuracy of potential viewsheds or ZVI. Wood (2000) found some ZVI to be accurate and others inaccurate due to not accounting for screening vegetation and structures. Several authors have commented on the degree of accuracy and error in such visibility analyses (Felleman 1986, Fisher 1992).

To summarize, key issues for visibility analysis are:

1. Distance zones specified,
2. Data sources to be utilized for sensitive viewpoints, elevation (DTM) and screening vegetation and/or structures,
3. Screening land forms and shadow zones to be included or not, and
4. Levels of accuracy and metadata required.

2.6 Visual impact thresholds

Once we have determined whether a proposed development is visible, the next question is, does it impose a significant visual impact on the landscape? One of the key considerations in any assessment of visual impact is at what point – that is, at what 'threshold' of distance, contrast, movement or simple annoyance – does the introduction of renewable energy equipment pose a 'problem' for people. We refer to this as the 'visual threshold', and it applies to both onshore and offshore developments. Studies of visual thresholds and causal factors of visual impact related to offshore windfarms have been done in Australia (Bishop & Miller 2007, Fenton & Syme 1989, Lothian 2006, Shang & Bishop 2000), Denmark (Ladenburg 2009, Ladenburg and Dubgaard 2007, Moller 2006), the UK (Sullivan et al. 2013), and Sweden (Johansson and Laike 2007).

2.6.1 Offshore wind impact perception

There have been no perception studies of offshore wind turbines in the US. The only related work of interest is Joan Nassauer's (1983) study of oil and gas development in the coastal landscape of Louisiana.

The Bishop and Miller study (2007), using photographic simulations of wind turbines presented via the web, assessed wind turbine visibility and impact at different distances, different lighting intensities and angles, and different atmospheric conditions, both for moving and stationary blades. The study identified distance and visual contrast as good predictors of perceived impact, and it showed a preference for moving blades. A related field study by Sullivan et al. (2013), assessed the visibility of built wind facilities in the UK from shore, and found that moderate-sized installations (25–140 turbines in 11 windfarms) with turbines up to 90 meters (295 feet) tall at blade tip height of 126 meters (413 feet) were visible to the unaided eye at distances greater than 42 kilometers (26 miles), and that turbine blade movement was visible up to 39 kilometers (24 miles). The observed wind facilities were judged to be a major focus of visual attention at distances up to 16 kilometers (10 miles), were noticeable to casual observers at distances of almost 29 kilometers (18 miles), and were visible with extended or concentrated viewing at distances beyond 40 kilometers (25 miles).

2.6.2 Land based windfarm impact

In a study conducted for BLM, Sullivan et al (2012b) assessed visibility of five wind facilities in Wyoming and Colorado under various lighting and weather conditions. They found these facilities to be visible to the naked eye at 58 km (36 miles) under optimal viewing conditions, much farther than the offshore wind turbines in the U.K. study. Turbine blade movement was faintly visible as far as 47 km (29 miles) and routinely visible at a distance of 39 km (24 miles) (Sullivan et al 2012b). Wind facilities were judged to be major foci of visual attention at up to 19 km (12 miles) and unlikely to be missed by casual observers at up to 32km (20 miles). Based on these findings, Sullivan recommends an appropriate radius for wind energy facility visual impact analyses would be 48 kilometers (30 miles) in western landscapes.

Visual impacts are not confined to daylight hours. Lighting is a necessity because the Federal Aviation Administration (FAA) recommends it for most structures taller than 200 feet to ensure aviation safety, and these lights are even more noticeable (and annoying) in dark-sky areas. Wind turbine lighting can be visible for very long distances at night, especially when turbines are located on ridgelines or mesas. Sullivan et al.'s Colorado and Wyoming wind turbine visibility study (2012b) found lighting on wind turbines to be plainly visible at a distance of 58 kilometers (36 miles). A typical adjustment has been to place lights on a selection of turbines, rather than on them all, and to have them synchronized rather than randomized, which is the mandatory FAA guidance in the US.

2.6.3 Solar array visual impact

Solar installations can also create a landscape signature. Sullivan et al. (2012c) assessed the visual characteristics of several parabolic trough PV and power tower facilities in the desert southwestern landscape. Observed sources of visual contrast from parabolic trough facilities included glare from heat transfer tubes, geometric patterns of reflected light that created scintillations, plumes associated with cooling towers, and reflections from supports and ancillary facilities. In certain cases glare was bright enough to cause visual discomfort and temporary afterimages when observed at distances greater than 6 kilometers (4 miles). A 1.6 square kilometer (400 acre) parabolic trough facility was found to be visible at distance of greater than 23 kilometers (14 miles), in both daytime and nighttime observations.

A thin-film PV facility was visible at a distance of 35 kilometers (22 miles) in daytime conditions. Other visual effects observed included dramatic and rapid changes in apparent color and reflectivity of solar collector arrays of both parabolic trough and thin-film PV facilities (Figure 2.10).

Sullivan et al. (2012c) found that reflected light from a 5-MW and a 20-MW power tower was easily visible at a distance exceeding 32 kilometers (20 miles) but did not cause visual discomfort. Sullivan and Abplanalp's more recent study (2015) of the much larger Ivanpah solar power tower facility (392 MW) in California showed that the three towers of the facility were easily visible from a mountain top 56 kilometers (35 miles) away, and that glare from the heliostat array could cause visual discomfort at a distance of 32 kilometers (20 miles) (Figure 2.11, Plate 4). This glare has precipitated substantial public attention (see Carullo et al. 2013, Chiabrando et al. 2009, 2011, Sullivan et al. 2012c, Sullivan & Abplanalp 2013).

From these studies, we can conclude that *distance* and *visual contrast* are two key variables needed to assess visual impact of renewable energy facilities, and second, *familiarity* and *experience of such development*

FIGURE 2.10 Glare off Solnova 1 and 3, two of the parabolic trough plants in operation at the Solucar Solar Complex, Seville, Spain.
(Photo: Business Wire.)

FIGURE 2.11 Unit #2 at the Ivanpah Central Receiver station. View is toward the east. Ivanpah Dry Lake is in the center of the image.
(Photo: R. Sullivan.)

will be key when assessing cumulative visual impacts. Cumulative impact issues will be covered in Chapters 8, 9 and 10. Studies in Europe have documented that familiarity with renewable energy development is key to negative or positive reactions (Graham et al. 2009, Johansson & Laike 2007, Ladenburg 2009, Torres-Sibille et al. 2009). More detailed coverage of social and visual impacts of renewable energy facilities is provided in Chapter 5.

2.7 Summary

Given the recent surge in renewable energy development, there has been a 'slight surge' of VIA methodology development sponsored mainly by the USDI Bureau of Land Management and the National Park Service. Such work was highlighted in recent meetings of the National Association of Environmental Professionals (NAEP):

- Visibility assessment of renewable energy projects (Palmer 2012, Rooney et al. 2012, Sullivan & Abplanalp 2013, 2015, Sullivan et al. 2012a, 2012b, 2012c, 2013, 2014),
- Air quality visibility impacts near and within national parks (Molenar et al. 2012),
- Limited visual perception research in the US (Ribe 2012, Palmer 2000, Palmer & Hoffman 2001), Germany (Roth 2012) and Australia (Bishop & Miller 2007),
- New mitigation measure effectiveness (Schwarzler & Cramer 2012, Sullivan et al. 2014),
- New set of best practices for renewable energy development visual impact mitigation on BLM lands (USDI BLM 2013),
- New guidance for assessing renewable energy projects near or on National Park Service lands (Sullivan and Meyer 2014).

References

Bacon, W. (1979). The visual management system of the Forest Service, USDA. In G. Elsner and R. Smardon (Eds.), *Our National Landscape*, pp. 660–665. Available at www.esf.edu/via and www.treesearch.fs.fed.us/pubs/27530

Berry, R., G. Higgs, M. Langford, and R. Fry. (n.d.). *An Evaluation of Online GIS-based Landscape and Visual Impact Assessment Tools and their Potential of Enhancing Public Participation in the UK*. Wales Institute of Social and Economic Research, Data and Methods & GIS Research Centre, University of Glamorgan, Wales UK. Available at www.academia.edu/2441090/An_evaluation_of_online_GIS-based_landscape_and_visual_impact_assessment_tools_and_their_potential_for_enhancing_public_participation_in_wind_farm_planning_in_

Bishop, I.D. (2002). Determination of thresholds of visual impact: The case of wind turbines. *Env. and Planning B: Planning and Design 29*(5), 707–718. doi: 101068/b12854.

Bishop, I.D. (2003). Assessment of visual qualities, impacts and behaviors in the landscape by using measures of visibility. *Env. and Planning B: Planning and Design 30*(5), 677–688. doi: 10.1068/b12956.

Bishop, I.D. and D.R. Miller. (2007). Visual assessment of offshore wind turbines: The influence of distance, contrast, movement and social variables. *Renewable Energy 32*(5), 814–831. doi: 10.1016/j.renene.2006.03.009.

Blair, G.E., L. Isaacson, and G.R. Jones. (1979). A comprehensive approach to visual resource management for highway agencies. In G. Elsner and R. Smardon (Eds.), *Our National Landscape*, pp. 365–372. Available at www.esf.edu/via and www.treesearch.fs.fed.us/pubs/27530

Cape Wind Associates LLC and Env. Design and Research (EDR). (2006). *Seascape and Shoreline Visibility Assessment, Cape Wind Energy Project*. Env. Design and Research Syracuse, NY.

Carullo, L., P. Russo, L. Riguccio, and G. Tomascelli. (2013). Evaluating the landscape capacity of protected rural areas to host photovoltaic parks in Sicily. *Natural Resources 4*, 460–472. doi: 10.4236/nr.2013.47057.

Chiabrando R., E. Fabrizio, and G. Garnero. (2009). The territorial and landscape impacts of photovoltaic systems: Definition of impacts and assessment of glare risk. *Renewable and Sustainable Energy Reviews 13*(9), 2441–2451. doi: 10.1016/j.rser.2009.06.008.

Chiabrando R., E. Fabrizio, and G. Garnero. (2011). On the applicability of the visual assessment OAI spp tool to photovoltaic plants. *Renewable and Sustainable Energy Reviews 15*(1), 845–850. doi: 10.1016/j.rser.2010.09.030.

Chias, P. and T. Abad. (2014). Impact assessment of the renewable energies in the cultural heritage: The case of the way of St. James in Spain. *International J. Natural Resources 5*, 165–171. doi: 10.5194/isprsarchives-xl-5-165-2014.

Crowe, S. (1958). *The Landscape of Power*. London: The Architectural Press.

Department of Transportation and Industry (DTI). (2005). *Guidance on the Assessment of the Impact of Offshore Wind Farms: Seascape and Visual Impact*. Report of the Dept. of Trade and Industry, Publication No. 80666, London.

Depellegrin, D., N. Blazauskas, and L. Egarter-Vigil. (2014). An integrated visual impact assessment model for offshore windfarm development. *Ocean and Coastal Management 98*, 95–110. doi: 10.1016/j.ocecoaman.2014.05.019.

Elsner, G.H. and R.C. Smardon (Eds.). (1979). *Proceedings of Our National Landscape: A Conference on Applied Techniques for Analysis and Management of the Visual Resource*. Incline Village, NV, April 23–25 1979, USDA For. Serv. Gen. Tech. Rpt. PSW-35, Berkeley CA, Pacific SW Forest & Range Experimental Station. Available at www.esf.edu/via and www.treesearch.fs.fed.us/pubs/27530

Engstrom, S. and B. Pershagen (Eds.). (1980). Aesthetic factors and visual effects of large scale wind energy conversion systems. *Final report NE 1980-20 of Task A5 for the International Energy Agency*. National Swedish Board of Energy Source Development.

Ervin, S. and C. Steinitz. (2003). Landscape visibility computation: Necessary, but not sufficient. *Env. and Planning B: Planning and Design 30*(5), 757–766. doi: 10.1068/b2968.

Fabrizio, E. and G. Garnero. (2012). Visual impact, landscape and renewable energy plants, the case of PV. TSO6L – Implementation of Plans and Infrastructure Paper No. 5946. In *FIG Working Week 2012: Knowing to Manage the Territory, Protect the Environment, Evaluate the Cultural Heritage*. Rome, IT, May 6–10 2012.

Felleman, J. (1986). Landscape visibility. In R.C. Smardon, J.F. Palmer. and J.P. Felleman (Eds.), *Foundations for Visual Project Analysis*. New York: Wiley, pp. 47–63. Retrieved from www.esf.edu/via

Fenton, D.M. and G.J. Syme. (1989). Perception and evaluation of the coastal zone: implications for coastal zone planning. *Coastal Mgmt. 17*(4), 295–308. doi: 10.1080/08920758909362092.

Fisher, P. (1992). First experiments in viewshed uncertainty, simulating fussy viewshed. *Photogrammetric Engineering and Remote Sensing 58*(3), 345–352.

Garcia-Garrido, E., P. Lara-Santillan, E. Zorzano-Alab, M. Mendoza-Villena, P. Zarzano-Santillamtameria, L.A. Fernanadez-Jimenez, and A. Falces. (2012). Visual impact assessment for small and medium size PV plants. In H. Walter (Ed.) *Advances in Powers and Energy Systems*. Prague, CZ, Sept. 24–26 2012, pp. 57–61. Available at www.wseas.org/multimedia/books/2012/Prague/POWER.pdf

Graham, J.B., J.R. Stephensen, and I.J. Smith. (2009). Public perceptions of wind energy developments: case studies from New Zealand. *Energy Policy 37*(9), 3348–3357. doi: 10.1016/j.enpol.2008.12.035.

Henderson, J.E., R.C. Smardon, and J.F. Palmer. (1988). *US Army Corps of Engineers Visual Resources Assessment Procedure*. Transportation Research Record No. 1189, pp. 67–71.

Higgs, G., R. Berry, D. Kidner, and M. Langford. (2008). Using IT approaches to promote public participation in renewable energy planning: prospects and challenges. *Land Use Policy 25*(4), 596–607. doi: 10.1016/j.landusepol.2007.12.001.

Hill, M., J. Briggs, P. Minto, D. Bagnall, K. Foley, and A. Williams. (2001). *Guide to Best Practice in Seascape Assessment*. Dublin: Countryside Commission of Wales, Brady Shipman Marsh and University College.

Horner & Maclennan and Envision. (2006). *Visual Representation of Wind Farms: Good Practice Guidance*. Report for Scottish Natural Heritage, The South Scottish Forum and Scottish Society for Direction and Planning.

Hurtado, J.P., J. Fernández, J.L. Parrondo, and E. Blanco. (2004). Spanish method of visual impact evaluation in windfarms. *Renewable and Sustainable Energy Reviews 8*(5), 483–491. doi: 10.1016/j.rser.2003.12.009.

Johansson, M. and T. Laike. (2007). Intention to respond to local wind turbines, the role of attitudes and visual perception. *Wind Energy 10*(5), 435–451. doi: 10.1002/we.232.

Jombach, S., D. Drexler, and A. Salloy. (2014). Using GIS for visibility assessment of a windfarm in Perenye, Hungary. In U. Wissen Hayak, P. Fricker, and E. Buhmann (Eds.), *Peer Reviewed Proceedings of Digital Landscape Architecture 2014 at ETH Zurich*. Offenbach: Wichmann Verlag.

Jones, G., B. Gray, and J. Ady. (1975). *Visual Impact Study: Statement of Findings, Alternative Closed Cycle Cooling Systems, Indian Point Nuclear Generating Plant*. Jones and Jones with Battelle Northwest for the US Nuclear Regulatory Commission.

Jones G., D. Sorey, and C. Scott. (2012). Visual resource assessment for regional landscapes: a modern response to NEPA. In *Science, Politics and Policy: Environmental Nexus: Conference Proceedings*, National Association of Environmental Professionals, 37th Annual Conference. Portland, OR, May 21–24 2012, pp. 641–656.
Ladenburg, J. (2009). Visual impact assessment of offshore windfarms and prior experience. *Applied Energy 86*(3), 380–387. doi: 10.1016/j.apenergy.2008.05005.
Ladenburg, J. and A. Dubgaard. (2007). Willingness to pay for reduced visual amenities from offshore windfarms on Denmark. *Energy Policy 35*(8), 4059–4071. doi: 10.1016/j.enpol.2007.01.023.
Landscape Institute and Institute of Env. Mgmt. and Assessment. (2013). *Guidelines for Landscape and Visual Impact Assessment*. 3rd Edition. London: Routledge.
Lothian, A. (2006). Visual impact assessment of windfarms in South Australia. In *New Zealand Assoc. of Impact Assessment Conference*. Dunedin, NZ, Nov. 30 2006.
McCarty, J., R. G. Sullivan, and H. Hartmann. (2014). Where and how do visual resources fit into the regional mitigation of onsite natural resource impacts? *National Association of Environmental Professionals 2014 Annual Conference*, St. Petersburg, FL, April 8–10 2014.
Molenar, J. V., R. W. Paterson, and M. D. Goodstein. (2012). Seeking clarity: scenic, economic and legal views on visibility. In *Science, Politics and Policy: Environmental Nexus: Conference Proceedings*, National Association of Environmental Professionals, 37th Annual Conference. Portland, OR, May 21–24 2012, pp. 767–817.
Molina-Ruiz, J., M. J. Martinez-Sanchez, C. Perez-Sirvent, M. L. Tudela-Serrano, and M. L. Garcia Lorenzo. (2011). Developing and applying a GIS-assisted approach to evaluate visual impact in windfarms. *Renewable Energy 36*(3), 1125–1132. doi: 10.1016/j.renene.2010.08.04.
Moller, B. (2006). Changing wind-power landscapes, regional assessment of visual impact on land use and population in Northern Jutland, Denmark. *Applied Energy 83*(5), 477–494. doi: 10.1016/j.apenergy.2005.04.004.
Murray, T. J., D. J. Bisenius, and J. G. Marcotte. (1979). Northwest Montana/North Idaho transmission corridor study, a computer-assisted corridor location and impact evaluation assessment. In G. Elsner and R. Smardon (Eds.), *Our National Landscape*, pp. 470–476. Available at www.esf.edu/via and www.treesearch.fs.fed.us/pubs/27530
Nassauer, J. I. (1983). Oil development in a coastal landscape: visual preferences and management implications, *Coastal Management 11*(3), 199–217. doi: 10.1080/08920758309361945.
National Research Council (NRC). (2007). *Environmental Impacts of Wind-Energy Projects*. Washington, DC: National Academy Press.
Palmer, J. F. (2000). Reliability of rating visible landscape qualities. *Landscape Journal 19*(1/2), 166–178.
Palmer, J. F. (2012). Maine's experience evaluating when scenic impacts from wind energy development are reasonably adverse. In *Science, Politics and Policy: Environmental Nexus: Conference Proceedings*, National Association of Environmental Professionals, 37th Annual Conference. Portland, OR, May 21–24 2012, pp. 602–620.
Palmer, J. F. and R. E. Hoffman. (2001). Rating reliability and representation validity in scenic landscape assessments. *Landscape and Urban Planning 54*(1–4), 149–161. doi: 10.1016/S0169-2046(01)00133-5.
Pasqualetti, M. J. (2001). Wind energy landscapes: society and technology in the California desert. *Society and Natural Resources, 14*(8), 689–699. doi: 10.1080/08941920117490.
Pasqualetti, M. J. (2011). Social barriers to renewable energy landscapes. *Geographical Review 101*(2), 201–223. doi: 10.1111/j.1931-0846.2011.00087.x.
Pasqualetti, M. J. (2013). Reading the changing energy landscape. In Sven Stremke and Andy van den Dobbelsteen (Eds.), *Sustainable Energy Landscapes: Designing, Planning, and Development*. Boca Raton, FL: CRC Press, pp. 11–44.
Pasqualetti, M. J., P. Gipe, and R. W. Righter (Eds.) (2002). *Wind Power in View*. San Diego: Academic Press.
Ribe, R. 2012. How do people's demographic attributes and environmental attitudes relate to their perception of scenic beauty? The case of forests in the Pacific Northwest. In *Science, Politics and Policy: Environmental Nexus: Conference Proceedings*, National Association of Environmental Professionals, 37th Annual Conference. Portland, OR, May 21–24 2012, pp. 971–1025.
Rodrigues, M., C. Montanes, N. Fueyo. (2010). A method for the assessment of the visual impact caused by the large-scale deployment of renewable energy facilities. *Environmental Impact Assessment Review 30*(4), 240–246. doi: 10.1016/j.eiar.2009.10.004.
Rooney, S., P. Dascomb, and S. Korjeff with R. Smardon. (2012). Technical Bulletin 12-001, *Visual Impact Assessment Methodology for Offshore Wind Energy Conversion Facilities*. Barnstable, MA: Cape Cod Commission. Available at www.esf.edu/via

Ross, R.W., Jr. (1979). The Bureau of Land Management and visual resource management: an overview. In G. Elsner and R. Smardon (Eds.), *Our National Landscape*, pp. 666–671. Available at www.esf.edu/via and www.treesearch.fs.fed.us/pubs/27530

Roth, M. (2012). Visual landscape quality assessment in landscape planning in Germany: a validation approach. In *Science, Politics and Policy: Environmental Nexus: Conference Proceedings*, National Association of Environmental Professionals, 37th Annual Conference. Portland, OR, May 21–24 2012, pp. 632–640.

Roth, M. and D. Gruehn. (2014). Digital participatory landscape planning for renewable energy-interactive visual landscape assessment as a basis for the geodesign of wind parks in Germany. In U. Wissen Hayak, P. Fricker, and E. Buhmann (Eds.), *Peer Reviewed Proceedings of Digital Landscape Architecture 2014 at ETH Zurich*. Offenbach: Wichmann Verlag.

Schauman, S. and C. Adams. (1979). Soil Conservation Service landscape management. In G. Elsner and R. Smardon (Eds.) *Our National Landscape*, pp. 671–675. Available at www.esf.edu/via and www.treesearch.fs.fed.us/pubs/27530

Schwarzler, K. and G. Cramer. (2012). The art and science of camouflage for visual mitigation in the landscape. In *Science, Politics and Policy: Environmental Nexus: Conference Proceedings*, National Association of Environmental Professionals, 37th Annual Conference. Portland, OR, May 21–24 2012, pp. 693–715.

Shang, H.-D. and I.D. Bishop. (2000). Visual thresholds for detection, recognition and visual impact in landscape settings. *J. Env. Psych. 20*(2), 125–140. doi: 10.1006/jvep.1999.0153.

Smardon, R.C. (1982). *An Organizational Analysis of Federal Agency Visual Resource Management Systems*. Ph.D. Dissertation: Environmental Planning Program at the College of Environmental Design, University of California, Berkeley. Available from International Microfilms, Ann Arbor, MI.

Smardon, R.C. (1986). Historical evolution of visual resource management within three Federal agencies. *J. Environ. Mgmt. 22*, 301–317.

Smardon, R.C., J.F. Palmer, J. Knopf, and K. Grinde with J.E. Henderson and L.D. Peyman-Dove. (1988). *Visual Resource Assessment Procedures for US Army Corps of Engineers*. Instruction Report EL-88-1, USACOE waterways Exp. Stn, Vicksburg, MS. Available at http.www.esf.edu/via

Smardon, R.C., S.R.J. Sheppard, and S. Newman. (1984). Visual Impact Assessment Manual. Occasional Paper No. ESF84-009, School of Landscape Architecture, SUNY/ESF, Syracuse, NY. Available at www.esf.edu/via

Stremke, S. and A. van den Dobbelsteen. (2013). *Sustainable Energy Landscapes: Designing, Planning and Development*. Boca Raton, FL: CRC Press.

Sullivan, R.G. and J.M. Abplanalp. (2013). Utility-Scale Solar Energy Facility Visual Impact Characterization and Mitigation Study: Project Final Report. Argonne, IL: Environmental Science Division, Argonne National Laboratory.

Sullivan, R.G. and J.M. Abplanalp. (2015). Visibility and Visual Characteristics of the Ivanpah Solar Electric Generating System Power Tower Facility. Argonne, IL: Environmental Science Division, Argonne National Laboratory.

Sullivan, R.G., J.M. Abplanalp, S. Lahti, L.J. Beckman, B.L. Cantwell, and P. Richmond. (2014). Electric transmission visibility and visual contrast distances in western landscapes. *National Association of Environmental Professionals 2014 Annual Conference*, St. Petersburg, FL, April 8–10 2014.

Sullivan, R.G., J. Cothren, M. Williamson, P. Smith, J. McCarty, and L.B. Kirchler. (2012a). Visual impact risk assessment and mitigation mapping system for utility-scale wind energy facilities. In *Science, Politics and Policy: Environmental Nexus: Conference Proceedings*, National Association of Environmental Professionals, 37th Annual Conference. Portland, OR, May 21–24 2012, pp. 667–692.

Sullivan, R.G., L.B. Kirchler, J. Cothren, and S.L. Winters. (2013). Offshore wind turbine visibility and visual impact threshold distances. *Environmental Practice 15*(1), 33–49. doi: 10.1017/s1466046612000464.

Sullivan, R.G., L.B. Kirchler, T. Lahti, S. Roche, K. Beckman, B. Cantwell, and P. Richmond. (2012b). Wind turbine visibility and visual impact distances in western landscapes. In *Science, Politics and Policy: Environmental Nexus: Conference Proceedings*, National Association of Environmental Professionals, 37th Annual Conference. Portland, OR, May 21–24 2012, pp. 897–942.

Sullivan, R.G., L.B. Kirchler, C. McCoy, J. McCarty, K. Beckman, and P. Richmond. (2012c). Visual impacts of utility-scale solar energy facilities on southwestern desert landscapes. Paper presented at the National Association of Environmental Professionals, 37th Annual Conference, Portland, OR, May 21–24 2012. Available at http://visualimpact.anl.gov/solarvis/

Sullivan, R. G. and M. Meyer. (2014). *Guide to Evaluating Visual Impact Assessments for Renewable Energy Projects.* Natural Resource Report NPS/ARD/NRR-2014/836 US National Parks Service and Argonne National Laboratory.

Survey Research Laboratory. (1977). Public Reactions to Wind Energy Devices: Final Report. Prepared for National Science Foundation and US Dept. of Energy. University of Illinois, Urbana.

Therkelsen, B., B. Grant, D. Bain, T. Gray, D. MacIntyre, and S. Ugoretz (1998). *Permitting of Wind Energy Facilities.* Washington, DC: National Wind Coordinating Committee.

Torres-Sibille, A. C., Victor-Andres Cloquell-Ballester, Vincente-Agustin Cloquell-Ballester, and R. Darton. (2009). Development and validation of a multi-criteria indicators for the assessment of objective aesthetic impact of windfarms. *Renewable and Sustainable Energy Reviews 13*(1), 40–66. doi: 10.1016/j.rser. 2007.05.002.

US Department of Agriculture, Forest Service (USDA). (1975). *National Forest Management: Utilities,* Vol. 2, Ch. 2. USDA Handbook No. 478, Washington, DC: US Gov. Print Office.

US Department of Agriculture, Forest Service (USDA FS). (1995). *Landscape Aesthetics, A Handbook for Scenery Management.* Agricultural Handbook No. 701. Washington, DC: US Gov. Print. Office. Available at www.esf.edu/via

US Department of the Interior, Bureau of Land Management (USDI BLM). (2013). *Best Management Practices for Reducing Visual Impacts of Renewable Energy Facilities on BLM-Administered Lands.* Cheyenne, WY: USDI Bureau of Land Management.

US Department of Transportation, Federal Highway Administration (FHWA). (2015). *Guidelines for the Visual Impact Assessment of Highway Projects.* Publication No. FHWA-HEP-15-029. Washington, DC: US DOT, Federal Highway Administration. Available at www.environment.fhwa.dot.gov/guidebook/documents/VIA_Guidelines_for_Highway_Projects.asp

Wagstaff and Brady. (1982). San Gorgonio Wind Resources Study Draft Environmental Impact Statement, App. 3. Berkeley, CA: Wagstaff and Brady.

Williams, W. and R. Whitcomb. (2007). *Cape Wind: Money, Celebrity, Class Politics and the Battle for Our Energy Future.* New York: Public Affairs.

Wood, C., B. Dipper and C. Jones. (2000). Auditing the assessment of environmental impacts of planning projects. *J. Env. Planning & Mgmt. 42*(5), 671–689.

Wood, G. (1999). Post-development auditing of EIA prediction techniques: a spatial analytic approach. *J. Env. Planning & Mgmt. 42*(5), 671–689. doi: 10.1080/09640569910948.

Wood, G. (2000). Is what you see what you get? Post development auditing of methods used in predicting the visual zone of influence in EIA. *Env. Impact Assessment Review 20*(5), 537–352. doi: 10.1016/S0195-9255(00)00055-X.

3

MANAGING NEW ENERGY LANDSCAPES IN THE USA, CANADA, AND AUSTRALIA

Richard Smardon, Ian Bishop, and Robert Ribe

Highlights

- US federal renewable incentives exist for wind, solar, and biomass renewable energy but individual US states have variability for renewable energy guidance, incentives, authority, and siting guidance – much of it relates to wind energy.
- There is very little guidance for utility-scale solar development in the US.
- Guidance for offshore wind energy in the US is starting to be developed.
- In Canada, most of the renewable energy policies have been developed in two provinces – Ontario and British Columbia for wind, solar and hydro.
- Almost all of the renewable energy guidance in Australia is for wind energy with more detailed guidance developed by New South Wales, South Australia, Victoria and Western Australia.

3.1 Introduction

This chapter addresses regulatory and legal tools for siting renewable energy facilities for the US, Canada and Australia. As stated in the 2007 US National Research Council's report on *Environmental Impacts of Wind-Energy Projects*, such projects are "amazingly controversial." If one looks at the history of the Cape Wind Project off Cape Cod, this is a complex history of local groups versus environmental groups, full of regional to national politics (Williams & Whitcomb, 2007). One of the major siting issues with Cape Wind was aesthetic – can we see it and do we want it in our view from the beach or from the water? This project also had a tale of uncertainty of just who had jurisdiction over the environmental review of the proposed project – the first commercial scale offshore wind project in the US. We also had major regional interconnecting power lines corridors proposed by DOE – which were to move power generated from one power-rich area to a power-poor area. Some of these such as the Utica, NY to Orange County, NY were stopped because of local resistance with viewshed issues again plus economic reasons.

It is also stated in the preface to the NRC study that "Regulation at the state and local level is variable among jurisdictions, some with well developed policies and others with little or no framework, relying on local zoning ordinances. There are virtually no policy or regulatory frameworks at the multistate regional scale." (NRC, 2007, p. ix). These are the major issues within this chapter such as jurisdictional/guidance conflict between federal agencies, as well as state and local agencies in addressing new and developing renewable energy development, especially at the industrial scale. Many of the issues from this chapter will be addressed in the policy Chapter 11.

Although most of the renewable energy regulatory action and scenic issues focuses on wind energy siting and regional interconnecting power lines, we also address solar and geothermal energy generation on the landside and wind energy generation offshore. We start with federal legislation/regulation as it affects renewable energy generation and transmission, then move to public lands issues, then to offshore renewable energy development, and finally to private or mixed land ownership issues under state and local jurisdiction. Then renewable energy generation guidance for Canada and Australia are covered. The final section will be a brief synthesis.

3.2 US federal and state support for renewable energy

The *Energy Policy Act of 2005* is probably the key statute affecting renewable energy development and transmission at the US federal level. It was signed into law by President Bush on August 8, 2005 and was over four years in the making. Key provisions include nuclear energy, coal, and renewable energy development research and incentives plus vehicle energy development and standards as well as electricity development, transmission, and safety. The key aspects of the act covered here relate to renewable energy development and transmission. For renewable energy the key provisions are summarized below:

Title II: Renewable energy

- Requires the Secretary of the Interior to study the potential of developing wind, solar and ocean energy resources on available federal land. Also directs the Secretary of Energy to annually review the available assessments of renewable energy resources within the United States, including solar, wind, biomass, ocean (tidal, wave, current, and thermal), geothermal, and hydroelectric energy resources.
- Extends the Renewable Energy Production Incentive (REPI), which makes nonprofit electric utilities eligible to receive the renewable energy Production Tax Credit.
- Requires the federal government to purchase 7.5% of its electricity from renewable sources by 2011.
- Creates two grant programs to encourage the production of electric energy or heat from biomass and to improve biomass utilization technology on Indian lands.
- Reforms the hydropower licensing process of the *Federal Power Act*, revives a DOE program to develop small hydropower projects, and provides incentives for increased production of hydropower through efficiency improvements at existing facilities.
- Authorizes projects on federal lands, on a cost-share basis with local utilities, to reduce dependence on fossil fuels used in the generation of electricity.

Title XIII: Energy policy tax incentives

- Extension and modification of the 1.8 cent per kWh renewable Production Tax Credit. Extends placed-in-service date by two years (through December 31, 2007) for qualifying facilities: wind facilities; closed-loop biomass facilities; open-loop biomass facilities; geothermal facilities; small irrigation power facilities; landfill gas facilities; and trash combustion facilities. Qualifying facilities receive credits per kWh for electricity produced over a 10-year period. Hydropower and Indian coal are added as new qualifying energy resources. Allows eligible cooperatives – more than 50%

owned by agricultural producers or entities owned by agricultural producers – to elect to pass any portion of the renewable electricity production credit to their patrons.

- Clean Renewable Energy Bond (CREB). CREBs are defined as bonds issued by a qualified issuer if, in addition to other requirements, 95% of proceeds are used to finance capital expenditures incurred for facilities qualifying for the renewable production tax credit. Qualified issuers include governmental bodies (including Indian tribal governments) and mutual or cooperative electric companies. Provision is effective for bonds issued after December 31, 2005.
- Creates an investment tax credit for purchases of solar photovoltaic and solar water heating technologies. The credit covers 30% of the cost of the solar technologies and applies to items installed during 2006–2007. The credit is capped at $2,000 for each solar system for residential facilities, while commercial facilities have no cap.

A key aspect of the Act is that Congress instructed the US Department of Interior (USDI) and Department of Energy (DOE) to collaborate in order to place at least 10,000 MW of non-hydroelectric renewable energy on federal land. As Glennon and Reeves (2010) point out, this has led to a 'stampede' as solar and wind energy companies have rushed to obtain permits for projects in Arizona, California, Colorado, Nevada, New Mexico, and Utah. Further fueling the stampede is the *American Recovery and Reinvestment Act of 2009*, which has further incentives for renewable energy development. The issue here, for very large solar arrays and focused power generation, is to find land that is not sensitive habitat, does not displace agriculture, and is not affecting National Parks and Trails.

In addition the Act established the role for the DOE and the Federal Energy Regulatory Commission (FERC) in transmission siting. The act directed DOE to create *transmission corridors* in locations with adequate transmission capacity. The act also granted FERC secondary authority over transmission siting in the corridors. This new federal role in the decision-making process that had previously been the province of the states was met with resistance. Although the process of creating transmission *corridors* moved forward, the Ninth Circuit Court vacated the congestion study that led to the designation of two such corridors.

Executive orders

There have been three Presidential executive orders that have helped shape renewable energy policy: Executive Order 13212, *Actions to Expedite Energy-Related Projects,* issued May 2001; Executive Order 13604 *Improving the Performance of Federal Permitting and Review of Infrastructure Projects,* issued March of 2012; and *Transforming our Nation's Electric Grid through Improved Siting, Permitting, and Review*, issued in 2013. All three executive orders are aimed at expediting both renewable energy development and transmission of such energy. The last executive order presents specific principles for established energy corridors for western and non-western states, and improved transmission siting, permitting, and review processes. These executive orders are reflected in the specific procedures and guidance with siting procedures for the Bureau of Ocean and Energy Management (BOEM), the US Department of Interior Bureau of Land Management (USDI BLM), and FERC, as we will see in subsequent chapter sections.

3.2.1 Public lands siting regulation – a jurisdictional mess

So with the incentives for renewable energy development from the *Energy Policy Act of 2005* plus three executive orders, there has been a tremendous push for large-scale renewable energy development on the western public lands – primarily wind and solar. Since private energy companies need leases to develop these facilities on public lands, the USDI BLM has inherited major responsibility for assessing environmental impacts under the *National Environmental Policy Act* (NEPA). Such responsibility has included developing a programmatic impact statement for wind (USDI BLM, 2005) and solar development (USDI BLM, 2012) under NEPA, plus special regulations for scoping (USDI BLM, 2011b) and assessing (USDI BLM, 2011a) renewable energy impacts.

In 2005 the BLM issued a *record of decision* on the implementation of a wind energy development program and how this would affect 52 land use plans in nine western states. This decision established policies and best management practices (BMPs) for the administration of wind energy development activities and established minimum requirements for mitigation. The policies and BMPs were evaluated in the *Final Wind Energy Programmatic Environmental Impact Statement* or PEIS (USDI BLM, 2005) (Fig. 3.2). The amendments to the 52 land use plans in nine states includes: adoption of BLM's *Wind Energy Development Program* policies, best management practices and identification of specific areas where wind energy development could occur (Fig. 3.1). With the development of this new program the BLM interim policy was replaced by a new policy in 2006 (Wind Energy Policy IM 2006–16). On BLM lands wind project development usually proceeds in two phases: (1) site testing and monitoring; and (2) if viable, a project construction and operation phase – both under the jurisdiction of the *Federal Land Policy and Management Act* (FLPMA) 43 USC sections 1701–174.

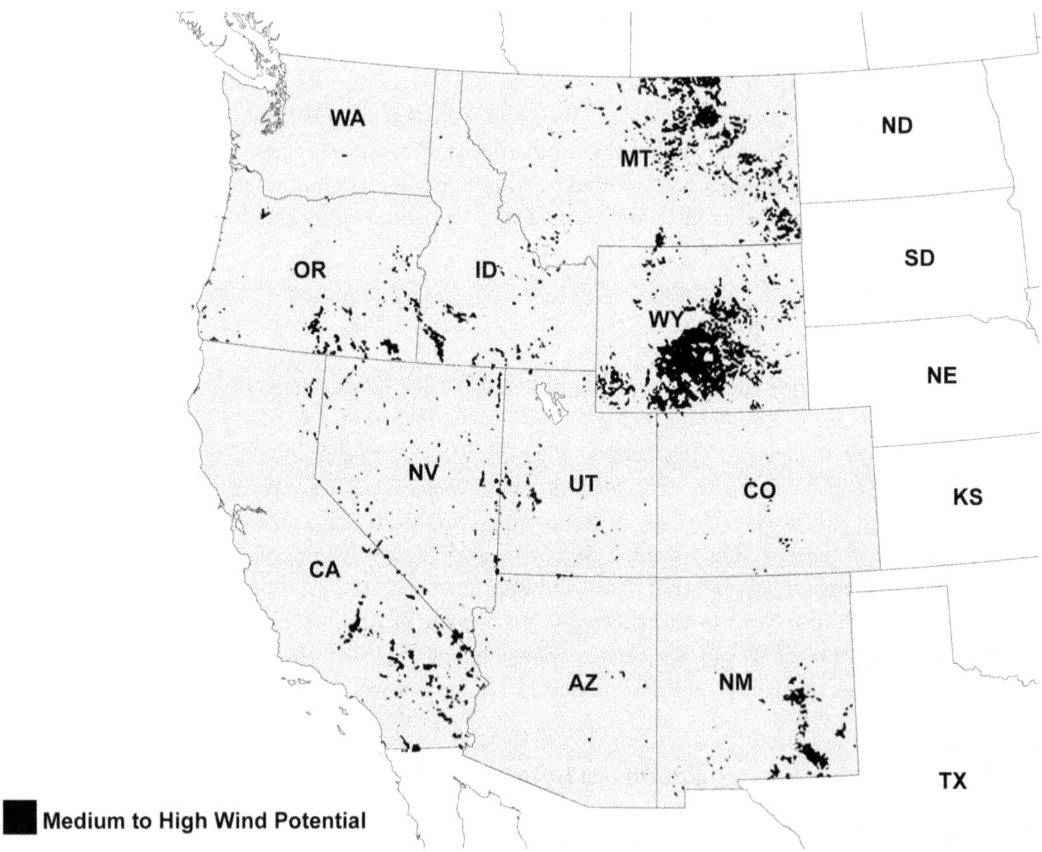

FIGURE 3.1 BLM-administered lands with medium or high potential for wind development.
(Redrawn by Mark Warfel Jr. and adapted from USDI BLM, 2005, pp. 2–4.)

FIGURE 3.2 BLM Final Programatic Environmental Impact Statement on Wind Energy Development. (Source: USDI BLM, 2005, cover.)

BLM issued a number of solar development policy memorandums post 2007. In 2010 BLM issued a *Draft Programmatic Environmental Impact Statement* (USDI BLM, 2010) to establish a new program and determine which lands are open for development of utility-scale solar energy projects as well as affect land use plans in six westerns states. The solar PEIS evaluated program development and impacts for areas in California, Nevada, Utah, Arizona, Colorado and New Mexico (Fig. 3.3). BLM identified 24 *solar development zones* within the six-state area with low-level environmental conflict. Within these zones there still would be site-specific and species-specific issues addressed during project reviews. The *Final Programmatic Environmental Impact Statement* with record of decision was published in 2012 (USDI BLM, 2012).

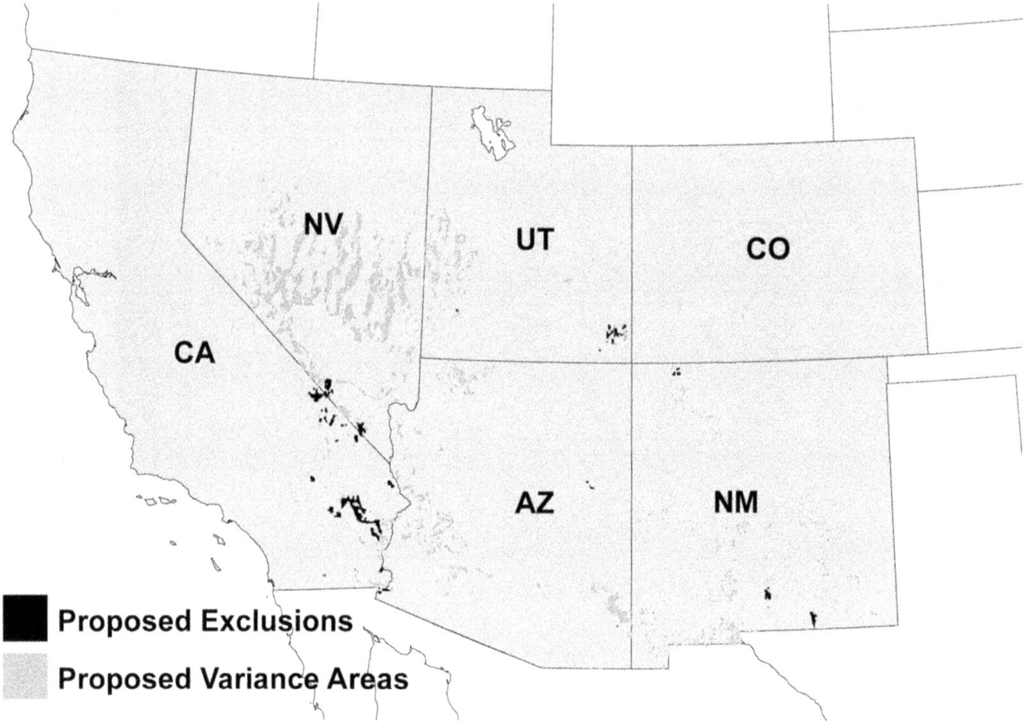

FIGURE 3.3 BLM PEIS solar exclusion zones.
(Redrawn by Mark Warfel Jr. and adapted from USDI BLM executive summary, 2012, p. ES-12.)

Other applicable laws as part of the environmental assessment process for specific renewable energy projects include the *National Environmental Policy Act* (NEPA) 42 USC sections 4321–4370, the *Endangered Species Act* (ESA) 16 USC section 1536(a)(2), the *Federal Noxious Weed Act* 7 USC section 2801 et seq., the *Migratory Bird Treaty Act* 16 USC 703–712), the *Bald and Golden Eagle Act* 16 USC sections 668–668c, the *Clean Water Act* 33 USC sections 1251–1387, and the *National Historic Preservation Act of 1966* (PL 102–575). Also see Smardon and Karp (1993), Chapter 11, pp. 182–208 for aesthetic project review under NEPA.

The other agency affected by these large-scale wind and solar projects is the National Park Service (NPS) whose parks and trails often abut BLM lands that are being targeted for development. The NPS is specifically concerned with aesthetic resource protection including: viewsheds and air quality; night skies; soundscapes; cultural/historic resources; wildlife, plants and habitat; water quality/quantity; as well as the overall visitor experience in parks and on trails. Given these large renewable energy projects, the NPS would like to use a *Smart from Start Program* including: (1) avoidance through appropriate siting; (2) minimization through use of design features; (3) offset impacts through mitigation; and (4) increased understanding through monitoring. From a legal perspective the DOI solicitor's office memorandum *Doe Run* (USDI 1993) states that the Secretary has a legal obligation to protect parks against an external development threat that is *direct, specific, and credible* and which *relates to a fundamental value or purpose to the park*. There also must be consideration of impacts to NPS resources and values through the NEPA process and the enabling legislation of park units outlines the purposes for which the park was created by Congress, and must be protected. See Smardon and Karp (1993), pp. 125–149 for a summary of sensitive land protection including wilderness and national parks.

Visual impacts of wind, solar and geothermal energy development are usually assessed utilizing the visual contrast rating system, which has been around since 1980 and has been used for linear phenomena in the landscape such as power lines. But renewable energy sources such as huge 200-foot-tall wind towers and turbines or large solar arrays create different kinds of visual impacts than what the agency is used to dealing with. So USDI BLM has engaged the Argonne National Laboratory to help develop appropriate tools for assessing visual impacts (Sullivan et al., 2012a, 2012c). The BLM Wyoming office has also commissioned Argonne National Laboratories to develop a comprehensive set of visual mitigation measures that can be used to mitigate geothermal, solar and wind commercial energy development (USDI BLM, 2013).

The other federal agencies have much less experience with commercial renewable energy development. The US Forest Service has a utility siting guide as part of its original Visual Management System (Smardon, 1986) and the National Park Service has limited guidance as covered above. The real issue as described in the 2007 NRC report is that some of these large windfarms and transmission corridors run through multiple federal and state agency jurisdictions in the mid-western to western states and there is no overarching policy for addressing aesthetic or other environmental impacts.

In terms of overall methodological guidance for assessing visual impact from renewable energy facilities, the most authoritative guidance can be found in Appendix D "A Visual Impact Assessment Process for Evaluating Wind-Energy Projects" in the 2007 NRC wind study. Basic steps include:

- Project description,
- Project visibility, appearance and landscape context,
- Scenic-resource values and sensitivity levels,
- Assessment of aesthetic impacts,
- Mitigation techniques, and
- Determination of acceptability or undue aesthetic impacts.

3.2.2 Interstate power transmission – who's on first?

In terms of power transmission we have an interesting and problematic situation. Under the *Energy Act of 2005* plus executive orders, DOE was charged with studying regional electric energy transmission congestion (NAERC, 2009), then recommend specific "national interest electric transmission corridors to develop new transmission capacity" – then share with the Federal Energy Regulatory Commission the review of such corridors as a kind of *backstop* if there is not adequate review at the state level. These corridors were accepted by DOE in 2007. But, in *California Wilderness v. US Department of Energy* in 2011 the Federal Appeals Court found against the US DOE. Key findings from the decision are:

> DOE failed to consult with the States in undertaking the Congestion Study as required by section 824p(a)(1) (of the Energy Act);
>
> DOE failed to properly consider the potential environmental consequences of its designation of national interest electric transmission corridors (NIETCs); and
>
> DOE's actual designations of the Mid-Atlantic Area National Corridor and the Southeast Area National Corridor are arbitrary, capricious, and not supported by the evidence; and
>
> We determine that DOE failed to properly consult with the affected States in conducting the Congestions Study and failed to undertake any environmental study for its NIETC Designation as required by the National Environmental Protection Act (NEPA) 42 USC ss. 4332(C).
>
> (*California Wilderness v. US Department of Energy* 2011)

So to date, the regional corridors are in a kind of 'limbo land' (Swanson & Jolivert, 2012; USDI NPS, 2010; Vann & DeBough, 2011) from a legal perspective and DOE still wants partial jurisdiction with FERC for review of regional power line interconnects. Even though DOE and FERC share responsibility for the review of interstate power corridors, only FERC (undated) has guidelines for considering aesthetic and environmental impacts and no specific methodology for doing such. So if states do not have specific guidelines under their respective utility or public service commissions for environmental and aesthetic review, this could pose major jurisdictional and substantive review problems.

3.2.3 Who's minding the shore?

In the Cape Wind case in Massachusetts there were major shifts of responsibility for who was the lead agency in permitting this large-scale windfarm. It moved from Mineral Management Service to the US Corps of Engineers to finally, the BOEM. The *Energy Policy Act of 2005* gave the USDI's Mineral Management Service (MMS) the authority to issue renewable energy easements, leases and rights-of-way on the Outer Continental Shelf (OCS). In addition the FERC, the National Oceanic and Atmospheric Administration's (NOAA) National Marine Fisheries Service, and the US Army Corps of Engineers are engaged in the permitting system. They are authorized to do such under the *Federal Power Act*, the *Endangered Species Act*, the *National Environmental Policy Act*, and the *Coastal Zone Management Act* respectively. Coastal state authority is also involved in matters over submerged waters up to 3 nautical miles from shore in regard to transmission lines and Coastal Management Zone consistency (Showalter & Bowling, 2011). In 2010 the MMS was replaced with the Bureau of Ocean Energy Management, Regulations and Enforcement (BOEMRE). In October 2011 BOEMRE further split into the BOEM and the Bureau of Safety and Environmental Enforcement (BSEE). The US Army Corps of Engineers retains its jurisdiction over permitting obstructions in navigable waterways while the BOEM is the lead agency with respect to issuing leases for project siting (Showalter & Bowling, 2011).

In 2007 the MMS completed a *Programmatic Environmental Impact Statement* on the effects of renewable energy projects on marine resources. In April 2009 the US Department of Interior published final rules on granting leases, easements and rights of way for renewable energy projects on the outer continental shelf, which is referred to as the *Framework*. The Framework has requirements for site assessment, construction and operation, payments and decommissioning. Applications for commercial leases require a Site Assessment Plan (SAP) and Construction and Operations Plan (COP). The Cape Wind project, for instance has completed all aspects except the COP. Specific guidelines for both SAP and COP have been published by the Bureau of Ocean Energy Management (2010). Leasing programs for renewable wind energy development have been executed in Massachusetts, Delaware and New Jersey. Three interim leases have been issued to commercial companies off New Jersey. States such as California, Oregon, Washington, Georgia, Hawaii, Florida, Maine, Maryland, New York, North Carolina, Rhode Island, South Carolina and Virginia have developed task forces or studies exploring possible renewable energy development (see Fig. 3.4) but no formal lease request have been made.

As we said above, Cape Wind is the only commercial scale renewable wind energy project to go through the entire process except the *Construction and Operations Plan*. In November of 2010 the Massachusetts Department of Public Utilities approved the 15-year purchase agreement with National Grid to buy Cape Wind's energy capacity and renewable energy credits. Cape Wind generated major controversy on Cape Cod among residents and different organizations with one of the major issues being visibility and aesthetic impact (Williams & Whitcomb, 2007). As a result of this controversy of just how to address scenic impact from siting such projects, the Cape Cod Commission was charged with developing visual siting guidance for renewable energy projects sited less than 3 nautical miles offshore. This

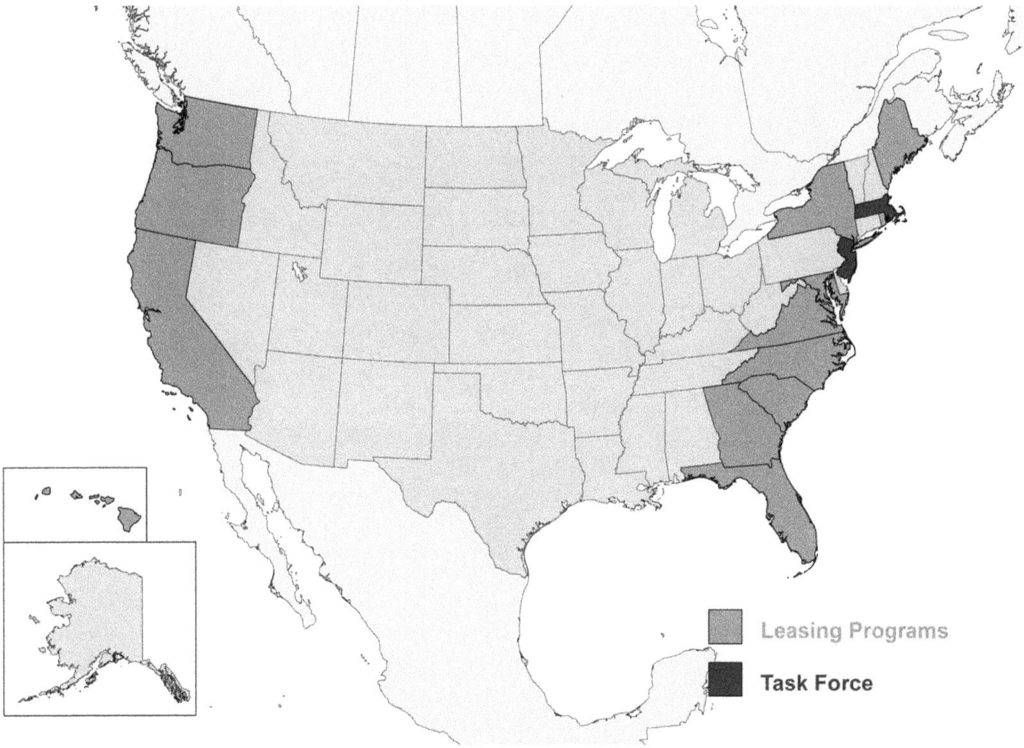

FIGURE 3.4 Status of BOEM offshore leasing activity.
(Redrawn by Mark Warfel Jr. and adapted from Showalter and Bowling, 2011, p. 5.)

is the first detailed visual resource guidance (Rooney et al., 2012) for renewable energy projects in the United States and drew from much work that had already been done for siting and analyzing offshore energy projects in Europe. The USDI BLM has also commissioned work by the Argonne National Laboratory (Sullivan et al., 2012b) to look at how visual resource evaluation and visibility assessment have been done in the UK.

So in summary we see a morass of jurisdictional issues with shore to offshore renewable energy development, with some siting guidance issues still not resolved. We see that European guidance for wind and biomass energy projects is much more developed than in North America.

3.3 State and local review of renewable energy projects – the crazy quilt

As we stated in the introduction, there is tremendous variation in state and local review of renewable energy projects. The Environmental Law Institute (2011), the National Wind Coordinating Committee (2006), and *Wind Power Siting Regulations and Wildlife Guidelines in the United States* produced by the US Department of Interior, Fish and Wildlife Service (US FWS 2007) have reviewed states that have renewable energy portfolios or standards and utility siting guidance (Fig. 3.5). A state and local comprehensive data source is the DIRE web site.[1] Also the American Planning Association (Rynne et al., 2011) has reviewed siting guidance in the US. As we will see, local governments exercise some authority over commercial scale wind facility siting in 48 of 50 states. In 34 states local government

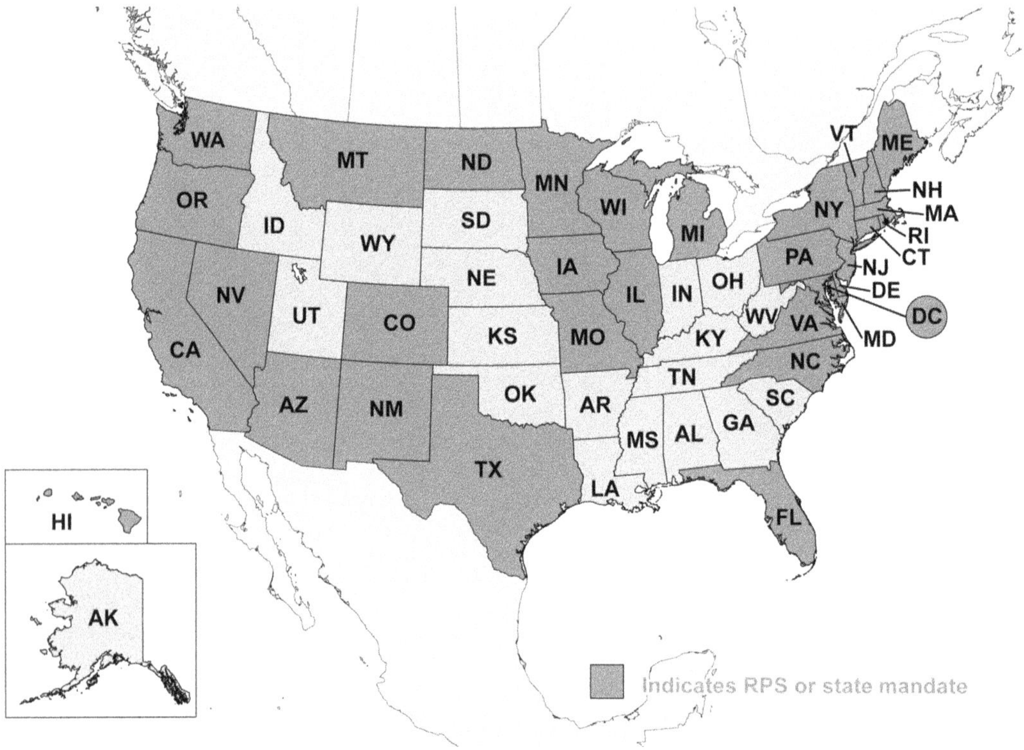

FIGURE 3.5 US states with renewable portfolio standards or other renewable energy mandates.
(Redrawn by Mark Warfel Jr. and adapted from Rabe, 2006, p. 4.)

have substantial autonomy. Eleven states set size thresholds for state regulatory involvement – usually with local governments regulating smaller facilities and state authorities larger ones. For one third of the states, siting of commercial scale wind facilities requires approval of both state and local government. Only a few states reserve all review rights at the state level. The following is a brief synopsis of some states in terms of their incentives for renewable energy development as well as siting and other provisions. Table 3.1 below includes an inventory of each states' renewable energy polices, incentives, regulatory authority and siting guidance.

Arizona has a Renewable Portfolio Standard (RPS) of 15% by 2025 plus green certification and tax incentives. The state also has wind energy development guidelines to recommend ways to minimize wildlife and construction impacts (AG&FD 2008).

Maricopa County of Arizona has a specific ordinance for siting renewable energy systems that are defined as any system other than fossil fuels or nuclear fission.

California has an RPS of 20% by 2017 plus a state-regulated Renewable Energy Certificate, which is a tradable instrument; net metering and tax credit enterprise zones. Siting is fairly complex and has to clear the *California State Environmental Quality Act* for environmental assessment, plus there are setback requirements for many counties. The California Energy Commission and the Department of Fish and Game have specific energy siting guidelines.[2]

Colorado has a Renewable Energy Standard set at 12% renewable by 2011 and 30% by 2012 and net metering offsets but not direct payments from utilities. The Colorado Public Utilities Commission (PUC) regulates renewable energy development for larger than 2 MW or structures over 50 feet in

TABLE 3.1 US state renewable energy policy guidance, incentives, authority and siting guidance

State	*RPS*	*Incentives*	*Authority*	*Siting Guidance*
Alabama	No	Few incentives	Local	None
Alaska	No	Tax, grants & rebates	Local	None
Arizona	15%	Green certification & tax incentives	Local	State guidelines
Arkansas	No	None	Local	None
California	20%	Renewable Energy Certificate	County State over 50 MW	Energy siting guidelines
Colorado	30%	Credit multipliers	Public Utilities Com. for net metered	State guidelines
Connecticut	10%	Grants & planning assistance	State Siting Council over 1 MW	Specific guidelines Certificates of Need
Delaware	10%	Grants & rebates	Local	None
Florida	No	Grants, rebates & tax incentives	Local	None
Georgia	No	Green certification & tax incentives	Local	None
Hawaii	20%	Tax incentives & industry recruitment	Local except special districts	None
Idaho	No	Tax incentives & loan program	Local	None
Illinois	25%	Tax abatement zones & property tax valuation	Local	County-level ordinances
Indiana	No	Financial incentives	Local	Biomass BMPs
Iowa	105 MW	Monetary incentives & wind tax credits	Dual permitting over 25 MW	Wind energy checklist
Kansas	20%	Property tax exemptions	Local	*Wind Siting Handbook*
Kentucky	No	Tax credit, loans, rebates & grants	Dual permitting over 10 MW	None
Louisiana	No	Tax incentives & loan programs	Local	*Coastal Zone Management Act*
Maine	10%	Grants & incentives	Local + dual > 20acres Land Use Reg. Com.	Wind project guidance
Maryland	7.5%	Multiple tax credits & incentives	Local + dual over 70 MW	Renewable guidance
Massachusetts	4% 1%/yr.	Rebates, tax incentives, loans, recruitment & production incentives	Local + state over 100 MW	Municipal model zoning + technical guidance
Michigan	10%	Grants, loans, incentives & recruitment	Local	Wind guidelines
Minnesota	17%	Loans, rebates, tax incentives & grants	Local + state over 25 MW	Siting standards bioenergy BMPs
Mississippi	No	Loan program	No	None
Missouri	15%	Noncompliance rules & incentives	Local	None
Montana	15%	Tax incentives & loan programs	Local	DEQ web site

(*Continued*)

TABLE 3.1 Continued

State	*RPS*	*Incentives*	*Authority*	*Siting Guidance*
Nebraska	No	Production tax credit	Local utility district	Wind energy guidelines
Nevada	20%	Rebates & tax incentives	Local	None
New Hampshire	No	Tax incentives & loan programs	Local + state over 30 MW	None
New Jersey	22.5%	Rebates, loans, tax incentives & recruitment	Local + state if in coastal zone	None
New Mexico	20%	Production credits	Local + dual over 300 MW	Wind guidelines
New York	24%	Property tax relief	Local + state over 25 MW	Wind guidelines
North Carolina	No	Tax incentives & loan programs	Local	Voluntary BMPs
North Dakota	Yes	Property tax credit wind lease provisions	Local + state over 60 MW	None
Ohio	25%	Tax exemptions	Local + state over 5 MW	Wind siting guidance
Oklahoma	15%	Income tax credits, green certification & industry recruitment	Local	None
Oregon	25%	Tax credits	Local + state over 35 MW	Energy siting standards
Pennsylvania	18%	Production tax credits	Local except net metered	Model ordinance
Rhode Island	15%	Rebates & tax incentives	Local + state over 1 MW preemption	Siting guidelines
South Carolina	No	Tax rebate & planning assistance	Local	Interconnection guidelines
South Dakota	10%	Tax rebate (voluntary)	Local + dual over 100 MW	Wind guidelines setback required
Tennessee	No	Grants, loans & tax incentives	Local	None
Texas	5,880 MW	Property tax, grants & recruitment	Local	Exception guidelines
Utah	No	Tax incentives	Local	None
Vermont	20%	Multiple tax credits, finance & rebates	State siting	Wind guidelines
Virginia	No	Tax incentives & recruitment	Local + dual 5–100 MW State over 100 MW	Solar ordinance
Washington	15%	Grants, tax & production incentives	Local or state developer choice	Wind guidelines
West Virginia	No	Tax incentives	Dual for commercial	None
Wisconsin	10%	State rebates	Local + state over 100 MW	Wind siting guidelines
Wyoming	No	Sales tax exemption for equipment	Local + dual over 30 turbines	Wind guidelines

height. The PUC rule requires specific consultation with the Colorado Division of Wildlife and the US Fish and Wildlife Service, plus specific surveys of potential impacts. For siting purposes landowners can create covenants to prohibit or restrict renewable energy devices and wind developers have to consult with Colorado agencies in order to obtain permits.[3]

Connecticut has an RPS of 10%, plus grant programs and planning assistance. The Connecticut Siting Council regulates renewable sources greater than 1 MW and the state addresses all renewables, not only wind. The Council has specific guidelines for filing Certificates of Environmental Compatibility and Public Need Certificates. Local municipal zoning and inland wetland agencies may regulate and restrict location of electric generating facilities.

Illinois has an RPS which requires 25% of energy from renewable sources by 2025, with 75% of that coming from wind. In December 2010 the Illinois State Power Commission gave approval to two utilities for long-term power purchase agreements for wind and solar power. There are also tax abatement enterprise zones in Illinois for wind projects, plus special property tax valuation. Illinois allows local regulation of renewable energy projects and many counties have adopted ordinances for such a purpose.

Indiana does not have a renewable energy standard but has 1,209 MW of wind power. There are some financial incentives and a utility must purchase alternative energy from a facility with less than 80 MW of generating capacity. Siting wind projects occurs at the local level with interaction with state agencies with state laws related to electric generation/transmission plus environmental requirements. Indiana has a guide for best management practices in harvesting biomass.

Iowa ranks second in the US with 3,675 MW of wind energy with an RPS of 105 MW. Iowa has a state-based monetary incentive program at $0.015 per kWh sold during the first 10 years of production and set a goal of energy independence by 2015. The state also has two production tax credits programs for eligible wind energy facilities. Iowa Utilities Board provides a certificate of public convenience, use and necessity for power plants over 25 MW, but projects are sometimes exempt. The Department of Natural Resources regulates all wind projects and has mapped areas of concern plus specific siting guidelines for energy developers to follow.

Kansas passed a *Renewable Energy Standards Act* in 2009, which requires 20% of renewable energy by 2020 including purchasing sub-targets and penalties for not making those targets. There are also property tax exemptions for renewable energy. There is a siting handbook with general guidance and most counties have permitting requirements. Windfarm development in the Flint Hills Tallgrass Prairie Region is discouraged and the Wabaunsee County ordinance prohibiting such development has been upheld by the State Supreme Court. There are web sites for the *Wind Siting Handbook* plus other guidance.[4]

Maine has an RPS of 10% by 2017 plus grant and incentives programs. Maine's Department of Environmental Protection regulates large structures over 20 acres. The Department of Inland Fisheries and Wildlife, the Department of Environmental Protection, and the Land Use Regulatory Commission (LURC) conduct environmental reviews of projects. The LURC just commissioned a study of cumulative impact of renewable energy development on unconsolidated territories in Northern Maine, which included aesthetic impacts. Specific guidance for wind projects can be accessed on the web.[5]

Maryland has an RPS of 7.5% from wind, solar and biomass by 2019. Maryland's Public Service Commission issues Certificates of Public Convenience and Necessity for projects over 70 MW and there are siting guidelines to mitigate avian and bat risks from wind power projects. Detailed guidance is available.[6]

Massachusetts has an RPS of 4% by 2009 plus a 1% increase each following year. The state also has rebates, tax incentives, grant and loan programs, industry recruitment and production incentives.

The Massachusetts Energy Facilities Siting Board is an independent state review board and has an *Energy Facilities Siting Handbook*. Agency requirements are not wind-specific and apply to generating plants of 100 kV or greater. Large projects also have to comply with the *Massachusetts Environmental Policy Act* as well as other agency review. Model zoning for municipalities was developed by the Division of Energy Resources and the Executive Office of Environmental Affairs, which is available on the web.[7]

Massachusetts also has a model *as-of-right zoning ordinance* or bylaw allowing use of large-scale solar energy facilities. Massachusetts's Cape Cod Commission has just issued technical guidance on siting renewable offshore energy facilities less than 3 nautical miles. This guide by Rooney et al. (2012) is also available on the web.[8]

Michigan has a proposed RPS of 10% by 2015 and 20% by 2025. Michigan has siting guidelines for wind systems over 3 MW plus local zoning. Guidelines are available on the web.[9]

Minnesota has two RPS requirements: a higher standard for the nuclear generating station and one for the other utilities by the year 2025 (12% by 2012, 17% by 2016, 20% by 2020, and 25% by 2025). Minnesota wind generators are required to pay a production tax of 0.012 cents to 0.12 cents per kWh in lieu of property taxes. The Minnesota Public Utility Commission preempts all local regulations and zoning for wind developments of 5 MW or more. Siting standards[10] include property line setbacks, internal spacing rules, noise standards, and setbacks from residences. Minnesota also has best management practices for bioenergy production and harvest.

Montana has an RPS of 15% by 2015 plus tax incentives and loan programs. Energy development is regulated by the Department of Environmental Quality if wetlands or water quality is impacted. Transmission lines greater than 69 kV need a certificate of Environmental Compatibility. The DEQ has detailed information on its web site.[11]

Nebraska has no mandate for RPS by law but is encouraging renewable energy sources. Electric facilities are all publically owned in Nebraska, so such facilities must be approved by the local utility district. Rather than pay property taxes a facility owner pays a 'nameplate capacity tax' equal to the total capacity of the facility. Nebraska also has special leasing arrangements for agricultural lands and renewable energy production tax credit for zero emission facilities. A Nebraska statute allows counties/municipalities with zoning to encourage solar and wind energy development within applicable local regulations.

New Jersey has an RPS of 22.5% plus rebates, tax incentives, loan programs, industry recruitment, and production incentives. Renewable projects would have to go through the Department of Environmental Protection and local planning and zoning commissions. The state is looking into offshore energy generation so state coastal management rules would apply. The state also has a special solar energy development program with special provision for protection of farmlands and a model local ordinance for solar facility siting. A New Jersey law was enacted in 2009 defining electric generating facilities using solar energy technologies, photovoltaic, and wind as permitted uses in industrial-zoned parcels.

New Mexico requires public utilities to generate 20% of their energy from renewable sources by 2020. Owners of such facilities can receive production credits of $0.01 per kWh for the first 400,000 MWh generated per year for 10 years. The New Mexico Public Regulations Commission requires state level certificates for 300 MW or larger facilities and there may be local permitting of smaller facilities. The web site has the guidelines.[12]

New York imposed an RPS in 2005 to increase renewables from the current 19% to 24% by 2015. The program is incentive based with relief from property taxes but local jurisdictions can choose not to provide such relief. Any facility of 80 MW or more is required to obtain a certificate from the New

York State (NYS) Public Service Commission and must include an environmental impact statement plus consistency with the state energy plan. The NYS Power Authority has examined offshore wind development in the Great Lakes but this proposal has been received with local opposition from shoreline communities and little economic interest for development.

North Dakota has a '25 by 25' initiative with a goal to produce 25% of all energy used in agricultural production from renewable sources by 2012 plus a general RPS objective by 2015. This is a voluntary program and there are 20-year property tax credit incentives and special provisions for wind leases. The North Dakota Public Service Commission regulates projects over 100 MW. A North Dakota law requires developers to obtain a permit from the state and have a transmission interconnection request in process within five years of entering into an easement or option agreement. This is designed to discourage developers from locking landowners into long-term commitments from speculative developments. Developers need to obtain a general permit from the public service commission and also comply with local zoning regulations.

Ohio requires utilities to provide 25% of their electricity from alternative energy by 2025 with half of this coming from renewable energy sources. Such development may be eligible for tax exemptions and an 'energy conversion facility' under the Ohio tax code. Ohio also has grant and loan programs plus industry recruitment. Wind facilities producing 50 MW or more are required to obtain a certificate from the Ohio Public Utility Commission and this satisfies all state and local agency requirements. There are two Ohio guidance documents: *Siting New Energy Infrastructure in Ohio: A Guidance Document* (available at the Ohio Power Siting Board web site[13]) and *Guidance for Siting* (Ohio SB 2005) and *Operation of Wind Generating Facilities in Ohio.*

Oregon utilities are required to achieve 5% renewable portfolio by 2011 and 25% by 2025 whereas smaller utilities only have to achieve a portfolio of 5% or 10% by 2025. There is a tax credit for facilities up to 50% of project cost with a maximum credit of $3.5 million for 2010 projects, $2.5 million for 2011 projects, and $1.5 million for projects after 2012. The Energy Facility Siting Council approves projects of 105 MW or greater. Oregon Public Utility Commission has formulated statewide siting requirements including use of existing roads, transmission facility consolidation, minimizing impact to raptors and bats, as well as visual and lighting impacts. There are still local zoning considerations but local jurisdictions are encouraged to support use of solar and wind systems. The Oregon legislature passed new laws in 2010 to incentivize offshore wind energy projects. The energy siting standards are at the Oregon Department of Energy web site.[14]

Pennsylvania's *Alternative Energy Portfolio Standards Act* requires utilities to meet 18% of electric need from renewable sources by 2020 and a portion of this is mandated to be solar. Renewable facilities may qualify for a production tax credit equal to 15% of development and construction costs up to $1 million. Siting such facilities is primarily controlled by local government, but there is a model ordinance that has been adopted by many Pennsylvania communities. This can be accessed at the Pennsylvania web site.[15]

South Dakota's RPS goal is to produce 10% of retail energy from renewable sources by 2015 and this is strictly voluntary. Instead of paying property taxes the facility must pay an annual tax equal to three dollars multiplied by the nameplate capacity of the windfarm. There is also a tax rebate equal to 50% of the cost of the transmission lines or collector systems. The South Dakota Utility Commission permits energy facilities over 100 MW. South Dakota imposes a statewide setback requirement for windfarms, e.g. turbines taller than 75 feet have a setback of 1.1 times their height or 500 feet. There is also provision for easements to ensure adequate exposure to wind. There is a web site for South Dakota's guidelines.[16]

Texas is the nation's leader in wind energy capacity at 10,085 MW. The Texas RPS requires energy generators to gradually increase the proportion of renewable energy to 2000 MW by 2009, 5,880 MW by 2015 and 10,000 MW by 2025. There is property tax abatement for wind energy developers as well as grant programs, planning assistance and industry recruitment. Texas passed a bill in 2005 creating *Competitive Renewal Energy Zones* in order to better coordinate the build out of renewable energy projects and transmission of such. A study was done and 25 such zones were identified. It is then up to the Texas Public Utility Commission to coordinate implementation within these zones. There have been some conflicts with local jurisdictions as well as financial commitment issues, which were clarified in 2010.

Vermont Public Service board has a requirement for petitions to construct electric and gas facilities and they have some specific requirements for wind facilities including cannot have undue adverse effects on aesthetics, historic sites, air and water purity, the natural environment, and public health and safety. Also aesthetics must be addressed for a 10-mile radius from the proposed project site. The Vermont Agency of Natural Resources has guidelines for the *Review and Evaluation of Potential Natural Resources Impacts from Utility-Scale Wind Facilities in Vermont* (2006). These guidelines are available at the Vermont Agency of Natural Resources web site.[17]

Virginia has no RPS but has tax incentives and industry recruitment. Virginia mandates that utilities must obtain a Certificate of Public Convenience and Necessity from the State Corporation Commission before constructing an electric generating facility. The Commission must consider the facility's potential environmental impact plus any mitigation needed and does not preempt local review authority. Virginia has a model ordinance for larger scale solar projects.[18]

Washington has an RPS of 15% by 2020 and a number of incentive programs including grant programs, planning assistance, tax and production incentives, industry recruitment and green certification (not a state agency program). The Washington Energy Facility Site Evaluation Council reviews all energy facilities greater than 350 MW. This includes a certification process and preliminary site study process. In addition some local jurisdictions have specific zoning provisions for windfarms. There is a web site for model guidelines.[19]

Wisconsin has the most detailed limitations on local decisions regarding wind siting. In 2009 the state Public Service Commission (PSC) promulgated rules for siting of wind facilities that are under local jurisdiction with capacities of less than 100 MW. The Public Service Commission provides a Certificate of Public Convenience and Necessity for projects larger than 100 MW. The local jurisdiction needs to enact a local ordinance. It cannot be more restrictive than the state law and local decisions can be appealed to the PSC. The state law and local ordinances include provision for setbacks and other environmental provisions. There are guidelines entitled *Considering Natural Resources Issues in Wind Farm Siting in Wisconsin,* which are available at the Wisconsin state web site.[20] Wisconsin also has *Sustainable Harvesting and Planting Guidelines for Non-Forest Biomass.*[21]

Wyoming has attracted wind development without an RPS or other incentives. The only incentive is a sales tax exemption for equipment purchases. There was a recent excise tax on wind-generated electricity. All industrial facilities that cost $96.9 million or more need a certificate from the Wyoming Public Service Commission and a siting permit from the Wyoming Department of Environmental Quality. There are new siting and permitting requirements as of 2010 which include setback and decommissioning. Six counties have specific zoning rules for wind projects.

3.4 What about utility-scale solar development in North America?

According to Paddock et al. (2009) there are several legal developments that relate to large-scale solar development in state and local jurisdictions. For utility-scale projects the major barriers involve zoning restrictions that may prevent siting projects where there is significant open space and more specifically prime agricultural land. For small-scale solar projects the major barriers are restrictive covenants or local zoning ordinances that prohibit or restrict installation or use of solar equipment. In the early 1980s states such as Arizona, California, Florida, Hawaii, New Mexico, North Carolina, Oregon and Wisconsin enacted laws that prohibited restrictive covenants that banned solar equipment. A number of states also prohibited local government from using local authorities to restrict solar equipment, including California, Indiana, New Mexico, Wisconsin and Wyoming. A few states enacted 'right to sunlight' legislation such as New Mexico and Wyoming. Most states have legislation protecting solar access or providing for easements. Very few states have laws relating to large-scale solar energy development. We found that Massachusetts, New Jersey and Virginia had such provisions.

The major issue with large-scale solar facilities is conflict with preservation of prime farmland and possibly visual impact in public protected areas such as BLM or National Park Areas. Canada, Germany, and Greece have enacted special measures to protect prime agricultural areas in the face of utility-scale solar development (Paddock et al., 2009). In the US this is just beginning to be an issue. In Michigan in 2009 legislation was proposed that would allow any renewable energy device on farmland that received a tax break under Michigan's Farmland Protection Program as long as farming continued. Otherwise PV arrays are not permitted on Michigan's agricultural preserves.

In New Jersey legislation was proposed that would designate solar panels as a permitted use on 20% of the land of agricultural areas. In California several large-scale solar projects have been proposed on agricultural land, but such development is limited by the *California Land Conservation Act of 1965*. The act authorizes cities and counties to establish agricultural preserves within which landowners may elect into contracts with minimum 10-year rolling terms that restrict land use to agricultural "compatible" uses in exchange for preferential property tax treatment and this affects 17% of California's agricultural lands (Paddock et al., 2009).

3.5 US legal issues with State and local renewable energy siting

Major sources for legal issues with renewable energy siting include Outka (2010), Salkin (2010), Salkin and Ostrow (2011), Helius (2011), and Stoel Rives (2010). The major aesthetic impacts that may or may not be dealt with in local ordinances are the impact on scenic views, flicker affect, and noise. Local ordinances may impose height restrictions, setback requirements, requirements for a visual impact statement and noise limit standards. Examples given by Salkin are local visual standards as part of a VIA for the towns of Bethany, Cohocton, and Ellington:

For the town of Ellington, NY there are special provisions for:

- Digital visibility map and computer rendered simulations,
- Non-reflective matte color or camouflage,
- Lighting to be limited to that required by the FAA,
- Transmission lines to be placed underground, and
- Ground level lighting to minimize lighting pollution.

For the town of Cohocton, NY the VIA must:

- Address impacts within a 5-mile radius,
- Submit scenic resource maps, viewshed maps and photographic simulations, and
- Suggest visual mitigation strategies.

For the town of Bethany, NY:

- They require blades to be painted in neutral, non-reflective colors,
- All wiring is to be underground or on existing wires, and
- That the shadow-flicker affect be no more than 30 hours per year and 30 minutes per day.

There are a few specific legal cases addressing renewable energy development and aesthetic impacts. Salkin again provides key information about specific court cases and has a blog for such cases (see the web site[22]).

In *UPC Vt. Wind LCC, 2009 WL279971, 2009 VT* "the [Vermont Service] board adequately considered aesthetic impact . . . the regional plan did not include aesthetic standards . . . the facility would not be shocking or offensive to the average person . . . views of it would be from a distance and intermittent" (Salkin, 2010, p. 38).

In *Rankin v. FPL Energy LLC 2008 WL3864829* (Tex App. 8/21/2008),

> the plaintiffs cannot assert a nuisance claim based on the windfarm's aesthetic impact . . . "successful nuisance actions involve an invasion of plaintiff's property by light, sound, odor or foreign substance" . . . "notions of beauty or unsightliness are necessarily subjective in nature and that giving someone an aesthetic veto over a neighbor's use of his land would be a recipe for legal chaos."
>
> (Salkin, 2010, p. 41)

These two court cases are an interesting contrast as in the Vermont case there are existing policy standards under *Vermont Act 250* when considering public related aesthetic impacts. In the Texas case there are no such standards for aesthetic impact and the plaintiff used a nuisance standard, which is individual-to-individual impact versus a public impact (see Smardon & Karp, 1993).

For many states the issue of consideration of aesthetics as part of local ordinances and regulation comes down to court case precedents that relate to defending local ordinances and zoning incorporating aesthetics. There is a state-by-state summary of such cases in Chapter 2 of Smardon and Karp (1993).[23]

3.6 Canadian laws, ordinances, regulations, and standards

Under Canadian law, an independent power producer (IPP) that proposes a project must obtain approvals from provincial and federal resource ministries. The legislative framework that governs renewable energy projects is the same for all industrial projects. Local zoning approval of a project was required until the spring of 2006. Passage of Bill 30, Miscellaneous Statutes Amendment Act, in May 2006 amended Section 121 of the *Utilities Commission Act* so that local government land use decisions (e.g. local zoning) cannot prevent public utilities from constructing a run-of-river facility. The amendment set out the following criteria to determine whether a clean energy project should be considered a *facility*: (1) it must be entirely located on provincial Crown land; (2) it has an electricity purchase agreement with BC Hydro, Powerex, or FortisBC; and (3) it has obtained necessary federal and provincial authorizations. Local governments retain the ability to participate in project reviews and provide information and feedback during the project permitting process. The Ministry of Energy and Mines has produced a guide entitled *Opportunities for Local Government and the Public Participation in Provincial Regulatory Processes for Independent Power Producers' Projects* (British Columbia, 2007).

3.6.1 Canadian environmental laws, ordinances, regulations, and statutes regulating renewable energy projects

The majority of renewable energy in Canada is from hydroelectric sources, but there has been development of wind power and tidal sources as well as solar energy, most of which is occurring in Ontario. Wind power is growing quickly – a 20 MW tidal plant sits in the Bay of Fundy at Annapolis, Nova Scotia and the first commercial solar project was built in Stone Mills, Ontario in 2009. There has been a national study on marine power development for Canada. Most of the renewable energy guidance for Canada is at the provincial level. See Table 3.2 below for an overview of Canadian renewable policy and guidance.

3.6.2 Canadian provincial guidance for renewable energy development

The **British Columbia** *Clean Energy Act of 2010* establishes a long-term vision for British Columbia (BC) to become a clean-energy powerhouse. It sets out 16 specific energy objectives that focus on three areas: ensuring electricity self-sufficiency at low rates, harnessing BC's clean power potential to create jobs, and strengthening environmental stewardship and reducing greenhouse gases; and is administered by the BC Ministry of Energy, Mines and Petroleum Resources. British Columbia's 2007 Energy Plan (www.energyplan.gov.bc.ca) is the provincial government's strategic plan for addressing energy issues. A key goal of the BC Energy Plan is to make BC energy self-sufficient by 2016. The province's Energy Plan has aggressive targets for zero net greenhouse gas emissions from electricity generation (zero emissions from coal-fired generation), new investments in innovation, and ambitious conservation goals aimed at reducing growth in electricity demand. Keeping electricity affordable for citizens and businesses is another goal of the Energy Plan.

While the Energy Plan sets a target for BC Hydro to meet half of its incremental resource needs through energy conservation, new supply will still be needed to ensure adequate electricity is available

TABLE 3.2 Canadian provincial renewable energy policy, incentives, authority and guidance

Province	*RPS*	*Incentives*	*Authority*	*Guidance*
British Columbia	No	See text below	Local	Clean Energy Production
Alberta	No	None	Utilities Comission	None
Saskatchewan	200 MW by 2013	Go Green Fund	Local	None
Manitoba	1000 MW	None	Dept. Science, Tech., Energy & Mines	None
New Brunswick	10%	None	Local	None
Newfoundland & Labrador	No	None	Local	None
Nova Scotia	25%	None	Local	None
Ontario	No	Small scale incentives	Ministry of Env., Ministry of Natural Resources	Municipal renewable guide
Prince Edward	15%	None	Local	None
Quebec	No	None	Ministry of Sustainable Development, Environment & Parks	

in the future. To accomplish this goal, the province is encouraging private sector businesses, First Nations, and other organizations to develop new supplies of electricity – and in particular, from clean and renewable sources. Clean energy producers are being invited to develop projects that generate electricity using water, wind, biomass, tidal and ocean geothermal, solar, and natural gas, with offsets. California has looked at developing or supporting run-of-river hydro in BC under the California RPS. BC has a *Clean Energy Production Guidebook*, outlining all the procedures for clean energy projects.[24]

Alberta has a market driven electricity supply industry, is deregulated, and has open access to transmission and distribution. There is no regulatory framework for development of renewable energy resources. There is potential for wind development in Southern Alberta. Construction and operation of renewable energy facilities requires approval of the Alberta Utilities Commission. Alberta has grid congestion issues, which restrict sale of wind power on their grid during many hours of the year.

Saskatchewan has set up a *Go Green Fund* for projects, which reduce greenhouse gas emissions. The intention is to use such funds to develop wind power and SaskPower is expanding wind power by an additional 200 MW by 2013.

Manitoba Hydro is a vertically integrated utility that supplies most of the Province's power needs and 95% is from hydropower. In 2005 the government released *Green and Growing* as a strategic framework, which includes developing 1,000 MW of wind power over the next decade. The Department of Science, Technology, Energy and Mines have a mandate to look at emerging energy technologies such as solar, geothermal, and wind.

Ontario passed the *Green Energy and Green Economy Act* (GEGEA) in 2009 in order to develop renewable energy as well as provide green jobs. The GEGEA takes a two-pronged approach to creating renewable energy commercialization. The first is to bring more renewable energy sources to the province and the second is the creation of more energy efficiency measures to help conserve energy. The bill led to the appointment of a Renewable Energy Facilitator to provide *one-window* assistance and support to project developers in order to facilitate project approvals.

The approvals process for transmission projects is streamlined and, for the first time in Ontario, the bill will cause development of standards for renewable energy projects. Homeowners have access to incentives to develop small-scale renewables such as low- or no-interest loans to finance the capital cost of renewable energy generating facilities like solar panels. The GEGEA is linked to the *Green Energy Act of 2009* (GEA), which in turn amended some 16 other provincial Acts. A major feature of the GEA in Ontario is a *Feed in Tariff Program* (FIT). This program has resulted in 1,800 small and large FIT contracts and about 25,000 micro FIT Programs, which total to some 4,800 MW of electricity generation capacity.

The Ontario Ministry of Environment (MOE) issues a single Renewable Energy Approval (REA) for all wind power faculties over 3 kW, with requirements established under the *Environmental Protection Act of 1999*. The Ministry of Natural Resources (MNR) may issue multiple approval and permits for such facilities on a site-by-site basis. Such MNR requirements are in the *Approval and Permitting Requirements Document for Renewable Energy Projects*.[25] Of specific interest are requirements for noise levels, setbacks, and mandatory public consultation with landowner and Aboriginal communities. Ontario has also published a guide for Municipalities in 2011, *Renewable Energy Development: A Guide for Municipalities*.[26]

In **Quebec,** Hydro-Quebec is a vertically integrated utility, which produces 92% of Quebec's electricity needs through hydropower. There is little incentive for other forms of renewable development. Every project over 10 MW is required to conduct and submit an environmental impact assessment to the Ministry of Sustainable Development, Environment and Parks (MDDEP). The EIA must include noise mitigation, setbacks from roads and buildings, compliance with forest management regulations, mitigation of impacts on farms and woodlands, plus landscape preservation measures for public lands. Based on the environmental impact assessment and consultation, the MDDEP makes a recommendation to the Quebec Council of Ministries, which, in turn, issues a favorable decree for the project to proceed. Hydro-Quebec is the sole purchaser of renewable energy and uses a bidding process to select the best projects. Prior to bidding, applicants need to obtain a Certificate of Conformity with the municipal use planning framework plus a construction permit from the relevant municipality.

The **Atlantic Provinces** are quite varied in their approaches to renewable energy. New Brunswick has an RPS of 10% renewable by 2016 and is working on deregulation of utilities. Power purchase agreements are in place for 330 MW of wind energy plus a 38 MW biomass energy project. Nova Scotia has an RPS standard of 25% from renewables by 2015 and is procuring 300 MW from independent renewable power producers. Prince Edward Island has an RPS of 15% renewable by 2010 and most of these are through procurement from windfarms. Newfoundland and Labrador Hydro owns and operates the province's electric generating facilities and over 80% of this is hydroelectric. The utility is now developing the lower Churchill Falls hydro primarily to export power.

So we can see great variation between US states and Canadian provinces that do or do not have RPS targets, and do or do not have incentives or penalties to achieve those targets, as well as a great diversity of siting jurisdiction and guidance rules. Most of the state and local siting issues so far have related to wind energy development but what about large-scale solar development? Little guidance exists for such development at the current time.

3.7 Renewable energy guidance for Australia

Almost all of the current renewable energy guidance in Australia is directed at windfarm development. No major new hydro schemes are under consideration and solar is highly distributed. There are draft National Guidelines for windfarms.[27] The document was produced by the Australian Environment Protection and Heritage Council in 2010 and begins:

> Burning coal and other fossil fuels for electricity generation accounts for more than one third of Australia's current greenhouse gas emissions. The expanded Renewable Energy Target (RET) requires that 20 per cent of Australia's electricity is generated from renewable energy sources by 2020 and is one of a number of strategies to reduce Australia's carbon emissions. The transition to a low carbon economy will require a significant transformation of the electricity sector.
>
> The 20/20 target is challenging; however, it is achievable and energy sources such as wind will have a key role in moving Australia to the clean economy of the future.
>
> (EPHC, 2010, p. 1)

In Australia each state has its own approach to windfarm regulation and the national guidelines were abandoned. This review therefore proceeds by considering the different guidelines in each state. These are quite varied.

The state guidelines can be considered in the context of existing installed and approved capacity. These figures and any stated objectives for each state are set out in Table 3.3. The table shows that South Australia is already exceeding the national target but that other states are a long way behind. The approved windfarms will raise capacity by 170% but will still leave wind energy production below 5% of the Australian total.

Meanwhile there is also expansion in solar energy production. Of the installed solar capacity of 2412 MW (December 2012) approximately 80% of this is from rooftop panels on around one million

TABLE 3.3 Wind energy by Australian state

STATE	*Installed (MW)*	*Approved and/or under construction (MW)*	*Production (GWh) 2011–2012*	*Proportion of current production (%)*	*Target*
New South Wales	282	1520	697.6	0.99	20% by 2020
Queensland	13	250	27.7	0.04	750 MW by 2020
South Australia	1223	582	3126.5	23.64	33% (all renewables) by 2020
Tasmania	308	548	423.1	3.99	No target identified
Victoria	947	1678	1416.2	2.49	20% (all renewables) by 2020
Western Australia	457	879	422.2	1.35	20% by 2025
Australian Capital Territory (ACT)	0	800 (situated outside the ACT)	0	0	90% (all imported) by 2020
Northern Territory	0	0	0	0	No target
AUSTRALIA (2011–2012)	3065	5209	6113	2.41	20% by 2020

(*Source:* Wind and total energy production figures from the Bureau of Resources and Energy Economics; installed and approved capacities from diverse online sources including Wikipedia.)

Australian homes. Capacity passed 3 GW in December 2013, meaning current production of over 3 GWh annually. In most states solar is around about 20% of the wind/solar totals.

3.7.1 Relevant legislation

Windfarms are typically regulated under state planning laws, for example, the *Victoria Planning and Environment Act*, and the *NSW Planning and Environmental Assessment Act*. Sometimes multiple Acts can be relevant to windfarm siting and approval processes and guidelines are typically issued based on the relevant Acts, together with contributing plans and policies.

State-by-state guidance

In **New South Wales** (NSW) the Department of Planning and Infrastructure (2011) released draft guidelines in December 2011. Three years later the guidelines are still considered 'draft'. At the same time the government was preparing a Renewable Energy Action Plan to support the national target of 20% renewable energy by 2020.[28]

> The guidelines outline the matters that will be considered in the assessment and determination of windfarm proposals, including:
>
> - Landscape and visual amenity
> - Noise impacts
> - Economic issues, including potential impacts on property values

- Ecological issues, including potential impacts on threatened species
- Auditing and compliance provisions
- Decommissioning and rehabilitation

(NSW Department of Planning and Infrastructure, 2011, p. iv)

The approval process depends, in the draft, on the capital investment value of the project. Under $5 million is a matter for local government, between $5 and $30 million the project goes to a Joint Regional Planning Panel, and over $30 million (or over $10 million if in an environmentally sensitive area) the project is assessed by the Department of Planning and Infrastructure with determination by the Planning Assessment Commission.

Key matters in the assessment process include:

- Proximity to existing dwellings: unless written consent is obtained from all residents within 2 kilometers, the proponent must apply for compatibility certification on the basis of–
  - Predicted noise levels (including low frequency),
  - Appearance, based on photomontages,
  - Impact on landscape values,
  - Blade glint and shadow flicker.
- Community consultation: through mandatory formation of a consultative committee, and 60 days of exhibition of the development application.
- Visual amenity: especially visual impacts on neighbours as determined through–
  - A well justified assessment methodology,
  - Descriptions of the 'worst case' layout and turbine heights; the landscape and its key features, significance and character,
  - Assessment of the visibility of the development (the zone of visual influence or ZVI – no less than 10 kilometers),
  - Photomontages from settlements and significant public view points,
  - Description of community and stakeholder values relating to existing visual amenity and perceptions of the project,
  - Assessment of cumulative impacts linked to transmission lines or other "approved or operational windfarms in the locality" (NSW Department of Planning and Infrastructure, 2011, p. 5).
- Noise: predicted equivalent noise (Leq. 10 minute) should not exceed 35 dB(A) or the background noise by 5 dB(A) whichever is greater, both day and night.

In **Queensland**, where the current level of wind energy is low, the emphasis seems to be on making it easier to develop rather than to implement specific protections. There are no guidelines but the State's Renewal Energy Policy (2009) says:

> A Renewable Energy Regulatory Taskforce will examine existing legislation and provide options to remove or reduce impediments and streamline planning processes for renewable energy projects. For example, the project will examine the best mechanisms for facilitating access to land for renewable energy, which may include acquisitions, land designations or declaration of State Development Areas.
>
> (Queensland DEEDI, 2009, p. 23)

It appears that there are no specific protections offered to Queensland residents.[29]

In **South Australia** the State-wide Wind Farms Development Plan Amendment (DPA) was introduced in 2012. This has specific provisions for the development plans of each of the local government areas in all but the very low population density regions of the state. In recent changes to windfarm visual impact management techniques it was determined to seek that wind turbines:

- "be set back at least 2 kilometers from defined and zoned township, settlement or urban areas (including deferred urban areas)"
- "be mounted on tubular, rather than lattice, towers"
- "avoid or minimize interference with low altitude aircraft movements associated with agriculture"
- "be setback from dwellings, tourist accommodation and frequently visited public places (such as viewing platforms) at a distance that will ensure turbine failure does not present an unacceptable risk to public safety."[30]

The amendment goes on to list a range of areas in which "the approved amendment does not introduce policy that explicitly envisages windfarms". These are primarily the main tourist areas of the State.

The Environmental Protection Authority – Wind Farms Environmental Noise Guidelines[31] provide that windfarm developments should be constructed and designed to ensure that noise generated will not exceed 5 dB(A) above the background sound level or 35 dB(A) using a 10-minute LAeq., whichever is the greater, at surrounding noise-sensitive premises.

The guidelines in **Victoria**[32] put local residents in a strong position. A site analysis must include:

> The location of all existing dwellings within two kilometres of the nearest turbine (adopting a precautionary approach, and accounting for micro-siting variation in final placement of turbines). Where the proposal includes any turbines within two kilometres of an existing dwelling, the application must be accompanied by evidence of the written consent of the owner of the dwelling. The application is prohibited under the planning scheme where evidence of the written consent is not provided.
>
> (Victoria DPCD, 2012, p. 23)

Noise follows NZ standard NZS 6808:2010:

> The Standard specifies a general 40 decibel limit for windfarm sound levels, or the sound should not exceed the background sound level by more than five decibels, whichever is the greater.
>
> Under section 5.3 of the Standard, a "high amenity noise limit" of 35 decibels applies in special circumstances. All windfarm applications must be assessed using section 5.3 of the Standard to determine whether a high amenity noise limit is justified for specific locations.
>
> (Victoria DPCD, 2012, p. 30)

Glint: "Blades should be finished with a surface treatment of low reflectivity to ensure that glint is minimised." (Victoria DPCD, 2012, p. 31)

Flicker: "The shadow flicker experienced immediately surrounding the area of a dwelling (garden fenced area) must not exceed 30 hours per year as a result of the operation of the wind energy facility." (Victoria DPCD, 2012, p. 31)

Visual: The visual impact guidelines are vague. Some significant visually attractive areas are explicitly protected. In other areas there are 'suggestions' for minimizing impact:

- siting and design to minimize impacts on views from areas used for recreation and from dwellings
- locating arrays of turbines to reflect dominant topographical and/or cultural features, such as ridgelines, the coastline, watercourses, windbreaks or transmission lines
- using turbine colour to reduce visual impacts from key public view points
- limiting night lighting to that required for safe operation of a wind energy facility and for aviation safety
- reducing the number of wind turbines with obstacle lights while not compromising aviation safety
- mitigating light glare from obstacle lighting through measures such as baffling
- selecting turbines that are consistent in height, appearance and rotate the same way
- spacing turbines to respond to landscape characteristics
- undergrounding electricity lines wherever practicable
- minimizing earthworks and providing measures to protect drainage lines and waterways
- minimizing removal of vegetation
- avoiding additional clutter on turbines, such as unrelated advertising and telecommunications apparatus.

(Victoria DPCD, 2012, p. 33)

Flora and Fauna: An assessment and (sometimes) an on-going monitoring plan are required.

> Where it is reasonably likely that species listed under the [relevant protection Acts] will be present on or near the site, or using the site as a migratory corridor, applicants for a wind energy facility permit should conduct surveys at the appropriate time for at least 12 months preceding the planning permit application. [The appropriate government department] should be consulted on the timing of the surveys. Survey work should determine the species present, any adverse impacts likely to arise from the proposed wind energy facility, and any appropriate mitigation measures.
>
> (Victoria DPCD, 2012, p. 26)

In addition, the removal of native vegetation should be avoided. If this is not possible losses should be minimized through expert input into project planning. The Victorian requirements, especially the 2-kilometer residential veto, are generally seen as the most restrictive in Australia.

Although the **Western Australian** guidelines[33] are the oldest among the states, they are among the most technically detailed and worth reporting in some detail.

Landscape and Visual Impact: Visual impact is based on a number of factors, which affect the perceived visual quality. The degree to which a windfarm development will impact on the landscape will depend upon:

- Siting, layout and design of the turbines, infrastructure, signage and ancillary facilities, including provision for tourism.
- Number, colour, shape, height and surface reflectivity of the towers and blades.
- Visibility of the development, having regard to the location, distance from which the development is visible, skyline and viewsheds.

- Significance and sensitivity of the landscape, having regard to topography, the extent and type of vegetation, natural features, land use patterns, built form character and community values.

(WAPC, 2004, p. 4)

Suggested approaches to reduction of visual impact include:

- Ensuring all turbines look alike, have a clean, sleek appearance and that the blades rotate in the same direction.
- Minimizing the number of turbines, as appropriate, by using the largest possible model (subject to the visual absorption capabilities and environmental considerations of the site) rather than numerous small ones.
- Siting the windfarm, ancillary buildings, access roads and transmission infrastructure to complement the natural landform contours and landform backdrop, including ridgelines.
- Ensuring the choice of materials and color (e.g. off-white and grey for turbines, low contrast for roads) for the development complements the skyline and the backdrop of the view sheds.
- Minimizing removal of vegetation and using advanced planting of vegetation screens as visual buffers where appropriate.
- Ensuring good quality vegetation and landform rehabilitation, onsite and off-site, where appropriate.
- Locating turbines to reflect landscape and topographical features (e.g. a random pattern may suit a rolling, varied landform and a linear pattern may suit a coastal edge, farm or industrial site).
- Avoiding clutter, such as advertisements and apparatus.

(WAPC, 2004, p. 4)

Noise:

To avoid adverse noise impacts on the amenity of the surrounding community, windfarm developments should include sufficient buffers or setbacks to noise sensitive premises. As a guide, the distance between the nearest turbine and a noise-sensitive building not associated with the windfarm, is likely to be 1km. The ultimate distance between sensitive uses and the wind turbine, may be determined on the basis of acoustical studies.

(WAPC, 2004, p. 4)

The South Australian criteria and approach were endorsed by Western Australia until a formal policy is adopted.

Other possible amenity effects: It was considered that a wind energy facility could also affect local amenity due to:

- Shadow flicker, which occurs when the sun passes behind the blades and the shadow flicks on and off, although in Australia this is uncommon.
- Glint, which occurs when the sun's light is at low angles and is reflected off the blades.
- Overshadowing, which affects adjacent developments.

(WAPC, 2004, pp. 4–5)

It was argued that "modelling can determine areas where these issues require further consideration. Careful siting and design, including the use of low reflectivity materials, can minimise or avoid any impact." (WAPC, 2004, p. 5)

Vegetation and fauna:

> The types, locations and significance of flora and fauna, particularly endangered or threatened species in the development area, should be mapped. Field surveys will help avoid highly sensitive areas of vegetation, including remnant native vegetation and enable roads and services to be placed appropriately. During construction, disturbance and vegetation clearance can be avoided or minimized, through careful siting and consideration of issues such as erosion, drainage run-off, habitat or food source destruction, dieback, weed hygiene, introduction of feral animals and contractor guidelines or penalties.
>
> Development issues to be addressed include controlling run-off, maintaining water quality, stabilizing topsoil and retaining existing vegetation, particularly in coastal areas where vegetation can be hard to re-establish. Any construction, particularly on slopes, should not cause degradation and careful attention will be required, especially in sensitive areas. As a general principle, steep slopes and ridgelines should be avoided.
>
> The impact of windfarms upon birds and bats should be considered. The cumulative effects of windfarms may have an impact on the migratory routes of certain bird species. A full avian study is recommended when a large wind facility is proposed. Solid towers and round nacelles prevent birds from nesting in the structure. The positioning of turbines away from migratory routes and the use of larger, slower turning turbines, may reduce the risk of avian strikes.
>
> (WAPC, 2004, p. 5)

Tasmania has extensive hydroelectric energy installations and has only seen very limited wind power development in fairly remote parts of the state. As their web site says: "Given the remote location of the current windfarm sites, issues at mainland windfarms (such as noise, visual etc.) have not been issues for the Tasmanian windfarms."[34] The document goes on to refer people to the now abandoned draft national guidelines.

Summary of Australian guidance

The diversity of approaches to wind energy in Australia reflects the availability of other energy options. South Australia, for example, has limited coal or hydro options and has therefore been more enthusiastic about wind energy than other states. Queensland has been slow to adopt the opportunities but now seems anxious to catch up with minimal controls. Victoria has seen considerable local opposition to windfarm development and has taken a very conservative approach. With a change of national government in 2013, Australia is currently intent on making the most of abundant coal resources and pushing climate change far down the list of public issues. The short-lived carbon tax was removed in 2014 and national renewable energy targets have been criticised by the conservative government as costing jobs by increasing energy prices. However, state-based targets seem to be more resilient with the Australian Capital Territory taking a very pro-active position of 90% of energy from renewable sources (mainly from outside the territory) by 2020.

3.8 Summary and conclusions

Most of the renewable energy guidance is for wind energy development in the US and Australian states plus some of the Canadian provinces. Guidance for commercial scale solar is still developing with the major exception being use of US federal lands in the southwestern US. There are also highly variable incentives across the US, Canada, and Australia for renewable energy development. Some regional jurisdictions such as British Columbia in Canada, Victoria in Australia, and Oregon and

Vermont in the US have detailed site assessment that includes aesthetic impacts. The US particularly has jurisdictional confusion over land and offshore wind development as well as regional transmission for such renewable energy.

As large-scale energy production facilities are often controversial, particularly with regard to aesthetics, it is reasonable to expect that many of the projects built so far are the 'low hanging fruit' both with respect to technical advantages and social perceptions, at least in the regions outside of Europe surveyed in this chapter. Early projects tend to be located in 'easier pickings' sites with lower perceived scenic impacts among the general population. They also often have easier pathways to approval in locations where the jobs and income are most needed, no new transmission lines are needed, and before other local projects have produced cumulative environmental impacts that provoke resistance. The state of policies, regulations and planning outlined above may be a reflection of these typically less environmentally and socially demanding circumstances. As new regions are brought on-line with large-scale renewable energy production, the people there may be greater in number, often less naïve about cumulative impacts, and generally more perceptive about potential damage upon sense of place, quality of life, or ecosystem services. If regional goals for renewable energy continue to grow, and many 'next best' sites prove to be aesthetically and otherwise valued by local populations, then the permitting policies are likely to evolve and become more demanding. The current diversity of regulatory regimes sketched above suggests several key dimensions and consequent potential pathways by which decision-making rules and procedures can become more demanding of project proponents.

Such evolution of more demanding permitting requirements may be restricted by a contradiction in the current state of renewable energy policies. Much regulation of large-scale renewable energy projects, if it exists at all, is limited to project attributes that are more readily measurable by technical means, and much of it is happening only at local jurisdiction levels. Local planners and decision-makers often have limited capacities to analyze, assess, communicate and make decisions related to perceptions of sense of place, quality of life, landscape identity, and social acceptability. Local regulators will face a sequence of forks in the road. They might take a path toward building capacities to support aesthetic decision-making and will need assistance to effectively and successfully incorporate such considerations into local planning. They might instead continue with current policies and procedures and rely on the ad-hoc judgments of local elected officials to decide aesthetic questions with attendant risk of court challenges. Or, regional and state agencies will fill the vacuum and adopt policies that largely 'take over' large scale renewable energy project permitting, at least with respect to complex social values such as those derived from landscape perceptions.

This contradiction is supported by results of a survey of local planners by the American Planning Association (2010). For wind energy projects, aesthetic issues were reported as very important by the most respondents (71%), with noise next (63%). In this context, ordinance writing (68%) and researching the issues (56%) were their dominant activities, as opposed to project review, plan writing and public engagement activities that dominate their normal work for other planning problems. Grappling with aesthetic considerations will likely also challenge state or regional planners, although they are more likely to afford staff specialists and/or landscape assessment and planning consultants. For planners at all levels, and for law-makers who give them their mandates, there are a few key regulatory choices that will come into play. These are sketched below with reference to some example jurisdictions where certain approaches are more prevalent. More detailed descriptions of planning options and methods are found in Rynne et al. (2011).

The most obvious way that regulations can become more challenging is when they shift up to state or national promulgation and enforcement, but this can obscure other more potent differences that can matter more. First among these is the distinction between regulations seeking to prevent or mitigate future specific, narrowly defined perceptual impacts or health risks versus those that seek to produce or co-optimize general social goals or public goods, such as scenic beauty or biodiversity. The first category includes glare from solar collectors or mirrors, or wind turbine impacts such as noise, shadow or blade-glint flicker affects, blade-throw risks, tower-light flashers at night, or reflections of glint or flashers off water. These problems mainly affect the sensory quality of life of nearby populations or those living, playing, and working quite close to the turbines or solar farms. These are more readily amenable to technical measurement, and are usually avoided or ameliorated by project location and design standards or sometimes by mitigation technologies. Regulations affecting these more technically measurable problems, creating perceptions of health risk or 'nuisance' among specific people, as opposed to more obviously aesthetic perceptions among larger populations, often prevail as the first step away from no regulation at all. These are common among existing regulatory regimes at all jurisdictional scales. A few of these may choose to conceptually, rather than operationally, define permitting standards, such as requiring a project to avoid 'undue' or 'unacceptable' noise impacts. Such conceptual standards serve to formulate discretionary regulations that provide decision-makers with the power to grant permits in spite of prospective nuisance or health risks. The alternative is to set measurable standards, such as a decibel limit or specified setback distances to places of residence, work or recreation.

Energy permitting policies that aim to optimize the mix of a project's more general public impacts upon those both near and far, such as landscape beauty, biodiversity, public health, historic preservation, recreation or overall economic development, are much less common, more complex and problematic. Such policies aim to provide planning procedures and/or standards or guidelines based on the definition of public goods and how they are to be measured and considered in decision-making. The weight of our review suggests that law-makers at all levels have thus far largely shied away from policies that require such considerations. Standards and procedures for decision-making based upon general public purposes seem only to arise when compelled by controversies and new case law; or they may develop when old policies designed to regulate smaller-scale projects – that predate large-scale renewable energy project approvals – must be borrowed, applied and adapted. Where this occurs, there are further dimensions along which project design and approval can become more demanding and about which current knowledge and methods may be found wanting.

3.8.1 Local regulation

In many US states local land use statutes form the basis for renewable energy project approvals. These zoning regulations are promulgated at the city, county or township level and often cannot consider impacts beyond these jurisdiction's boundaries, depending on the applicable state land use planning enablement or case law. Local zoning affecting energy project permits is not enabled in the commonwealth countries reviewed above, and is sometimes preempted by provincial or national law in European and other nations. Local land use permitting of renewable energy projects is largely the exclusive mode of regulation in many US states, such as Illinois, Virginia, Pennsylvania, Indiana, Kansas, and Nebraska. In these cases, the consideration of public goods – as opposed to particular perceptual nuisance impacts – via zoning ordinances is nominally founded upon the goals and values articulated in the relevant local comprehensive plan, particularly with respect to aesthetics. In common instances where aesthetic values are poorly, or not explicitly, articulated by such plans – which is likely for unprecedented large-scale renewable energy projects – decisions are largely guided by any local precedents, potentially meager case law from state courts, and the potentially ad-hoc judgments of planning commissioners and elected officials. This lack of clear legal guidance can lead to high levels of discretion in permitting and opportunities for project advocates to 'buy' approvals by offering to help fund local projects, hire local workers or suppliers, pay local taxes, or by proffering other local good-will favors. These advantages in gaining permits can be powerful even in locations with some state regulation and permitting. Many states, such as New Mexico, Oregon, Maryland, South Dakota, Michigan, New South Wales, Washington and Maine, only regulate renewable energy projects above, widely diverse, project size thresholds, and developers may often break their projects up into multiple adjacent projects with separate permit applications to keep approvals at the local level. Many states apply permitting standards to any proposed major land use change that likely impacts water quality, wildlife conservation, public health, wetland protection, and other public objectives. These usually apply to large-scale renewable energy proposals even if there is no statewide regulation specifically applied to such projects.

In the absence of state guidance or directives, if local governments wish to further strengthen permitting rules within zoning ordinances they often begin by adopting specific zones or overlay zones where large-scale energy projects are prohibited. They may also or instead adopt specific design or impact limiting standards within certain districts, or adopt moratoriums on further projects. The review above suggests that such localities rarely adopt separate statutes governing renewable energy projects with clear planning procedures and standards related to public goods or aesthetics. This is likely because such less rule-based policies – based upon analysis and deliberations aimed at complex ideas of non-market driven public goods – are politically unpopular at local levels. Such policies also tend to be expensive to administer and defend within local budgets and local capacities for planning and legal advocacy. By whatever means, stronger local permitting rules affecting renewable energy projects are more likely in places where the local economy is dependent upon amenities threatened by large-scale energy projects, e.g. tourism, traditional cultural landscapes, or recreational homes.

A growing number of US states continue to largely delegate large-scale renewable energy permitting to local land use powers, but provide standards or guidelines that local jurisdictions should or must follow, such as Arizona, Colorado, Michigan, Nebraska, New Jersey, New Mexico, North Carolina, Rhode Island, Wisconsin and Wyoming. Others provide such guidance but also overlay statewide regulations and/or permitting procedures, such as California, Maine, Minnesota, New York, Ohio, Oregon, Vermont and Washington. A few states provide model land use ordinances for localities to adopt with their own revisions and additions, such as Illinois, Massachusetts, Pennsylvania and Virginia. In all such instances, local ordinances typically have considerable discretion to apply many state guidelines, standards, or model ordinance language with considerable discretion. These may be written as local mandatory standards or as optional or recommended conceptual guidelines. In Maine, for example, a county can limit wind development only to a specified area, effectively excluding it from all other areas, or it can allow windfarms throughout the county but set more stringent conceptually defined habitat conservation or aesthetic impact standards, that are difficult for any project to satisfy, in portions of the county. New York allows the local description of conceptual guidelines for judging aesthetic impacts but at least one community (Ellington) has applied more specifically defined and measurable aesthetic standards as necessary to gain permits.

Regulation at the local level highlights another important dimension affecting the strength of new energy policies that cuts across all jurisdictional scales. Local zoning and permitting decisions are usually made upon weaker standards of evidence in compilation of the relevant factual record. That record is usually based upon the testimony and participation of engaged and interested members of the public and stakeholder representatives who come to public meetings, serve on advisory committees, or testify at hearings. There is typically little legal compulsion that these facts be interrogated by experts, be shown to be representative of the values and perceptions of the whole affected population, or be based upon the best available and sanctioned professional or scientific methods and knowledge. The factual record for local decisions is often held to account primarily through the re-election of the public officials that vote to grant or deny a permit, unless a decision is of sufficiently 'high stakes' that a developer or interest group appeals a local decision to a higher court, state agency or legislature.

Whenever such legal challenges, or the politics arising from controversial projects, produce state regulation and permitting of renewable energy projects, rules of evidence applicable to factual records usually become more demanding. Public comments remain relevant but permitting hearings and decisions tend to rest more on evidence that seeks to be acceptably valid and reliable. This kind of evidence arises from careful planning analyses employing best professional practices, expert scientific or professional opinions based upon current knowledge, and potentially more scientific surveys of public perceptions intended to validly sample whole affected populations. The challenges in gaining permits for renewable energy projects at state, provincial or national levels depend on exactly how these forms of evidence are required or interpreted in each jurisdiction. So far, some states with supra-local permitting requirements may not yet have staff planners or attorneys with the expertise needed to demand high quality visual impact assessment (VIA) assessments against different standards of acceptable impacts (i.e. visual quality objectives) and can be unable, unwilling, or not yet required, to interrogate on behalf of the public the aesthetic evidence submitted by applicants. This lack of capacity may create unnecessary controversies and legally protracted permitting processes as proposed projects move into more aesthetically sensitive areas.

3.8.2 Supra-local regulation

If supra-local permitting is required, a key choice is whether to employ guidelines, standards, or prescribed decision analysis procedures, or some combination of these. As every law-maker or lawyer knows, guidelines almost always provide the weakest regulatory permitting guidance inasmuch as they are not mandatory and are often broadly conceptual and non-specific in not prescribing operational outcomes, technologies or particular designs. Such guidelines usually provide permitting authorities with considerable discretion as long as project proposals seem to stay within a broad scope of the meaning and logic of the guidelines. This option largely prevails in New Mexico, Colorado, Connecticut, Michigan, Iowa, Rhode Island, Victoria, Washington, Wyoming, and perhaps Alberta.

In these cases permitting agencies typically require a factual record, usually generated by the applicant, which must follow provisions for analysis specified or implied in the permitting rules and/or guidelines. The agency then engages in 'considerations' of the applicant's analysis as well as other stakeholders' testimony and proffered counter-facts. If the guidelines are met the permit is issued but often with modifications negotiated among the contestants and decision-makers. In the absence of permitting standards, further contest is usually based upon such doctrines as 'undue impact,' 'necessity,' 'public trust,' or other such legal concepts, perhaps by reference to the guidelines, but adjudicated by case law or the judgment of hearing examiners and judges. In all these considerations and appeals, the rules of evidence that govern the admissibility of elements of the factual record are usually stronger than at the local level. Issues of validity and reliability and use of best professional analysis methods can come more into play at the state level.

In states and provinces where stronger permitting standards are applied they often only regulate nuisance impacts in the form of a few specified and operationally measurable sensory or environmental impacts, such as decibel levels, height limits, setbacks, land use settings, landscape positions, etc. In these cases, impacts related to the production of public goods, such as biodiversity, scenery, quality of life or overall regional economic prosperity, may not be considered at all or only via guidelines. This situation allows jurisdictions to stay comfortably within the less political and easier rule-based domain of impact abatement by technical means. This situation largely prevails in South Dakota, Victoria, Ontario,

Minnesota, Maryland, North Dakota, Ohio, Montana, California away from the Pacific coast, and South Australia outside of zones where projects are prohibited. Another option for minimal but strong permitting regulations is to simply apply statewide zoning in a manner to avoid or minimize 'nuisance' impacts, such as in Texas. Another is to direct projects mainly into low overall likely impact zones, such as in recent executive orders affecting the southwestern US and potentially in Western Australia or Victoria.

3.8.3 Development of siting standards

A few states and provinces have chosen to apply clear standards related to public goods in their permitting of renewable energy projects, albeit usually in combination with guidelines and nuisance standards. Law-makers generally shy away from clear regulatory standards because they tie the hands of agencies and legislatures, unless the courts compel clear standards. With this potentially stronger basis for regulation there are two paths these jurisdictions may follow, with potential mixing of the two. The first path is to employ a permitting process that solicits evidence of compliance with less discretionary, more quantifiable and more operationally defined standards of socially acceptable impacts, rather than optional or vague guidelines. The permitting agency arguably must engage in 'deliberations' rather than 'considerations' inasmuch as they must judge the extent to which renewable energy projects' performance measures and assessments achieve desired levels of public good production, and assess the acceptability of unavoidable tradeoffs among these. The application of such project standards can compel even stronger standards of admissible evidence for the factual record than the cases sketched above. Ad-hoc testimony of stakeholders or members of the public tend to carry little ultimate weight. If there is a standard related to landscape aesthetics, more valid and reliable measures of scenic impacts must be performed consistent with the language of the standard. States where such permitting of renewable energy projects is arguably significantly based upon standards judged against a factual record, usually in the form of an analysis formulated by the applicant, include Ontario, Massachusetts, Wisconsin, and Oregon, and those with an explicit standard related to aesthetics seem to be Vermont, Maine, and New South Wales.

Arguably the most challenging regulatory regimes affecting renewable energy projects are found in states and provinces that pursue the general policy intention of some kind of 'net maximization of public goods' by means of required, formal, comprehensive impact assessment methods. These usually analyze a list of impact assessments to produce a factual record across a variety of public goods, often including landscape aesthetics. In these situations where aesthetic values must be included, widely applied methods, as informed by research, of VIA are the best practice. More valid and reliable professional or scientific methods than VIA might be employed instead but these are usually more costly. Such visual assessments will form the basis for minimizing the scenic impact of renewable energy projects, or their denial based upon unacceptable aesthetic impacts. These impact assessments may prove inadequate if these methods fail to assess general public perceptions of renewable energy projects. This may occur if methods are poorly informed with respect to how their 'green' attributes interact with their foreign and industrial presence in certain highly valued landscapes, and the extent to which people become hedonically adapted to built projects.

In some states, the whole impact assessment process may sometimes be applied in-lieu of tests against a few specific project performance standards. Such environmental impact assessments rarely determine a permitting decision by an algorithm or decision analysis method, such as cost-benefit analysis, linear programming, or multi-attribute decision analysis. Instead, the robust impact assessment is simply a factual record to inform the discretionary judgment of the permitting authority, within the bounds set by any project performance standards or directives that may constrain such decisions. Impact assessment procedures usually must address a list of issues and be applied using the best available science or professional methods. If challenged, standards of evidence for the factual record are normally strong and minimize the value of ad-hoc testimony. States with permitting guidelines but where such formal and comprehensive impact assessments may sometimes be required include New York, Western Australia, Quebec, British Columbia, New South Wales and California near the Pacific coast. It is important to note that such assessments will be required anywhere in the US where a nexus of federal involvement compels the application of the US *National Environmental Policy Act*, such as on federally owned lands or where substantial federal funding or permitting applies, or perhaps in the future near US national parks. In such instances, these comprehensive impact assessments will likely become part of the factual record that informs state permitting.

Planning and permitting of large-scale renewable energy projects is still in a relatively nascent and fluid state. New projects will likely encounter growing political challenges as more controversial project sites are proposed. These challenges will likely be much more related to general public perceptions related to landscape aesthetics and other complex and broad public values, as opposed to the site-specific and more technically measurable impacts previously emphasized in permitting processes. These political challenges may instigate the evolution of more demanding law, analysis requirements and evidentiary quality. In these conditions, project planners and designers will need to become more sophisticated and creative, to become highly skilled in public engagement and project simulations (Manyoky et al., 2015), and can expect more project applications to fail.

Notes

1 At www.direusa.org
2 The web site for guidelines is at www.energy.ca.gov/2010publications/CEC-600-2010-007/CEC-600-2010-007.pdf
3 The guidelines are available at www.sos.state.co.us/CCR
4 At www.kansasenergy.org/Kansas_Siting_Guidelines.PDF
5 At www.fws.gov/mainefieldoffice/PDFs/MEFO%20wind%20guidance%2011.12.09.pdf
6 At http://energy.maryland.gov/govt/Documents/MSEC_Renewable_Energy_Guidelines_and_Resources_2-5-14_002.pdf
7 At www.mass.gov/eea/docs/doer/gca/wind-not-by-right-bylaw-june13-2011.pdf
8 At www.esf.edu/via
9 At www.mcecc.org/documents/Siting_Wind_Systems.pdf
10 These can be seen at http://mn.gov/commerce/energyfacilities/#tabs=3
11 At http://deq.mt.gov/Energy/Renewable/WindWeb/indexWindinMT.asp
12 At www.emnrd.state.nm.us/ECMD/RenewableEnergy/documents/NMwinddevelopmenthandbook.pdf
13 At www.puco.ohio.gov/emplibrary/files/media/OPSB/OhioSitingManual.pdf
14 At www.dfw.state.or.us/lands/docs/OR_wind_siting_guidelines.pdf and the model ordinance is at www.oregon.gov/ENERGY/SITING/docs/ModelEnergyOrdinance.pdf
15 At www.pawindenergynow.org/pa/Model_Wind_Ordinance_Final_3_21_06.pdf
16 At www.sdgfp.info/Wildlife/Diversity/windpower.htm
17 At www.fishwildlife.org/files/Vermont.pdf
18 At www.deq.virginia.gov/Programs/RenewableEnergy/ModelOrdinances.aspx
19 At http://wdfw.wa.gov/publications/00294/wdfw00294.pdf
20 At https://psc.wi.gov/renewables/documents/windSitingReport2014.pdf
21 At http://datcp.wi.gov/uploads/About/pdf/WI-NFBGuidelinesFinalOct2011.pdf
22 www.lawoftheland.wordpress.com
23 This is available at www.esf.edu/via
24 At www2.gov.bc.ca/assets/gov/farming-natural-resources-and-industry/natural-resource-use/natural-resource-major-projects/major-projects-office/guidebooks/clean-energy-projects/clean_energy_guidebook.pdf

25 https://dr6j45jk9xcmk.cloudfront.net/documents/2714/277097.pdf
26 http://www.energy.gov.on.ca/en/files2016/01/muncicipla_guide_english_web_2016.pdf
27 At www.scew.gov.au/system/files/resources/8e446a1a-ab93-5f84-99d0-12d3422d2a23/files/draft-national-wind-farm-development-guidelines-july-2010.pdf
28 See www.planning.nsw.gov.au/Policy-and-Legislation/~/media/1C3284EB49E244FEA7539B8FFFD3D9BA.ashx
29 See http://rti.cabinet.qld.gov.au/documents/20090/jun/qld%20renewable%20energy%20plan/Attachments/Qld%20Renewable%20Energy%20Plan.pdf
30 At www.lga.sa.gov.au/webdata/resources/files/2012.32%20-%20Windfarm%20Development%20Guidelines%20-%20Final%20Report.pdf. All quotes are in the section of document labeled "Approval Stage" which is without page numbers.
31 www.epa.sa.gov.au/files/47788_windfarms.pdf
32 See www.dtpli.vic.gov.au/__data/assets/pdf_file/0009/231768/Policy-and-planning-guidelines-for-development-of-wind-energy-facilities-in-Victoria-July-2012.pdf
33 See www.planning.wa.gov.au/dop_pub_pdf/pb67May04.pdf
34 At http://epa.tas.gov.au/regulation/wind-farms

References

American Planning Association. 2010. Wind Energy Planning: Results of the American Planning Association Survey. Retrieved from www.planning.org/research/wind/surveyreport.htm

Arizona Game and Fish Department (AG&FD). 2008. Guidelines for Reducing Impacts to Wildlife from Wind Energy Development in Arizona, Phoenix: Arizona Fish and Game Department. Retrieved from www.azgfd.gov/hgis/pdfs/WindEnergyGuidelines.pdf

British Columbia. 2007. *Opportunities for Local Government and the Public Participation in Provincial Regulatory Processes for Independent Power Producers' Projects.* Victoria: British Columbia Ministry of Energy, Mines and Petroleum. Retrieved from www.empr.gov.bc.ca/PoerDev/EAED_IPP_Mini-guide_for_Local_Government_enclosure.pdf

Environment Protection and Heritage Council (EPHC). 2010. National Wind Farm Development Guidelines DRAFT. Adelaide: Australian Environment Protection and Heritage Council. Retrieved from www.scew.gov.au/system/files/resources/8e446a1a-ab93-5f84-99d0-12d3422d2a23/files/draft-national-wind-farm-development-guidelines-july-2010.pdf

Environmental Law Institute (ELI). 2011. *State Enabling Legislation for Commercial Scale Wind Power Siting and the Local Government Role.* Washington, DC: ELI.

Federal Energy Regulatory Commission (FERC). Undated. *A Guide to the FERC Electric Facilities Permit Process.* Washington, DC: FREC.

Glennon, R., & Reeves, A.M. 2010. Solar energy's cloudy future. *Arizona Journal of Environmental Law and Policy, 1*, 91. Retrieved from www.ajelp.com/articles/solar-energys-cloudy-future/

Helius, L. 2011. The law of solar energy. Stoel Rives LLP. Retrieved from www.stoel.com/renewableenergy

Manyoky, M., Ribe, R., Grêt-Regamey, A., & Wissen, U.H. 2015. Visual-acoustic simulations for integrating soft landscape values into wind park assessments. In: E. Buhmann, S. Ervin, & M. Piethc (Eds.), *Peer Reviewed Proceedings of Digital Landscape Architecture 2015 at Anhalt University of Applied Sciences*, pp. 135–140. Berlin, Germany: Herbert Wichmann Verlag.

National Research Council (NRC). 2007. *Environmental Impacts of Wind-Energy Projects.* Washington, DC: National Academy Press.

National Wind Coordinating Committee (NWCC). 2006. *State Siting and Permitting of Wind Energy Facilities.* Washington, DC: National Wind Coordinating Council & National Conference of State Legislatures.

North American Electric Reliability Corporation (NAERC). 2009. *2009 Long-Term Reliability Assessment: 2009–2018.* Washington, DC: NAERC.

NSW Department of Planning and Infrastructure. 2011. NSW Planning Guidelines; Wind Farms. State of New South Wales: Dept. of Planning and Infrastructure. Retrieved from www.planning.nsw.gov.au/~/media/Files/DPE/Guidelines/draft-nsw-planning-guidelines-wind-farms-a-resource-for-the-community-applicants-and-consent-2011–12.ashx?la=en

Ohio Power Siting Board. 2005. Siting New Energy Infrastructure in Ohio. Columbus: The Ohio Power Siting Board. Retrieved from www.puco.ohio.gov/emplibrary/files/media/OPSB/OhioSitingManual.pdf

Outka, U. 2010. Siting renewable energy: land use and regulatory context. *Ecology Law Quarterly, 37*(4), 1041. Retrieved from http://scholarship.law.berkeley.edu/cgi/viewcontent.cgi?article=1934&context=elq

Paddock, L., Grinlinton, D., Burton, S., Bruenjes, A., Heaven, G., & Zubair, S. 2009. *Legal Framework for Solar Energy*. Washington, DC: Law School Environmental Law Program, The George Washington University. Retrieved from http://solar.gwu.edu/file/442/download

Queensland Dept. of Employment, Economic Development and Innovation (DEEDI). 2009. The Queensland Renewable Energy Plan: A Clean Energy Future for Queensland. Queensland Dept. of Employment, Economic Development and Innovation. Accessed at http://rti.cabinet.qld.gov.au/documents/2009/jun/qld%20renewable%20energy%20plan/Attachments/Qld%20Renewable%20Energy%20Plan.pdf

Rabe, B.G. 2006. *Race to the Top: The Expanding Role of US State Renewable Portfolio Standards.* Prepared for the Pew Center for Global Climate Change by the University of Michigan, Ann Arbor.

Rooney, S., Dascomb, P., Korjeff, S., Prahm, G., & Smardon, R. 2012. *Visual Impact Assessment Methodology for Offshore Wind Energy Conversion Facilities.* Technical Bulletin 12-001. Barnstable, MA: Cape Cod Commission. Retrieved from www.esf.edu/via

Rynne, S., Flowers, L., Lantz, E., & Heller, E. (Eds.). 2011. *Planning for Wind Energy*. Planning Advisory Service Report No. 566. Chicago: American Planning Association.

Salkin, P. 2010. Renewable energy and land use regulation (part 2). *ALI-ABA Business Law Course Materials Journal, 2010* (April), 27–43. Retrieved from http://papers.ssrn.com/sol3/papers.cfm?abstract_id=1588766

Salkin, P., & Ostrow, A.P. 2011. Cooperative federalism and wind: a new framework for achieving sustainability. *Hofstra Law Review, 37*(4), 1049.

Showalter, S., & Bowling, T. 2011. *Offshore Renewable Energy Primer.* University, MS: National Sea Grant Law Center, University of Mississippi. Retrieved from http://nsglc.olemiss.edu/offshore.pdf

Smardon, R.C. 1986. Historical evolution of visual resource management within three federal agencies. *Journal of Environmental Management, 22*, 301–317.

Smardon, R.C., & Karp, J. 1993. *The Legal Landscape: Guidelines for Regulating Environmental and Aesthetic Quality.* New York: Van Nostrand-Reinhold. Retrieved from www.esf.edu/via

Stoel Rives. 2010. The Law of Wind. Stoel Rives LLP. Retrieved from www.stoel.com/renewableenergy

Sullivan, R.G., Cothren, J., Williamson, M., Smith, P., McCarty, J., & Kirchler, L.B. 2012a. Visual impact risk assessment and mitigation mapping system for utility-scale wind energy facilities. In *Science, Politics and Policy: Environmental Nexus.* Proceedings of the 37th Annual National Association of Environmental Professionals Conference, May 21–24, 2012, Portland, OR, pp. 667–692.

Sullivan, R.G., Kirchler, L.B., Cothren, J., & Winters, S.L. 2012b. Preliminary assessment of offshore wind turbine visibility and visual impact threshold distances. In *Science, Politics and Policy: Environmental Nexus.* Proceedings of the 37th Annual National Association of Environmental Professionals Conference, May 21–24, 2012, Portland, OR, pp. 943–969.

Sullivan, R.G., Kirchler, L.B., Lahti, T., Roche, S., Beckman, K., Cantwell, B., & Richmond, P. 2012c. Wind turbine visibility and visual impact distances in western landscapes. In *Science, Politics and Policy: Environmental Nexus.* Proceedings of the 37th Annual National Association of Environmental Professionals Conference, May 21–24, 2012, Portland, OR, pp. 897–942.

Swanson, D., & Jolivert, M. 2012. DOE transmission corridor designates and FERC backstop siting controversy: has the Energy Policy Act of 2005 succeeded in stimulating the development of new transmission facilities? *Energy Law Journal, 30*, 415–466. Retrieved from www.felj.org/sites/default/files/docs/elj302/17swanstrom-and-jolivert.pdf

USC 7, Sections 28101 et seq., The Federal Noxious Weed Act

USC 16, Chapter 12, The Federal Power Act

USC 16, Sections 470 et seq., The National Historic Preservation Act

USC 16, Sections 668–668c. The Bald and Golden Eagle Act

USC 16, Sections 1456 et seq., The Coastal Zone Management Act

USC 16, Section 1536 (a)(2), The Endangered Species Act

USC 33, Sections 1251–1387, The Clean Water Act

USC 42, Sections 4321–4370, The National Environmental Policy Act

USC 42, Sections 15801 et seq., Energy Policy Act of 2005
USC 43, Sections 1701–1704, The Federal Land Policy and Management Act
US Court of Appeals, Ninth Circuit. 2011. California Wilderness Coalition v. US Department of Energy Case Numbers 08-71074, 08-71823, 08-71829, 08-71831, 08-71845, 08-71870. 08-71870, 08-71872, 08-71884, 08-71908, 08-72644, and 08-72835.
US Department of Interior (USDI). 1993. Options Regarding Applications for Hardrock Mineral Prospecting Permits on Acquired Lands near a Unit of the National Park System, Memorandum M-36993 (so called Doe Run memo). US Department of the Interior, Office of the Solicitor, Washington, DC. https://solicitor.doi.gov/opinions/M-36993.pdf
US Department of Interior, Bureau of Land Management (USDI BLM). 2005. *Final Programmatic Environmental Impact Statement on Wind Energy Development on BLM-Administered Lands in the Western United States.* Washington, DC: USDI, BLM.
US Department of Interior, Bureau of Land Management (USDI BLM). 2006. Wind Energy Policy, IM 2006-16. Washington, DC: USDI, BLM.
US Department of Interior, Bureau of Land Management (USDI BLM). 2010. *Draft Programmatic Environmental Impact Statement for Solar Energy Development in Six Southwestern States.* Washington, DC: USDI BLM & US DOE.
US Department of Interior, Bureau of Land Management (USDI BLM). 2011a. *National Environmental Policy Act Compliance for Utility Scale Renewable Energy Right-of-Way authorizations.* EMS TRANSMISSION 02/08/2011 Instruction Memorandum No. 2011-059. Washington, DC: USDI, BLM.
US Department of Interior, Bureau of Land Management (USDI BLM). 2011b. *Solar and Wind Applications: Pre-Application and Screening.* EMS TRANSMISSION 02/08/2011 Instruction Memorandum No. 2011-061. Washington, DC: USDI, BLM.
US Department of Interior, Bureau of Land Management (USDI BLM). 2012. *Final Programmatic Environmental Impact Statement (PEIS) for Solar Energy Development in Six Southwestern States.* FES 12-24 DOE/EIS-0403. Washington, DC: USDI BLM & US DOE.
US Department of Interior, Bureau of Land Management (USDI BLM). 2013. *Best Management Practices for Reducing Visual Impacts of Renewable Energy Facilities on BLM Administered Lands.* Cheyenne, WY: USDI, BLM.
US Department of Interior, Bureau of Ocean Energy Management (BOEM). 2010. *Guidelines for Information Requirements for a Renewable Energy Construction and Operations Plan (COP).* Washington, DC: USDI, BOEM.
US Department of Interior, Fish and Wildlife Service (US FWS). 2007. *Wind Power Siting Regulations and Wildlife Guidelines in the United States.* Washington, DC: Association of Fish and Wildlife Agencies & USDI, FWS.
US Department of Interior, Minerals Management Service (MMS). 2007. *Programmatic Environmental Impact Statement for Alternative Energy Development and Production and Alternate Use of Facilities on the Outer Continental Shelf, Final Environmental Impact Statement.* OCS EIS/EA MMS 2007-046, October 2007. Washington, DC: USDI, MMS.
US Department of Interior, National Park Service (USDI NPS). 2010. Director's memo, *Implementation Guidance for the Interagency Transmission Memorandum of Understanding.* Washington, DC: NPS.
US Supreme Court. 2011. *California Wilderness v. USDOE.*
Vann, A., & DeBough, J. 2011. The Federal Government's Role in Electric Transmission Facility Siting. *Congressional Research Service,* 7-5700. Washington, DC.
Vermont Agency for Natural Resources. 2006. Draft Guidelines for the Review and Evaluation of Potential Natural Resources Impacts from Utility-Scale Wind Energy Facilities in Vermont. Montpelier: Agency for Natural Resources. Retrieved from www.fishwildlife.org/files/Vermont.pdf
Victoria Department of Planning and Community Development (DPCD). 2012. Policy and Planning Guidelines for Development of Wind Energy Facilities in Victoria. Melbourne: Victoria Dept. of Planning and Community Development. Retrieved from www.dtpli.vic.gov.au/__data/assets/pdf_file/0009/231768/Policy-and-planning-guidelines-for-development-of-wind-energy-facilities-in-Victoria-July-2012.pdf
Western Australian Planning Commission (WAPC). 2004. Guidelines for Wind Farm Development. Perth: Western Australia Planning Commission Dept. for Planning and Infrastructure. Retrieved from www.planning.wa.gov.au/dop_pub_pdf/pb67May04.pdf
Williams, W., & Whitcomb, R. 2007. *Cape Wind: Money, Celebrity, Class, Politics and the Battle for Our Energy Future.* New York: Public Affairs.

Key web sources of federal agency renewable energy development regulation guidance

For BLM: www.blm.gov/wo/st/en/prog/energy/wind_energy.htm
For BOEM: www.boem.gov/Renewable-Energy-Program/index.aspx
For EPA: www.epa.gov/renewableenergyland
For FERC: www.ferc.gov/industries/electric/indus-act/siting.asp
For NOAA: www.energy.noaa.gov/renewable-energy/

Web sources for state renewable energy development guidance

Arizona at www.azgfd.gov/hgis/pdfs/WindEnergyGuidelines.pdf
California at www.energy.ca.gov/2010publications/CEC-600-2010-007/CEC-600-2010-007.pdf
Colorado at www.sos.state.co.us/CCR
Kansas at www.kansasenergy.org/Kansas_Siting_Guidelines.PDF
Maine at www.fws.gov/mainefieldoffice/PDFs/MEFO%20wind%20guidance%2011.12.09.pdf
Maryland at http://energy.maryland.gov/govt/Documents/MSEC_Renewable_Energy_Guidelines_and_Resources_2-5-14_002.pdf
Massachusetts at www.mass.gov/eea/docs/doer/gca/wind-not-by-right-bylaw-june13-2011.pdf
Michigan at www.mcecc.org/documents/Siting_Wind_Systems.pdf
Minnesota at http://mn.gov/commerce/energyfacilities/#tabs=3
Montana at http://deq.mt.gov/Energy/Renewable/WindWeb/indexWindinMT.asp
New Mexico at www.emnrd.state.nm.us/ECMD/RenewableEnergy/documents/NMwinddevelopmenthandbook.pdf
Ohio at www.puco.ohio.gov/emplibrary/files/media/OPSB/OhioSitingManual.pdf
Oregon at www.dfw.state.or.us/lands/docs/OR_wind_siting_guidelines.pdf and the model ordinance is at www.oregon.gov/ENERGY/SITING/docs/ModelEnergyOrdinance.pdf
Pennsylvania at www.pawindenergynow.org/pa/Model_Wind_Ordinance_Final_3_21_06.pdf
South Dakota at https://gfp.sd.gov/wildlife/docs/wind-power-siting-guidelines.pdf
Vermont at www.anr.state.vt.us/site/html/plan/DraftWindGuidelines.pdf
Virginia at www.deq.virginia.gov/Programs/RenewableEnergy/ModelOrdinances.aspx
Washington at wdfw.wa.gov/hab/engineer/windpower/index.htm
British Columbia at www2.gov.bc.ca/assets/gov/farming-natural-resources-and-industry/natural-resource-use/natural-resource-major-projects/major-projects-office/guidebooks/clean-energy-projects/clean_energy_guidebook.pdf
New South Wales at www.planning.nsw.gov.au/Policy-and-Legislation/~/media/1C3284EB49E244FEA7539B8FFFD3D9BA.ashx
Queensland at http://rti.cabinet.qld.gov.au/documents/2009/jun/qld%20renewable%20energy%20plan/Attachments/Qld%20Renewable%20Energy%20Plan.pdf
South Australia at www.lga.sa.gov.au/webdata/resources/files/2012.32%20-%20Windfarm%20Development%20Guidelines%20-%20Final%20Report.pdf
Victoria at www.dtpli.vic.gov.au/__data/assets/pdf_file/0009/231768/Policy-and-planning-guidelines-for-development-of-wind-energy-facilities-in-Victoria-July-2012.pdf
Western Australia at www.planning.wa.gov.au/dop_pub_pdf/pb67May04.pdf

4

ADJUSTING TO RENEWABLE ENERGY IN A CROWDED EUROPE

Simon Bell

Highlights

- Summarizes landscape impact assessment related to wind energy development in Europe.
- Provides a detailed discussion of windfarm development visual design standards in the UK, particularly in Scotland.

4.1 Introduction

This chapter aims to broaden the discussion of the field of impacts and effects of renewable energy in the landscape by looking at the European experience in general and the UK experience in particular. Following an overview of the situation some methods used for strategic planning and for site-level impact assessment will be presented. These in part reflect the accepted methods for making project-level assessments but also incorporate some modifications by the author and also some developmental work in strategic planning.

In Scotland, where windfarm development has been a focus of Scottish government policy for a number of years, the agency responsible for protecting and conserving landscape and nature, Scottish Natural Heritage (SNH), has been at the forefront of developing guidance, especially at the impact assessment level. This is generally considered to be some of the best developed and tested methodology in Europe and beyond. The *Guidelines for Landscape and Visual Impact Assessment* (Landscape Institute and Institute for Environmental Management and Assessment, 2013), now in its third edition is also held up internationally as being the best of its kind currently available in English. It covers all types of development which require a landscape and visual impact assessment. There are aspects and factors specific to renewable energy and to windfarms which have led to more specific guidance, such as that provided by SNH.

While I make reference to the published documents and guidance, it is not my aim to repeat what is already available in detailed form and can be found in publications or on various websites. Instead, I will summarise the main aspects which are taken into account, illustrated with some examples. The European dimension is difficult to summarise completely as national legislation and impact assessment guidance tends to be written in the native language. However, there are some specific Europe-wide aspects which in theory overcome this and should ensure a reasonably common approach across the continent. I say "in theory" because while the same general rules should apply based on European Union-level directives (see p. 80) the implementation of this is not always as strict as it should be or taken as seriously in all countries within the EU.

Wind energy obviously needs wind to make it work and for this reason those countries of Europe with the windier climates, such as those on the Atlantic seaboard, have also tended to be those with some of the largest scale developments. In addition, the political climate has had a major impact. Countries which have historically favoured left of center governments and have a strong Green political element have also been front runners in wind energy development. Denmark was an early leader and developed its own manufacturing industry for wind turbines, such that it is now the market leader in Europe. The Netherlands, used to having windmills to power the pumps to keep Holland dry as well as having an engineered landscape, has found it easier to adopt modern forms of wind power. Spain and Portugal, before the economic crisis which hit them in recent years, also adopted aggressive wind power

(A) (B) (C) (D)

FIGURE 4.1(A)–(D) Examples of windfarms in different European landscapes: (a) is a group of turbines located in western Spain but viewed from a natural park in Portugal – an international visual impact; (b) shows a small group located on the summit of a ridge in Portugal – a typical location which has a high impact from all around; (c) is a scene in eastern Germany where there are many inter-visible groups of turbines into the distance; (d) is an array in the flat open landscape of the Netherlands.

(Photos: Creative Commons.)

development. Germany is not so windy but its politics are Green and, depending on the individual state within the federation and local laws, has also been a leader (see p. 81). The UK, while a slow starter, has caught up quickly and, owing to its windier climate than most of the continent, has installed a huge capacity, particularly in Scotland, as noted above.

This chapter starts with an overview of the European policy context – not only in terms of renewable energy policy but also environmental policies and environmental impact regulations, planning policy and landscape policy. These four aspects are not always harmonious or harmonized and not always implemented to the same degree in all countries. One country where all forms are strong legally and implemented conscientiously is the UK, which makes it a good country to focus on. Following this the nature and character of the European landscape – or rather landscapes – and the factors which set it apart from the landscapes of the USA, for example, will be considered. Next will be a look at strategic landscape planning aspects, illustrated with a case study, after which windfarm site layout principles, design factors and the general process of landscape and visual impact assessment will be presented and discussed.

4.2 Policy context

The European Union, comprising the majority of the states of Europe, has a political as well as economic dimension. While there is a European Parliament, the main decision-making body is the European Commission, an executive body comprising commissioners appointed by each EU member state and divided into different Directorships General (DGs), for example for energy, environment, transport, agriculture, research, etc. The Commission is funded by contributions from member states according to their economic status, and feeds back huge sums in grants and subsidies to help develop and strengthen individual member states and regions (especially economically and socially disadvantaged ones) and to support certain industrial sectors.

4.2.1 EU and member states energy policy

The EU acts as an international signatory in its own right in many international conventions and was an advocate for, as well as an early signatory to, the Kyoto protocol on climate change and reduction in carbon dioxide emissions. Within the EU membership, individual member states also made commitments, often with rather ambitious targets and time scales. While in general, carbon dioxide emissions have only reduced where de-industrialization has taken place, such as in eastern Europe following the collapse of the Communist system, the need to divert energy sources away from "conventional" fuels such as hard coal and brown coal to try to reduce emissions has led to the adoption of different energy sources such as natural gas, hydroelectricity, nuclear power (not popular in countries with strong Green movements, especially Germany), solar (especially possible in the Mediterranean, sunny countries), biomass and, of course wind.

The EU Directive on Electricity Production from Renewable Energy Sources promotes renewable energy use in electricity generation. It is officially named 2001/77/EC, commonly known as the RES Directive and took effect in October 2001. It sets national indicative targets for renewable energy production from individual member states. The EU as a body does not strictly enforce these targets but the European Commission monitors the progress of EU member states and will, if necessary, propose mandatory targets for those who miss their goals. Regulators wanted a 12 per cent share of gross renewable domestic energy consumption by 2010 – and aim for a 20 per cent share by 2020. Wind power is of course one of the main types of renewable energy generation. As examples of aggressive policies on wind energy development, two member states, the UK and Germany are presented below.

In the UK, the Climate Change Act came into force in 2008. The act commits the UK to reducing emissions by at least 80 per cent in 2050 from 1990 levels. This target was based on advice from the

Committee on Climate Change report, *Building a Low-carbon Economy* (CCC, 2008). The 80 per cent target includes greenhouse gas emissions from the devolved administrations (Scotland, Wales and Northern Ireland), which currently account for around 20 per cent of the UK's total emissions. The UK legislation is designed to be harmonious with the EU so as to reduce any potential legal conflicts. The UK has its own oil and gas reserves, although peaking and declining. It experienced a "dash for gas" to build gas-powered generating stations in the 1990s and also has many ageing nuclear power stations (with plans to build some replacements). The country relied primarily on coal until quite recently, having large coal reserves which had been exploited since the Industrial Revolution but which proved to be increasingly expensive to mine, leading to the use of imported coal. The sheer scale and visual impact of large-scale construction of windfarms in all parts of the UK has led to a public backlash in a number of areas and the focus now is moving offshore (see Figure 4.2).

Germany is the most populous and economically powerful country in the EU and has promoted renewable energy at both federal and state level. Germany's renewable energy sector is among the most innovative and successful worldwide. Net-generation from renewable energy sources reached about 30 per cent in 2014. For the first time, wind, biogas, and solar combined accounted for a larger portion of net electricity production than brown coal. Germany has been called "the world's first major renewable energy economy". More than 23,000 wind turbines are spread across the country's 357,000 km^2 land area (see Figure 4.3). The recent focus is on offshore wind production, the major challenge being development of sufficient network capacity for transmitting the power generated in the North Sea to the rest of the country. With so many turbines they have become a common sight, especially in the more open and less industrialised parts of the east.

4.2.2 EU environmental policy

The EU has many environmental policies set out in a number of directives, such as the Habitats Directive, the Water Framework Directive and so on. However, at present there is no specific recognition of landscape as a subject and no directive to safeguard it (although it is covered in a different way via the European Landscape Convention, see p. 85).

(A) **(B)**

FIGURE 4.2(A) AND (B) (a) A windfarm at Ardrossan in Scotland, showing the size of the turbines which dominate an otherwise rather small-scale landscape; (b) an example of an offshore windfarm at Scroby Sands in England which can be seen from the shore but whose scale is not as dominating.

(Photo: Creative Commons.)

FIGURE 4.3 Wind turbines in Bernberg, East Germany viewed from the train, where, as can be seen, the landscape is dominated by turbines. What is not obvious from the photo is the movement which at this distance can be quite disturbing. The red bands also make the turbines more noticeable.
(Photo: S. Bell.)

A major directive of interest for this chapter is the Environmental Impact Assessment Directive (85/337/EEC) which has been in force since 1985, and amended several times. This requires all member states, through the adoption of the regulations into national legislation, to consider the impact of developments on a range of resources, of which the landscape and people are two, along with many others: physical, ecological and cultural. As with most EU legislation, this is applied more stringently in some countries than others and, as noted above, the approach developed and applied in the UK is considered to be one of the best and most worthy of imitation by other countries.

4.2.3 EU planning policy

There is no specifically EU-wide coherent planning policy – whether regional planning or development planning. There are a number of different urban planning paradigms across the continent and the force with which they are (or can be) applied in practice also varies considerably. In the UK development planning is well-developed and strictly applied with some variations across the constituent nations but in general with a strong harmony between them. Germany has a particular form of landscape planning which is mainly focused on biodiversity protection but which is also strictly applied at the federal state level. In France there is a more centralized system. Table 4.1 gives an overview of the strength of control over land use change across Europe. The variation leads to differences in how landscape is treated.

TABLE 4.1 Strength of degree of public control over land use change policies, 7 being the highest and 1 the lowest

Value	*Countries*
7	–
6	Denmark, The Netherlands, Portugal, UK
5	Belgium, Cyprus, France, Germany, Greece, Ireland, Lithuania
4	Italy, Spain, Sweden
3	Austria, Bulgaria, Finland
2	Estonia, Latvia, Luxembourg, Malta, Poland, Slovenia
1	Czech Republic, Hungary, Romania, Slovakia

(*Source:* Piorr *et al*, 2011.)

The issue is raised here because as a form of development windfarms require some form of planning permission which is granted or withheld by the relevant regional or local planning authority. Governments may make a special case for strategic developments to be handled at a higher level to prevent local opposition from denying permission, but these applications may be turned down. In the countries where the system is harder to enforce this is often due to economic and business interests having more political power and so permissions for developments may be easier to obtain.

4.3 The European landscape

Europe as a whole has been settled and modified by human activity since the earliest times, and this is reflected in the landscape. With a range of climate from oceanic to continental, Mediterranean to boreal and with mountain ranges, hills, plains and islands, it is incredibly diverse in its basic physiographic and ecological fundamentals. Cultural aspects also lend diversity in the form of crops, vernacular architecture, field patterns, building materials and settlement forms. As well as having a long history the continent is on average densely populated and urbanised, although there are wide variations (see Figure 4.4). The most densely populated areas are those with concentrations of resources, industry, commerce, ports and locations of strategic importance. The south of England, the Ruhr Region of Germany, the Netherlands and Belgium and parts of France and Italy have the densest population concentrations and the largest metropolitan areas. Parts of the north of Scotland, the Iberian peninsula, together with eastern, south-eastern and northern Europe, are sparsely populated and in fact are depopulating, with, in many places, forest returning to once settled landscapes. The majority of the continent south of the Nordic countries is settled agricultural landscape with fields, farms and villages, small towns and dense road networks.

As well as the overwhelmingly settled nature of the continent, much of it is also rich in cultural history and valued for its scenic qualities. This leads to many areas being protected for cultural and landscape reasons, with a network of national and regional parks, UNESCO world heritage sites and spectacular coastlines popular for tourism. Nowadays these values are frequently recognized as a package and form a brand identity used for marketing regionally specialized products, landscapes, cultural monuments and tourism. This can lead to tensions when faced with accommodating landscape changes which threaten these values, especially in areas where tourism and related activities form the principal means of earning a living.

At the other end of the landscape quality spectrum within Europe is a significant amount of post-industrial landscape, urban sprawl and intensive industrialized agriculture. This coincides with the urban

FIGURE 4.4(A)–(F) A series of photos capturing something of the range of European landscapes showing their diversity and distinct character as long-settled: (a) is a landscape in southern Finland, dominated by forest and lakes but with little relief; (b) is in the Austrian Alps, where alpine pastures lie among the forest; (c) shows vineyards in the Bordeaux region of France, famous for its wines; (d) Northern Portugal is rugged with much planted forest and white-painted houses; (e) the Peloponnese in Greece is mountainous, with isolated villages and terraces of olive and other trees; (f) is an area in Herefordshire in England, where the strong pattern of enclosed fields and nucleated villages is typical of many parts of the countryside.

(Photos: S. Bell.)

areas where some 80 per cent of the European population live and which are already heavily influenced by urban infrastructure, electricity power lines, motorways, railways, industrial buildings and port facilities (see Figure 4.5).

In order to come to terms with this diversity of landscape as well as to conserve, develop and manage it as a whole, the European Landscape Convention (ELC) was signed in 2000 (Council of Europe, 2000). This is not a European Union convention but is organised by the Council of Europe (CoE), an international organisation which includes a number of non-EU member states. This convention, which has been signed and ratified by a majority of CoE countries, provides a standard definition of landscape and requires signatories to undertake certain tasks. It is worth quoting some key parts of this, as it helps to understand what is expected and the fact that it is not a simple case of identifying and protecting the best landscapes but that all landscapes are important.

The landscape, according to the ELC (Council of Europe, 2000), is "an area, as perceived by people, whose character is the result of the action and interaction of natural and/or human factors" (Article 1). This is an extremely important definition because it explicitly identifies perception as the key, converting land into landscape, and this leads to the importance of how people perceive it and value it on aesthetic grounds.

FIGURE 4.5 The urban edge next to heavily urbanized countryside is a common aspect of European, frequently relatively low-rise, cities. There is often little or no separation of one settlement from another in the most densely populated areas. In this example the center of Brussels can be seen from the outskirts.
(Photo: S. Bell.)

The tasks to be undertaken by signatory states to the Convention includes:

> With the active participation of the interested parties, as stipulated in Article 5.c, and with a view to improving knowledge of its landscapes, each Party undertakes:
>
> (a) i to identify its own landscapes throughout its territory;
> ii to analyse their characteristics and the forces and pressures transforming them;
> iii to take note of changes;
>
> (b) to assess the landscapes thus identified, taking into account the particular values assigned to them by the interested parties and the population concerned.
>
> (Council of Europe, 2000, Article 6)

One of the key aspects in the convention is the role of the public. They are supposed to become involved in identifying and placing values on their own local landscapes. However, public participation is not very well developed in most places.

In addition, Resolution 128 (2002)1 on the problems of Europe's countryside by the Congress of Regional and Local Authorities states:

> 9. Europe's countryside, and the people who live in it, are a highly valued and varied asset for the whole population of the continent: the largest part of rural Europe is covered by agricultural land and forests, which have a strong influence on the character of European landscapes;
> 10. . . . It is our duty to understand, protect and enhance this heritage;
> 11. At present, in many parts of Europe, the rural heritage is being rapidly eroded and even destroyed by social or technological changes, modern agriculture, urban growth, neglect and other forces;

One of these technological changes is the spread of windfarms, using a standard, more or less identical industrial structure which appears the same wherever it is placed.

One of the main ways in which the signatory states implement the ELC is under item (a) noted above. This usually means that something generally called a "landscape character assessment" (LCA) is undertaken, which is an exercise in mapping and describing a territory in terms not only of land use but also of landscape. This has been applied in many countries using similar approaches and represents an incredibly valuable resource for use in both strategic planning and landscape and visual impact assessment, as will become clear later in the chapter.

A landscape character assessment is often carried out at a national or regional level, possibly by national agencies, with more local and detailed assessments often being prepared in addition by local authorities. The process is essentially the same everywhere and is summed up in the excellent guidance published jointly by the English Countryside Agency (now known as Natural England) and Scottish Natural Heritage (Countryside Agency and Scottish Natural Heritage, 2002).

The basic procedure follows several steps and essentially involves looking at a series of map layers, for example solid and superficial geology, hydrology, soils, ecology, land use, settlement, transport infrastructure and cultural monuments. Overlaying these maps reveals areas with different combinations of characteristics which separates them from other, adjacent areas, and which also give a sense of specific character or identity. Field work and visual-aesthetic assessment confirms these units and a report of the finalized landscape character types is prepared. Normally each separate unit, or a cluster of several of the same type in different locations around the study area, is described in terms of the physical and visual characteristics, condition or state and the identifiable forces for change.

Many countries in Europe (the UK constituent nations among them), have national maps, often set up in an electronic database accessible from the Internet. In Scotland, which will be the focus of a case study later in the chapter, the national assessment was compiled from a series of separate ones covering the country and completed in 1996. The dataset is accessible free of charge as a set of files suitable for use in a geographic information system (GIS) and can be used in any planning study or impact assessment. Of course, if these assessments were carried out 20 or more years ago and there have been continual changes to the landscape, more in some areas than others, the use of such material as a baseline may not be completely reliable and may need updating when used for a project. There have also been attempts to try to harmonize assessments from different countries into a Europe-wide map, with some limited success. This might be important when cross-border impacts of proposed developments may occur.

In comparison with the USA, Canada or many other countries, Europe in general is in a strong position in terms of having data about landscapes and a legal framework within which to ensure that it is taken into account in a proper way at both strategic and project scales.

4.4 Overview of methods and approaches to considering landscape in windfarm development

There are three main aspects which need to be considered in ensuring that wind power developments have the least practical impact on valued landscape. The first of these is strategic – determining, prior to development proposals, where windfarms should best be located and where they should be avoided (see also Chapter 7). This is essentially a regional planning task and may well be carried out as a multi-factorial process where a number of key aspects are considered, not only landscape and visual concerns but other environmental factors such as archaeology – of which there is a great deal in Europe and the UK – as well as nature conservation and biodiversity. In addition there are numerous technical aspects which may also act at a regional scale, such as the pattern of windiness, the location of the main electricity grid (into which the generated power must feed), road access and civil and military aviation factors. Together these may be used to determine broad areas of suitability or to guide the type of windfarm, such as small-scale community developments or large-scale commercial developments, as well as to suggest key areas to be protected, such as large complexes of Natura 2000 sites (these are sites protected for biodiversity under EU legislation), scenically important views from National Parks or from important heritage sites.

The next logical step is to use such strategic plans to direct the potential locations for developments, taking into account the critical factors, and to design the extent and layout of the development, to select the appropriate type of turbine and to consider the associated features such as access roads, anemometer masts (for monitoring wind speeds) and routes for electricity lines, each of which adds impact and affects the landscape. There are a number of aspects which need to be taken into account in the landscape and visual design of windfarms which will be considered below.

Following on from the design stage is the environmental impact assessment (EIA), the landscape and visual impact assessment being an important and integral part of any EIA (though not always taken as seriously as it should be in some countries). Since any specific windfarm development may not be the only one in the area, the cumulative impact of all such developments may also be an important task.

The next sections will consider each of the above aspects in turn.

4.5 Strategic planning: locational aspects and landscape capacity

For the regional-scale locational aspects of windfarm planning as far as landscape and visual aspects are concerned, one valuable approach is to use the concept of landscape and visual sensitivity and capacity. These are two related concepts but are usually treated separately (as is also the case in landscape and visual impact assessment – see p. 93).

4.5.1 Landscape character sensitivity and capacity

The concept of landscape character sensitivity was developed some years ago in a special Topic Paper to support the implementation and use of landscape character assessments (Countryside Agency and Scottish Natural Heritage, 2003). The main concepts were defined as follows:

1 Overall landscape sensitivity: This term refers primarily to the inherent sensitivity of the landscape itself, irrespective of the type of change that may be under consideration. It is likely to be most relevant in work at the strategic level, for example in preparation of regional and sub-regional spatial strategies.

Relating it to the definitions used in landscape and visual impact assessment, landscape sensitivity can be defined as embracing a combination of:

- the sensitivity of the landscape resource (in terms of both its character as a whole and the individual elements contributing to character);
- the visual sensitivity of the landscape, assessed in terms of a combination of factors such as views, visibility, the number and nature of people perceiving the landscape and the scope to mitigate visual impact.

2 Landscape sensitivity to a specific type of change: This term covers assessment of the sensitivity of the landscape to a particular type of change or development. It should be defined in terms of the interactions between the landscape itself, the way that it is perceived and the particular nature of the type of change or development in question.

3 Landscape capacity: This term describes the ability of a landscape to accommodate different amounts of change or development of a specific type. This should reflect:

- the inherent sensitivity of the landscape itself, but more specifically its sensitivity to the particular type of development in question, as in (1) and (2). This means that capacity will reflect both the sensitivity of the landscape resource and its visual sensitivity;
- the value attached to the landscape or to specific elements in it.

Essentially, the formula can be summed up as: the higher the sensitivity, the lower the capacity to accommodate development, with more specific recommendations being determined by key aspects affecting particular parts of the landscape.

The same Topic Paper considered how to assess visual sensitivity and capacity as being part of an overall assessment of landscape sensitivity and capacity as follows:

4.5.2 Visual sensitivity

In a comprehensive study of landscape sensitivity, an account would be taken of the visual sensitivity. This requires careful thinking about the way that people see and perceive the landscape. This depends on:

- the probability of change being highly visible, based particularly on the nature of the landforms and the extent of tree cover, both of which have a major bearing on visibility;
- the numbers of people likely to perceive any changes and their reasons for being in the landscape, for example as permanent residents, as temporary residents staying in the area, as travelers passing through, as visitors engaged in recreation, or as people working there;
- the likelihood that change could be mitigated, without the mitigation measures in themselves having an adverse effect (for example, planting trees to screen development in an open, upland landscape could have as great an effect as the development itself).

Since these definitions were established, many assessments of landscape sensitivity to windfarm developments have been undertaken, and the concept has developed considerably through this practice. As an example, a project to assess both landscape character sensitivity and visual sensitivity as part of supplementary planning guidance to Highland Council and Scottish Natural Heritage undertaken by the author and colleagues in the then Macaulay Land Use Research Institute in Aberdeen Scotland (now the James Hutton Institute) will be presented.

In this study we carried out a combination of desk and field analysis. There were two independent parts as follows:

Assessment of landscape character sensitivity

For this work we used the SNH LCA database which is linked to the map of LCA units together with files suitable for use in a GIS (Plate 5(a)). We examined the LCA types in order to extract the main characteristics which we judged to be significant for contributing to landscape character sensitivity out of the many possible ones in the database. On the basis of earlier work and following consultation with the clients we defined four aspects: landform scale, landform complexity, land cover complexity and land cover naturalness which we believe are most relevant. Using the GIS we could give a scale of each value to each LCA unit and then combine them so as to produce an overall score, appropriately weighted (Plate 5(b)). This approach reflects the need as discussed in the Topic Paper referred to earlier, about the need for sensitivity assessments to be tailored according to the development type under evaluation.

Following the GIS analysis a draft set of maps for each category and a draft composite map were taken out into the field and each LCA type was visited and checked to see that the computer database was correct. From this ground-truthing a final set of maps and an accompanying report were compiled. Each LCA unit was given a final grading of landscape character sensitivity as high, medium or low (see Plates 5(a) and (b)).

Assessment of visual sensitivity

To measure visual sensitivity we used the factors discussed in the Topic Paper such as how people would see a landscape, from where and how much would be visible. We identified a set of candidate viewpoint types such as major tourist routes, settlements, visitor attractions, hotels and other accommodation. We also selected hiking routes and mountain summits (Munros, Scottish mountains over 3,000 ft (914.4 m) originally listed by the late Sir Hugh Munro, and Corbetts, Scottish mountains over 2,500 ft (762 m) and below 3,000 ft (914.4 m) originally listed by the late John Rooke Corbett). These offer different kinds of viewing experiences. Roads and settlements are often in lower-lying areas with views up towards summits and are often quite limited by landform, while views from mountain summits can be very extensive and open. A network of viewpoints of different types was identified and the extent of visibility from each was assessed using a landform model (digital elevation or terrain model) and a computer program to calculate the extent of visibility (viewshed) to certain radii. This produced a series of maps which showed different types of visibility depending on the category of viewpoint (see Plate 6). Each viewpoint type and individual viewpoint was given a specific degree of sensitivity so that individual or composite maps could be produced in the GIS. The maps were not registered to the LCA units although the degree of each LCA type within a viewshed was used to help to ascribe the sensitivity. These were added to the report to the clients.

The result was a digital product which could be used in different ways in order to keep it more flexible and open to testing of different types of sensitivity in relation to a proposed development rather than a fixed map. It could be combined with other digital layers so as to give planners the means of weighing different factors in their decision making.

The capacity of the landscape was not given as a final output owing to the complexity of adding the landscape character and visual sensitivities together. Consequently the study left the clients with the sensitivity information instead. A capacity map had been produced for an earlier study but became rather dated owing to the speed of development of the wind turbine technology. For example, a zone with a high capacity for an anticipated windfarm development in groups of 10–25 100 m high turbines could substantially change when the average development becomes 50–60 turbines each 150 m high. Thus we felt our approach offered greater flexibility and resilience to the planners.

4.6 Site level planning and design

Once a locational strategy has determined which areas can best accommodate a specific type of development the next stage is to consider how to plan and design the layout, the choice of turbine type – especially its height and rotor diameter, turbine position and spacing. Work has been carried out on the design aspects by Scottish Natural Heritage (2009) and some of the aspects are presented briefly below.

(A) (B)

FIGURE 4.6(A) AND (B) Perception of the size of the turbines varies according to landscape characteristics: (a) shows turbines in a very open landscape with no elements to enable the viewer to judge the size and therefore the scale; (b) is an example where the landscape contains elements which allow the size of the turbines to be assessed and their scale impact evaluated by the viewer.

(Source: Caroline Stanton.)

Since turbines are a standard industrial product they cannot be designed uniquely to fit a specific landscape. There are few alternative designs and the only factor tending to change is their overall height and rotor width in proportion to this height. Over time turbines have developed to have rather large diameter rotors in proportion to the towers and this can have an impact in terms of their visual balance and proportion.

One of the main factors affecting design is the size of the turbine in relation to the scale of the landscape. In many places, as the turbines of whatever size tend to look the same and possess no features which enable a viewer to detect their size, their scale can be very deceptive (see Figure 4.6). Since landscapes vary in scale, the effect of a standard-looking turbine can be such that it seems to dominate the landscape or can be lost in it. Thus a large turbine on a small hill can have a much larger visual impact than the same turbine on a large mountain of similar shape and profile (see Figure 4.7). This is a really complex issue which can affect the way people perceive the development.

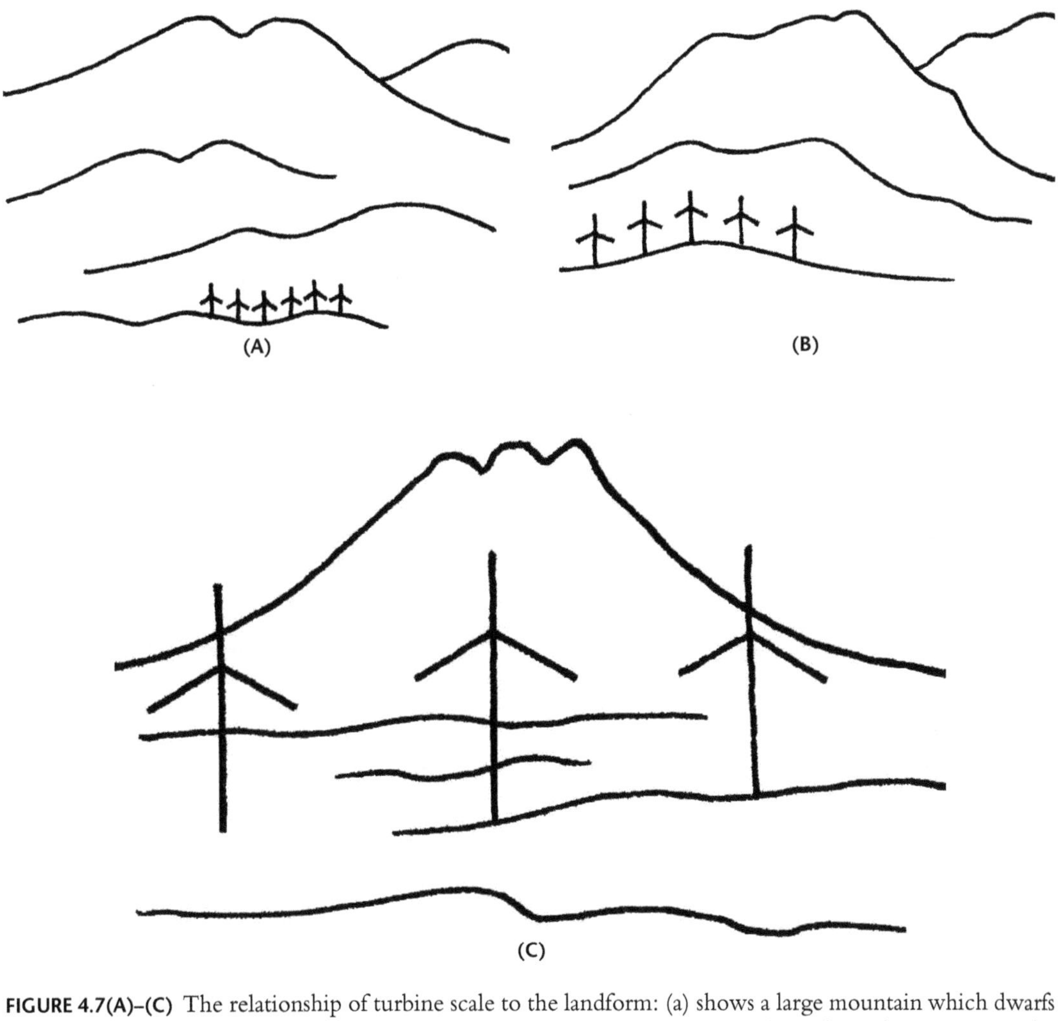

FIGURE 4.7(A)–(C) The relationship of turbine scale to the landform: (a) shows a large mountain which dwarfs the size of the turbines and so reduces their visual impact; (b) is a situation where the turbines are a similar size to the mountain; and (c) shows turbines appearing visually dominant in comparison to the mountain. (Source: Caroline Stanton.)

Another important aspect is the position of each turbine according to the landform. In irregular convex topography such as is typical on a hill or mountain it is important that turbines sit comfortably on a summit of a knoll or in the bottom of a hollow, not half way between. In irregular topography the spacing of turbines should not be on a grid but vary to take account of the micro-landform (see Figure 4.8).

The vegetation or other land cover patterns can also have an effect – the simpler the land cover the easier it is to position turbines, while in complex and diverse landscapes they may clash with other objects and patterns such as roads, field boundaries, patches of woodland and so on (see Figure 4.9). If the landscape is dominated by or at least has many other artificial and technical structures within it such as masts, chimneys, pylons or large buildings then turbines are likely to become part of such a scene more readily than when they are the only man-made elements in an otherwise natural landscape.

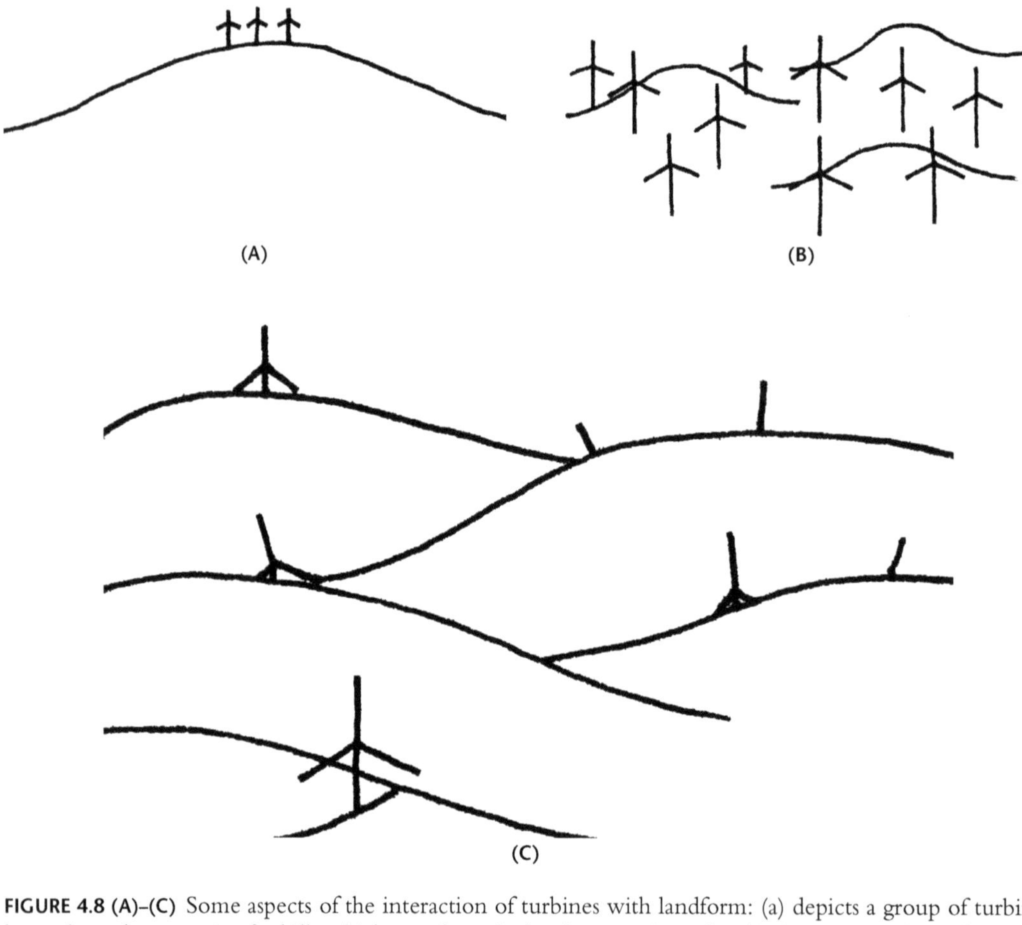

FIGURE 4.8 (A)–(C) Some aspects of the interaction of turbines with landform: (a) depicts a group of turbines located on the summit of a hill, which are also sub-dominant to its scale; (b) shows a number of turbines whose position does not relate well to landform and which are big enough to dominate it; (c) shows the common problem of turbines sited in hollows being only partly visible and their rotating blades coming and going from view can be very intrusive.

(Source: Caroline Stanton.)

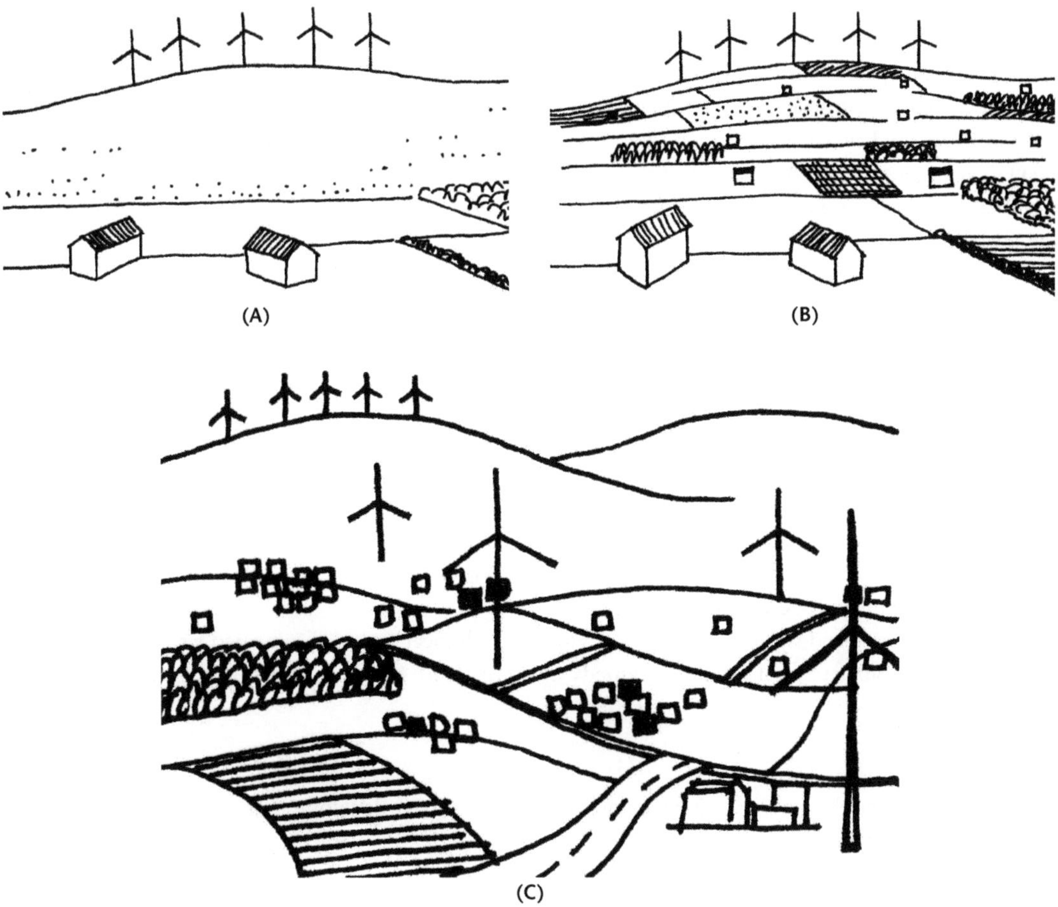

FIGURE 4.9 (A)–(C) The relationship of turbines to the overall pattern of the landscape has visual implications: (a) shows some turbines on part of a fairly simple landscape pattern where they appear fairly neutral; (b) shows the same turbines in a landscape with a stronger pattern where they tend to compete for attention; (c) shows the potential effect of the same type of turbines located in various parts of a landscape where issues of pattern and scale cause a severe visual conflict.

(Source: Caroline Stanton.)

4.7 Landscape and visual impact assessment

Once a design has been developed which meets the technical requirements and takes into account a range of constraints as noted above, the proposal will be submitted for planning consent and an environmental impact assessment will be completed (by the developer).

Article 3 of the EU Environmental Impact Assessment Directive states:

> The environmental impact assessment shall identify, describe and assess . . . the direct and indirect effects of a project on . . . :
>
> - *Human beings*, fauna and flora
> - Soil, water, air, climate and the *landscape*

- Material assets and the *cultural heritage*
- The *interaction between the factors* . . . (author's emphasis)

The relevant part of an EIA covering the aspects of landscape and effects on people is known as a landscape and visual impact assessment (LVIA). The approach presented here will be mainly based on the *Guidelines for Landscape and Visual Impact Assessment* (Landscape Institute and Institute for Environmental Management and Assessment, 2013) and therefore reflects UK practice. However, the approach is similar to other European countries which have decided to follow the UK practice as noted earlier in the chapter.

It is current practice to assess two very closely related effects:

- on the landscape (landscape impact),
- on people (visual impact).

These cannot in reality be totally separated because they are not independent, but in practice the assessment of each is separate.

The aim of an LVIA is to:

- identify systematically the likely effects of the development;
- estimate the magnitude of the effects;
- assess the nature and significance of these effects in a logical and well-reasoned fashion; and
- indicate measures to avoid, reduce, remedy or compensate for those effects (mitigation measures).

Effects can be:

- negative (adverse) or positive (beneficial), although for most developments it is the adverse effects that are of concern;
- direct or indirect, secondary or cumulative;
- permanent or temporary (short, medium or long term);
- arise at different scales (local, regional or national).

4.8 Assessment methodology

The methodology includes the following tasks:

1. Establishment of the extent of the study area.
2. Description of the setting and context of the development including land use and local development and landscape policies.
3. Identification of the landscape resources likely to be affected occurring within the study area and their sensitivity (which can be related to the factors affecting sensitivity at the strategic level described earlier).
4. Identification of the visual resources likely to be affected occurring within the study area and their sensitivity (which can be related to the factors affecting sensitivity at the strategic level described earlier).
5. Assessment of the magnitude of effect and significance of effect on the landscape resources.

6. Assessment of the magnitude of effect and significance of effect on the visual resources.
7. Identification of mitigation measures to reduce any assessed effects.
8. Cumulative impact assessment with other developments in the study area or overlapping with it.

Tasks 1, 2, 3 and 4 together form the baseline assessment. Tasks 5 and 6 form the landscape and visual impact assessment.

The site and surrounding area should be visited to obtain familiarity with the landscape both in winter and in summer in order to assess the way that seasonal differences affect landscape character and visual amenity. Field studies of the landscape character together with desk studies of existing data from maps and reports and field assessment of potential viewpoints allow the existing state of the landscape character and other resources and of the existing visual amenity of the area to be evaluated.

4.8.1 Establishment of the extent of the study area

The study area should be identified by establishing the zone of theoretical visibility (ZTV) of the development. The most common method is to use a GIS to calculate zones of theoretical visibility based on an accurate digital terrain model. Visibility can be analyzed for the whole of the turbine (from the base of the tower to the rotor tips), the rotor hub and blades, or just the blade tips. The study area is usually defined as the extent of the ZTV to the blade tips, in order that the worst case, or maximum visibility, scenario is used.

The ZTV for windfarm developments is usually calculated to a range of 35 km from the perimeter of the proposed development, ie a line following the outer perimeter of the proposed turbine layout, not the center point of the development. Such a ZTV, based solely on topography, presents the potential maximum visibility of the landscape since it does not account for any screening effects provided by vegetation, buildings or other constructions (such as earthworks) not contained within the digital elevation model (see Plate 7). Thus it overstates the actual visibility, which is taken into account during field work and final assessment. No allowance is made for the effects of atmospheric haze or pollution which can also reduce actual visibility. A 35 km limit for assessing potential visibility in the ZTV has been established as representing the maximum extent at which the naked eye can see a turbine with tower diameter and rotor blade width typical for a 100–120 m height design. At that distance it is usually rather difficult to see a turbine unless the light is good and the atmosphere free of haze. Within the overall radius of a ZTV it is normal to define different radii expanding outwards from the proposed development site. These reflect the fact that both landscape and visual impacts tend to be greater the nearer the landscape or visual resource is to the site. It is typical to show radii of 0.5, 1.0, 5.0, 10.0, 15.0, 20.0, 25.0, 30.0 and 35.0 km distant. As will become clear, proximity is one factor when defining criteria for assessing magnitude and significance.

4.8.2 Description of the setting and context of the development including land use and landscape policies

Site visits and desk studies are used to establish the site context and its wider setting within the study area. Any relevant local government plans are consulted to establish the range of policies associated with land use, development and landscape that could have an effect on the proposed development or be affected by the proposed development. Plans for housing development, for example, may result in many more viewers or housing being too close to the development.

4.8.3 Identification of the landscape resources likely to be affected occurring within the study area and their sensitivity

The landscape resources considered for assessment fall under several categories in the methodology. They represent the way in which aspects of landscape are classified and evaluated in national and local policies. They usually include most or all of the following categories:

- *Physical resources* on site such as trees, hedges, other vegetation or structures which may be affected by removal or alteration due to the development design and layout and which may form a significant part of the local landscape.
- The *landscape character types or areas* occurring within the study area as defined by the ZTV, identified from assessments carried out by national or local authorities. These may occur at several levels – larger-scale character units defined at a national or regional level with subdivisions made at a local assessment. The extent to which each type occurs within the ZTV and the degree of visibility from it to the proposed development as well as the general proximity according to the different radii from the proposed development need to be assessed on site. Any LCA units with no actual visibility within the ZTV for whatever reason – such as being heavily forested with no views possible – are then eliminated from the assessment. In addition, the LCA unit within which the development would be located is likely to be an especially important one.
- Any *historic gardens and designed landscapes* located within the study area should be identified and assessed for their visibility. The most valued examples are those which are recognized as having a high value, being laid out by renowned landscape architects or gardeners, which are listed in national, regional or local registers or inventories. These are an especially important landscape element in the UK, which has a heritage of such parks and gardens forming the grounds of stately homes and country houses. Many urban parks and other designed landscapes also occur and need to be identified. Once identified each must be visited to see if the proposed development will be visible. Often these sites have strong enclosure by trees, blocking external views. In other cases, such as those in the picturesque tradition, the external and often distant landscape has been incorporated into vistas and in these cases the site visits should establish whether such vistas and views may be compromised by the proposed development.
- *Conservation areas* designated by local authorities. These include historic town and village centers, for example, replete with old vernacular architecture. They are affected by special regulations within the UK planning system and their siting and background form important aspects of the whole ensemble. Once again, the actual extent to which the proposed development would be visible from such areas needs to be ascertained through site visits. It is unlikely that a wind turbine development would be sited within such a place unless it was a very small one, in which case it would need special planning permission.
- *Nationally designated landscapes* (National Parks, Areas of Outstanding Natural Beauty, National Scenic Areas) form the top tier of nationally and internationally recognized valuable landscapes in the UK. While turbine developments may not be proposed to be within any of these, views of windfarms from such areas may be a major concern.
- *Locally important landscapes* designated by local authorities (Areas of Great Landscape Value, Special Landscape Areas or other terminology may be used) form a lower tier of landscape value and status compared with the nationally designated areas and the degree to which a proposed development may be visible needs to be checked on the ground as for every other resource. The local status is reflected in the way the criteria for landscape sensitivity are taken into account (see p. 97).

Given the time that may have passed since many of these resources were initially assessed, described or designated, it is necessary to check them through field work to ensure that the descriptions given are up to date and correct and to fill in any gaps in coverage. For example, if the landscape character assessment was prepared some time ago the character of some LCA units may have altered as a result of various developments and boundaries of descriptions may need to be updated.

Assessment of the sensitivity of the landscape resources

Landscape effects arise from changes as a result of developments which may affect its features, character and quality. The scale and significance of the potential effect must be examined with regard to a number of factors associated with the sensitivity of the landscape resource, such as its importance, intactness, quality and capacity to accept change. For presentation, it is often helpful to present the sensitivity assessment of each resource in turn, classified by type, and then to summarise them in a table together with the assessment of magnitude and significance, so that the general pattern can be seen as a whole and the reader can check the way the assessment was carried out.

Landscape sensitivity: criteria used to assess this

The sensitivity of the landscape resources to change is defined for each category – landscape character type, garden and designed landscape, nationally designated landscape or locally designated landscape – as being high, medium, low or negligible, based on professional interpretation, aided by the site visit, of a combination of all or some of the criteria given in Table 4.2.

TABLE 4.2 Factors contributing to the sensitivity of the landscape[1]

	Sensitivity of the landscape resource			
Criteria	*High*	*Medium*	*Low*	*Negligible*
Landscape designations	Landscape designated for its national landscape value	Landscape designated for regional or county-wide landscape value	Landscape designated for local value or valued locally, for example as an important open space	No designations present
Landscape quality	Distinctive landscape with strong sense of place and integrity	Distinctive landscape with strong sense of place but with some detractors	Landscape with relatively ordinary characteristics, some detractors	Featureless, spoiled or mundane landscape with weak sense of place
Cultural heritage interests	Contains features or sites of national importance	Contains sites of regional importance	Contains some sites of local importance	Few sites or features of importance
Landscape characteristics such as pattern, scale, form, tranquillity, wildness	Landscape with characteristics that are highly sensitive and highly affected by windfarm development	Characteristics moderately sensitive to change from windfarm development	Characteristics not greatly affected by windfarm development	Characteristics relatively unaffected by windfarm development

Owing to the number of landscape resources and the cumulative effect caused by the presence of, for example, a designated landscape in a given LCA unit, it can useful to combine these and to produce a synthesised sensitivity for each LCA unit most directly affected. This is often limited to those within a 10 km radius of the site as beyond this range the effect is reduced by the number of intervening elements and their proportionately greater influence on landscape character.

4.8.4 Identification of the visual resources likely to be affected occurring within the study area and their sensitivity

The term "visual resources" normally means how people are likely to see a proposed development (i.e. places from where people would be able to see the development). In practice, since the number of potential viewing locations is theoretically vast, a sample number of viewpoints is usually chosen which best represent the way a development would be seen. Potential viewpoints should be scoped initially from the ZTV map, favouring as potential viewpoint locations places accessible by both residents and visitors. The rationale for their selection is usually to establish a broad sample ranging around the area and at different distances from the site. Typical viewpoint locations included residential areas (ranging from small villages, or sometimes individual houses close to the site, to the outskirts of larger towns), roads carrying national, regional or local traffic, railways, canals, rivers used for traffic, ferries, long distance footpaths and cycle ways and, finally, specific viewpoints, such as from major cultural sites or visitor attractions. The initial selection, normally made by a consultant on behalf of the developer, is often submitted to the local authority or government agency such as SNH for approval. Once approved, no party can claim that a specific and important viewpoint has been missed or that the viewpoints have been selected to minimise the potential impact. In fact it is best to use viewpoints which show as much of the development as possible.

Following the scoping of potential viewpoints as a desk study using the ZTV map, and the revision of the preliminary list, site visits should be undertaken, with each potential viewpoint visited. Because the ZTV is based on landform only it takes no account of the presence of screening elements such as trees, woodland or buildings, of which there are often many. This may mean that while the windfarm is potentially visible from a specific location, in fact it may not actually be so. This can only be assessed on the ground. It may also be the case that due to local topographic variation or the height of buildings and other constructed viewing locations (bridges, elevated sections of road, river embankments) other viewpoints emerge as being more significant or else a small shift in position may reveal a greater degree of visibility. All candidate viewpoints are then photographed in both winter and summer in order to assess the effects of different seasons and the visibility with and without leaves on the trees.

Viewpoint photography

The viewpoint photography must be taken with specific purposes in mind: to provide a typical panorama of the area consistent with the normal human field of view from that location when focusing on the site of the proposed development, and also to provide a base image for use in photomontage work later on. This subject has been extensively studied and tested in Scotland and detailed recommendations on how to take photographs have been published and should be followed (see also Chapter 9).

Assessment of the sensitivity of the effect on the visual resources

Visual effects result from the changes in character and quality in people's views arising from the development. The significance of the impact is determined by the sensitivity of the visual receptor and the magnitude of the visual effect. The assessment is made from a combination of wireframe renderings of the proposed development from each viewpoint and also using photomontages based on panoramic photographs taken from each viewpoint (see Plate 8).

Criteria used to assess visual sensitivity

The degree to which people are sensitive to and concerned about landscape change depends on several factors:

- the visibility of the landscape and the development within it;
- the number of people who see the landscape;
- the nature of the viewing experience by those people.

Landscape visibility: The visibility of the landscape depends mainly on the topography, the presence of elements that block or screen views and the amount of the landscape accessible to potential viewers. The viewpoints scoped within the study area may range from open and unobstructed to those heavily affected and partly screened by fore- or middle-ground trees, buildings or other features. Landform is the major landscape element to screen views and the use of the visibility analysis to create the ZTV means that this is already taken into account, except for identifying how many turbines or how much of a turbine is visible from a given point. The other factors that affect visibility are the distance to the site from the viewer and the viewing direction in relation to the lighting direction. Up to 5 km away a site can be considered as foreground and highly visible, possibly dominating views, while from 5 km to 15 km the site will be seen as part of the general landscape. Beyond 15 km it is more likely to be seen as part of the background and attention is easily diverted from it. The use of a 35 km ZTV radius removes potential viewpoints that are too far away and the lighting direction varies from view to view, so should be considered separately. In addition, the prevailing wind direction is likely to affect how turbines are seen – views against the prevailing wind will tend to see the turbines face-on, while those at right angles will see less of the turbines by comparison.

Numbers of viewers: There is usually little or no hard data available on the number of viewers expected to see the proposed wind turbine development, so this is often an estimate and may reflect local characteristics and be more of a relative than an absolute calculation. The number of viewers can, in practice, usually be inferred from information on population and from observation of the study area, the strength of the settlement pattern of towns and villages and the number and importance of transport routes and places used for recreation in a given area. Some viewpoints are likely to be used by fewer people since they are on less important roads or at smaller settlements.

The nature of the viewing experience: People who live in a particular area experience the landscape all year round together with its changing moods. They are used to seeing the landscape as it is and may not react favourably to changes taking place. Visitors to the area may see the landscape primarily from the roads and other transport routes, although footpaths and cycle ways also provide a limited but significant type of experience. Travellers see the landscape as a moving experience and may spend greater or lesser times travelling through or around the landscape seeing the development. Local people driving to and from work or using local services are likely to be more sensitive than purely business or commercial travellers passing through. Tourists travelling around specifically to experience the landscape will also be more sensitive. Table 4.3 categorizes these criteria into different levels of contribution to sensitivity. Judgement is needed in order to decide on the average or weighted average of all three factors into the overall level of sensitivity for each viewpoint.

TABLE 4.3 Factors affecting visual sensitivity

Degree of contribution to sensitivity	*Factors affecting sensitivity*		
	Visibility	*Numbers of people*	*Nature of the viewing experience*
High	• Development is clearly visible from the viewpoint • Landscape is open and unobstructed • Turbines normally face the viewpoint due to the prevailing wind • Viewing distance up to 5 km from the site	• Large numbers of residents • High volumes of travellers	• Residential viewing • Local travellers frequently using the area • Recreation and tourism visitors spend time in the area
Medium	• Development is mostly visible from the viewpoint • Landscape is partly open and unobstructed • Turbines normally face away from the viewpoint due to the prevailing wind • Viewing distance from 5–15 km from the site	• Moderate numbers of residents • Moderate numbers of travellers	• Residential viewing • Local travellers • Some recreation and tourism visitors to the area
Low	• Development is partly visible from the viewpoint • Landscape is mostly obstructed by objects in the view • Viewing distance 15–35 km from the site	• Small numbers of residents • Small numbers of travellers	• Some residential viewing • Travellers mainly on business • Few if any recreation or tourism visitors

4.8.5 Assessing the magnitude and significance of the landscape effect

The magnitude of change to the landscape resources likely to be caused by the proposed development is assessed as being large, medium, small or negligible, based on the interpretation of a combination of criteria, presented in Table 4.4. The magnitude of effect is affected by the size of the development, the distance of the landscape resource from the site, the scale of the landscape, the degree of openness and the presence of features that provide a measure of scale such as electricity pylons.

Significance of landscape effects

A combined evaluation of the landscape sensitivity and the magnitude of the effect give the significance of the landscape effect, which may be adverse or beneficial, as noted earlier, although here we are only concerned with negative effects. The definitions of the degrees of significance of the landscape effect are presented in Table 4.5. The method of assessing the combined effects of sensitivity and magnitude is given in Table 4.6.

Although there is a strong relationship between magnitude and sensitivity, it should not be used in a formulaic way: each situation needs more subtlety and there are unique aspects which may affect the final outcome. Thus professional judgement is also needed.

TABLE 4.4 Criteria used to assess the magnitude of the landscape effect

Class	*Name*	*Description*
Large	Dominant	• Very extensive, highly noticeable change, affecting most key characteristics and dominating the experience of the landscape • Introduction of highly incongruous elements • Changes over a long period of time
Medium	Conspicuous	• Noticeable change to a significant proportion of the landscape, affecting some key characteristics and the experience of the landscape • Development with some uncharacteristic elements • Changes that will affect the landscape over the medium term
Small	Apparent	• Minor change, affecting some characteristics and the experience of the landscape to an extent • Introduction of elements that are not uncharacteristic • Changes that are of short duration
Negligible	Inconspicuous	• Little perceptible change • Introduction of elements that are not uncharacteristic • Changes that are of very short duration

TABLE 4.5 Definition of the degrees of significance of the landscape effect

Significance	*Definition*
Severe adverse	The proposed project would result in effects that: • are at complete variance with the landform, scale and pattern of the landscape; • would permanently degrade, diminish or destroy the integrity of valued landscape features, elements and/or their setting; • would cause a very high quality landscape to be permanently changed and its quality diminished.
Major adverse	The proposed scheme would result in effects that: • cannot be fully mitigated and may cumulatively amount to a severe adverse effect; • are at a considerable variance to the landscape degrading the integrity of the landscape; • will be substantially damaging to a high quality landscape.
Moderate adverse	The proposed scheme would: • be out of scale with the landscape or at odds with the local pattern and landform; • leave an adverse impact on a landscape of recognised quality.
Minor adverse	The proposed scheme would: • not quite fit into the landform and scale of the landscape; • affect an area of recognised landscape character.
Negligible	The proposed scheme would: • complement the scale, landform and pattern of the landscape; maintain existing landscape quality.

TABLE 4.6 Assessment of significance of landscape and visual effects

Magnitude of change	*Landscape sensitivity*			
	High	*Medium*	*Low*	*Negligible*
Large	**Severe**	**Major**	**Moderate**	Moderate/Minor
Medium	**Severe/Major**	**Major/Moderate**	Moderate/Minor	Minor
Small	**Moderate**	Moderate/Minor	Minor	Minor/Negligible
Negligible	Moderate/Minor	Minor	Minor/Negligible	None

TABLE 4.7 Criteria used to assess the magnitude of the visual effect

Magnitude	*Description*	*Description – appearance in field of vision*
Large	Dominant	• Major change to the view, striking, sharp, unmistakeable, easily seen • Creation/Removal of dominant visual focus • Highly incongruous elements or pattern introduced • High proportion of the view affected • Long-term change
Medium	Conspicuous	• Noticeable change to the view, distinct, clearly visible, well defined • Creation or removal of a visual focus • Some elements of the development fit underlying visual composition • Significant proportion of the view affected • View changed over the medium term
Small	Apparent	• Minor change to the view but still evident • Little change to focus of the view • Fits intrinsic visual composition • Small proportion of the view affected • Short-term change to the view
Negligible	Inconspicuous	• No real change to perception of the view • Weak, not legible, near limit of acuity of human eye

4.8.6 Assessing the magnitude and significance of the visual effect

The magnitude of change in the view is determined partly by the degree of intrusion and obstruction of views in relation to the person viewing it and partly by the degree to which the nature and scale of the proposals assimilate with or alter the character and quality of the existing views as demonstrated using wirelines and photomontages. The criteria used to determine the magnitude of the visual effect are presented in Table 4.7. They are largely quantifiable and therefore relatively objective to assess.

Several factors modify the visual effect. Some of these are related to human perception and some related to the physical environment. They have been taken into account in the assessment of magnitude of visual effect.

Significance of the visual effect

The degree of significance of the visual effect (adverse or beneficial, though beneficial is not considered a likely outcome and is omitted from the criteria) is judged from the combined evaluation of the visual sensitivity of each viewpoint and the magnitude of the visual effect in a similar fashion to that of the landscape effect. The criteria used to determine the significance of the visual effect are presented in Table 4.9. Table 4.10 shows how the sensitivity and magnitude are combined, largely following the same formula as for the significance of the landscape effect. The assessment is made from the wirelines and photomontages prepared for each viewpoint (see Plate 8).

TABLE 4.8 Factors that affect apparent magnitude

Tend to reduce apparent magnitude	*Tend to increase apparent magnitude*
Static	Movement
Backgrounding	Skylining
Cloudy sky	Clear sky
Low visibility	High visibility
Absence of visual clues	Visual clues
Mobile receptor	Static receptor
Windfarm not focal point	Windfarm as focal point
Complex scene	Simple scene
Low contrast	High contrast
Screening	Lack of screening
High elevation of viewpoint	Low elevation

TABLE 4.9 Criteria used to assess the significance of the visual effect

Degree of impact	*Description*
Major adverse impact	Where the scheme would cause a significant deterioration in the existing view
Moderate adverse impact	Where the scheme would cause a noticeable deterioration in the existing view
Minor adverse impact	Where the scheme would cause a barely perceptible deterioration in the existing view
No change	No discernable deterioration or improvement in the existing view

TABLE 4.10 Assessment of significance of visual effects

Magnitude of change	*Visual sensitivity*			
	High	*Medium*	*Low*	*Negligible*
Large	**Major**	**Major/Moderate**	**Moderate**	Moderate/Slight
Medium	**Major/Moderate**	**Moderate**	Moderate/Minor	Minor
Small	**Moderate**	Moderate/Minor	Minor	Minor/Negligible
Negligible	Moderate/Minor	Minor	Minor/Negligible	None

Where the landscape or visual impact has been classified as major, major/moderate or moderate, this is deemed to be equivalent to significant effects referred to in the *Town and Country Planning (Environmental Impact Assessment) Regulations 1999.*

4.8.7 Identification of mitigation measures to reduce any assessed effects

Once the assessment of landscape and visual impact has been made and before the Environmental Statement is finalised there is an opportunity to consider what mitigation might be possible in order to reduce any major, major/moderate or moderate landscape and/or visual impact. This may include reducing the size of turbines, reducing the number of turbines, altering their position on the site, their colour or design. This is an iterative process and involves professional judgement depending on the main factors that affect the impact in relation to specific combinations of receptors. Following the identification of mitigation measures the revised proposals are reassessed to check the extent of the reduction of impact achieved.

4.8.8 Cumulative impact assessment with other developments in the study area or overlapping with it

At this point the impact assessment ceases to involve only the proposed development but examines the cumulative impact of it with other proposed, planned or constructed windfarms within a combined or "back-to-back" ZTV of 70 km. The approach examines the combined ZTV of the proposed development with each of the other sites in turn and possibly with all together (depending on the number to be included and their spatial distribution). Cumulative wirelines and photomontages showing the expected appearance are used for the assessment of the visual impact.

The selection of viewpoints for the cumulative impact assessment is based on an analysis of the draft cumulative ZTVs. They should be chosen to represent the following fixed or moving position cumulative visual impact scenarios:

- *Combined or simultaneous visibility* occurs where the observer is able to see two or more developments from one viewpoint, without moving his or her head. A 90 degree arc of view should be shown and the effects represented as described below.
- *Successive or repetitive visibility* occurs where the observer is able to see two or more windfarms from one viewpoint but has to move his or her head to do so. Visualizations, such as 180 or 360 degree arc of view wirelines, will be useful in assessing these effects.
- *Sequential effects on visibility* occur when the observer has to move to another viewpoint to see other developments or a different view of the same development.

The study of such sequential effects is a relatively new and emerging field of EIA/LVIA and this method follows guidance developed by SNH. The framework to allow such effects to be consistently described and the significance of impact determined is as follows:

- Routes to be assessed should be defined and agreed with the planning authority. The extent of these study routes should be informed by the 70 km base plan and the combined cumulative ZTVs.
- The assessment should clearly describe the baseline conditions and then describe to what extent the proposal would add additional visual impacts.

- The description should be informed and depicted by supporting wireline drawings and, where relevant, photomontages.
- Computer generated moving images or video-montage techniques may also be appropriate. Alternatively, a series of static images could be produced and viewed in time sequence.

The "journey scenario" clearly describes the notable points along the route where impact occurs and should be described and assessed in terms of:

- direction of view ("direct", "oblique", "aligned on route", or "looking NW of route", etc);
- distance from nearest turbine;
- the number of turbines visible at each windfarm development; and
- which parts of the turbines are visible at each development (eg blade tips, hubs, upper towers or full towers).

The duration of effect should also be described. For example, "assuming an average driving speed of 'x', this effect will be apparent for approximately 10 minutes between 12 and 8 km from the nearest turbine". Whether views are aligned on direction of travel or oblique to the development also needs to be made clear.

4.9 Taking account of public perceptions and opinions

The process described above is primarily a technical/expert approach and takes no heed of local preferences and/or opinions which may be in favour of or against the proposed development. While at the planning application stage the proposal will be publicly available and there are processes for comment by any and all interested parties, this is always after the proposal has been submitted and there is little opportunity to modify it. Participatory planning always works best if carried out at an earlier stage. If both developers and local planning authorities wish to improve the public acceptance of landscape change then it is a good idea to involve local residents in the process so that they can check the potential impacts from locations they deem to be important (as opposed to the consultants of local authorities) and also have some say in the layout and scale of the proposal.

While photomontages from set viewpoints can show something of the impact, because turbines move, the static effect of such an image tends to reduce the real impact. This is because the human eye and brain have developed to spot movement and a set of randomly rotating turbine blades, reflecting the light as they turn, for example, can be much more conspicuous than it appears in a photomontage. In addition, since there are many moving viewpoints – from cars, buses, trains, bicycles or from foot traffic – this also complicates the viewing experience for many people.

Using virtual reality, where an accurate computer three-dimensional model complete with rotating turbines, accurate lighting and the capacity to simulate travel through the modeled area is available, presents a different range of possibilities. Using a virtual landscape theatre, with a curved screen upon which it projects the virtual landscape (see Figure 4.10), viewers can obtain an immersive experience, can drive along a road, go to personally important viewpoints or see the difference between different layout options. If the scenarios are presented by a knowledgeable but neutral facilitator, it is also possible to discuss the pros and cons of different scenarios and to vote for preferred options (or to vote against all).

FIGURE 4.10 Use of a virtual landscape theatre to show the effects of a development in an interactive way and with turbine rotation. Members of the public can discuss the pros and cons of a specific development proposal or of different alternatives.

(Photo: Peeter Vassiljev.)

4.10 Conclusions

This chapter has summarised the situation for taking landscape and visual factors into account in the development of wind energy in Europe in general and the UK (and Scotland) in particular. It has demonstrated that there is a fairly strong European and national legislative and policy framework governing the development of windfarms. At present these requirements may be overridden if the energy policy priorities are considered to be more important, and this can lead to frustration among local communities. Although the mechanisms for public participation are in place as outlined above, the main elements of the strategic and impact assessment processes are technical and do not take public views into account. This has led to something of a backlash in some areas, where people consider that there are enough or more than enough turbines in their area and that other values such as landscape and visual impact need to be given greater weighting than energy policy.

Note

1 Tables 4.2–4.10, containing criteria describing sensitivity, magnitude and significance used in the method present, here represent the author's version of factors presented in various forms in a variety of published and unpublished reports and project environmental statements.

References

Committee on Climate Change (2008) Building a Low-carbon Economy. CCC, London.

Council of Europe (2000) European Landscape Convention, ETS No. 176, Florence, Italy.

Countryside Agency and Scottish Natural Heritage (2002) Landscape Character Assessment: Guidance for England and Scotland. Countryside Agency, Cheltenham.

Countryside Agency and Scottish Natural Heritage (2003) Landscape Character Assessment: Guidance for England and Scotland: Topic Paper 6. Countryside Agency, Cheltenham.

European Commission (1985) Environmental Impact Assessment Directive (85/337/EEC). EC, Brussels (amended 1997, 2003, 2009).

European Commission (2001) Directive on Electricity Production from Renewable Energy Sources (2001/77/EC). EC, Brussels.

Landscape Institute and Institute for Environmental Management and Assessment (2013) *Guidelines for Landscape and Visual Impact Assessment* (3rd edition). Routledge, Abingdon.

Piorr, A., Ravetz, J. and Tosics, I. (eds) (2011) Peri-urbanisation in Europe: Towards European Policies to Sustain Urban-rural Futures. PLUREL Consortium, Copenhagen.

Scottish Natural Heritage (2009) Siting and Designing Windfarms in the Landscape (Version 1) SNH, Perth.

5

SOCIAL ACCEPTANCE OF RENEWABLE ENERGY LANDSCAPES

Richard Smardon and Martin J. Pasqualetti

Highlights

- The successful introduction of renewable energy is fundamentally linked to its social acceptance.
- Visual perceptions of renewable energy equipment dominate public acceptability but differ between visitors and residents.
- Land-based wind installations are generally accepted, except in areas with sensitive landscape values, including recreation and tourism.
- Acceptance of offshore wind installations increase with distance offshore.
- Large-scale solar is generally acceptable but is so new that little research has been done in this regard.
- Geothermal development is generally accepted if air quality and indigenous landscape values can be addressed.

5.1 Introduction

Whether we are considering France, Greece, Germany, Italy, Portugal, the US, or just about any other country, a preponderance of survey respondents say they favor renewable energy over conventional forms of generation (e.g. Jobert et al. 2007, Kaldellis et al. 2012, Oikonomou et al. 2009, Cicia et al. 2012, Ribero et al. 2013, Klick and Smith 2010). But the table turns when it comes to actually siting large-scale renewable energy facilities – people rarely like them 'in their back yards' or even within their viewshed (Pasqualetti 2011a, 2011b, 2012) (Figures 5.1–5.3). Many organizations are dedicated to opposing them.[1]

Public responses to renewable energy are complicated, varying by country, location, technology, policy, and a host of demographic factors. Our goal is to unpack these reactions, by examining both those surrounding social acceptance and those surrounding opposition to large-scale renewable energy facilities in several countries.

We begin by summarizing, country-by-country, public views of renewable energy development. We follow this by summarizing socio-cultural attributes affecting acceptability. Then we drill down to public perceptions for specific renewable development for wind on land and offshore, solar, and

FIGURE 5.1 Opposition to wind power in western-most Virginia.
(Photo: M. Pasqualetti.)

geothermal. We include table summaries by country, using available publications. Later, we break down these findings by the themes we identify as most salient.

5.2 General public reactions to renewable energy

While reactions to prospective renewable energy development are generally favorable, a review of the international literature on acceptance reveals that the public has a more nuanced view, and it is broadening. Over the past 20 years, reactions typically have targeted wind energy, but other types of

FIGURE 5.2 Opposition to wind power in France.
(Photo: Paul Gipe. Used with permission.)

FIGURE 5.3 Opposition to solar power in California.
(Photo: Miriam Raftery. www.EastCountyMagazine.org. Used with permission.)

renewables are attracting attention as well. These include hydrokinetic energy (wave and tidal) (Devine-Wright 2011), solar (Heras-Saizarbitoria et al. 2011, Sovacool and Ratan 2012), and large-scale biomass (Williams 2011) energy production. Again, we find support in a general sense that the 'idea' of renewable energy development is attractive, but this support does not always carry over to specific projects.

Wind projects offer several good examples of what we are getting at. For example, Bohn and Lant (2009) teased out further insights that general community acceptability depends on: (1) procedural legitimacy in siting decisions, and (2) perceived aesthetic fit with the local landscape. Similarly, Toke et al. (2008) compared wind power deployment in six places – Denmark, Germany, Scotland, the Netherlands, England and Wales – and found several key institutional factors contributed to positive public opinion: (1) more effective planning support systems in some countries, (2) better financial support (e.g. Demark, Germany and Spain), (3) strong landscape protection organizations, as in England, and (4) local ownership.

Wolsink's research solidified the concern for aesthetic intrusion that had been raised by others such as Pasqualetti, Gipe, and Righter (2002), concluding that "visual evaluation of the impact of wind power on landscape values is by far the dominant factor in explaining why some are opposed to wind power and others are supporting it" (Wolsink 2010, p. 188). Wolsink (2010) and Pasqualetti (2011b), working 5,000 miles apart, found that opinions about renewable energy are liable to change once developments are initiated, shifting over time from conceptual acceptance to specific opposition to pragmatic support.

In general, populations support renewable energy development, even while recognizing some of the inherent drawbacks. Support is especially evident in Germany, France, Spain and the UK, where there has already been substantial wind energy development. While similarities in the public reaction to wind power exist from place to place, there are also differences, as summarized in Table 5.1. A review of the literature cited in this table also reveals the acceptance of renewable energy decreases with age, and in most countries, acceptance appears to rise with higher levels of education. We anticipate reactions to other forms of renewable energy would be in a similar vein.

TABLE 5.1 General public reactions to renewable energy

Country	*Summary Findings*	*Sample Size/Year*
Australia	95% supported building windfarms 94% supported renewable energy target Source: Australian Research Group for AusWEA (2003)[2]	1,027/2003
Austria	86% favor solar 55% favor hydro 50% favor wind Source: Gallup Institute + Wind Direction Focus (Oct. 2003)[3]	1,500/2003
Denmark	68% support continued wind development Source: Wind Direction Focus (Oct. 2003)[4]	random national/2001
Finland	Positive understanding of renewable energy Related to age: older people are less supportive Supported if cost savings Source: Moula et al. 2013	N/A

(*Continued*)

TABLE 5.1 Continued

Country	*Summary Findings*	*Sample Size/Year*
France	93% consider wind non-polluting 89% consider wind as waste-free 82% support renewables as energy independence 63% acknowledge countryside loss of attractiveness 28% wind is disturbing, 21% noisy 52% support further wind development	2,800/2002
France – Aude region	55% consider wind as clean energy 51% consider it economical 63% considered as tourist appeal Source: Wind Direction Focus (Oct. 2003)[5]	2,090/2003
Germany	67% support further wind development 88% support windfarms if planning criteria are met 86% agreed to increase wind in renewable power mix Source: EMIND 2002 and 2003, Bielefeld University 2003[6]	National/2003
	Some coastal regions' residents disapprove Replacing turbines or building new one Source: Meyerhoff et al. 2009	2008
Greece	51% positive for existing wind parks 39% positive for new wind parks Source: Kaldellis 2005	417/2001–2002
	62.3% consider as clean energy 32.8% consider as clean energy Source: Kontogianni et al. 2014	183 households/2005
New Zealand	Wide range of factors affecting support and opposition to windfarms Source: Graham et al. 2009	2006–2007
Scotland	82% support increase in wind energy 50% increase in turbines in local areas Source: Braunholtz 2003	1,800/2003
Scotland – Tourism	43% residents consider windfarms equally positive or negative	307/2002
	75% visitors either positive or neutral Sources: Tourists Attitudes Toward Wind Farms MORI Scotland 2002 and Investigation into the Potential Impact of Wind farms on Tourism in Wales[7]	180/2002
South Africa	85% support windfarm development on the west coast region Source: Lombard and Ferreira 2014	98/2012
Spain – Navarre	85% in favor of wind development	2001/region
	75% preferred wind previously Source: CIES + Wind Direction Focus (Oct. 2003)[8]	1998/region

TABLE 5.1 Continued

Country	*Summary Findings*	*Sample Size/Year*
Spain – Taragonna	80% favor wind in Catalonia 62% advantages of wind outweigh negatives Source: APPA Spanish Renewable Energy Assoc.[9]	2001–2004
Spain – Albacete	79% see wind energy as a benefit 79–91% benefits compensate for negative effects Source: CIES (2002) + Wind Direction Focus (Oct. 2003)[10]	2002/region
Sweden	64% prefer wind for renewable energy 39% prefer hydro Source: SIFO (2002) + Wind Direction Focus (Oct. 2003)[11]	2002/region
	49% positive, 19% somewhat positive 23% neutral and 10% negative toward wind power support decreasing with age Source: Ek 2005 Generally positive for wind power but 27% see landscape impacts Source: Söderholm et al. 2007	1,000/2002
Turkey	66% residents endorsed investment in renewable energy Source: Ertör-Akyazı et al. 2012	2,422/2007
	Solar and then wind are highest ranked clean energy sources Source: Erbil 2011	National/2011
UK	77% from 42 surveys favor wind energy Source: IPSOS for British Wind Energy Association of Surveys[12]	2,600/1990–2002
	80% support wind energy Sources: Bell et al. 2005, Toke et al. 2008	2000
USA	Public understanding of wind energy is low and there are both positive and negative factors involved Source: Klick and Smith 2010	National/2008

5.3 National public response to renewable energy

Taking the general public acceptability to renewable energy as a starting point, we now examine social factors that affect the acceptability of renewable energy development, with greater granularity, at the national level. We have two goals. First, we wish for the first time to summarize the findings of a significant sample of surveys of public opinion into land-based wind development. Second, we aim to identify within this quickly growing literature the key similarities and differences that cross international borders. We later summarize those findings in Table 5.2.

We begin with surveys in Ontario and Nova Scotia, Canada; and Michigan, Indiana, and Texas, United States.

5.3.1 North America

Canada and United States

Canada (10,000 MW installed) and the United States (66,000 MW) have been among the leading countries in wind development (Figures 5.4 and 5.5). Both are also steeped in strong support from environmental protection. In Ontario, the key concerns about wind development are health, risk perception, community benefits, community control, fair distribution of community benefits, general community enhancement, and a stronger support for small-scale deployment, from one to ten turbines (Baxter et al. 2103). To the east, along the Atlantic coast in Nova Scotia, Corscadden et al. (2102) found residents looking to wind power as a sustainable option that, yes, maximizes environmental benefits, but also helps control utility bill increases.

Across the border to the south, in Presque Ile County, Michigan, Bidwell (2013) identified the single largest perceived drawback as the effect on the local economy within a community altruism value linked to 'positive energy' support. Wind enthusiasm tended to diminish with age. Another survey in Michigan (1,500 MW installed), this time of 1,000 households in the eastern part of the state, identified the strongest concerns as visual aesthetics, noise, health, and unwanted change (Groth and Vogt 2014). Social concerns were more influential than environmental benefits within a specific township, while social and environmental factors were more influential than economic within a larger county area.

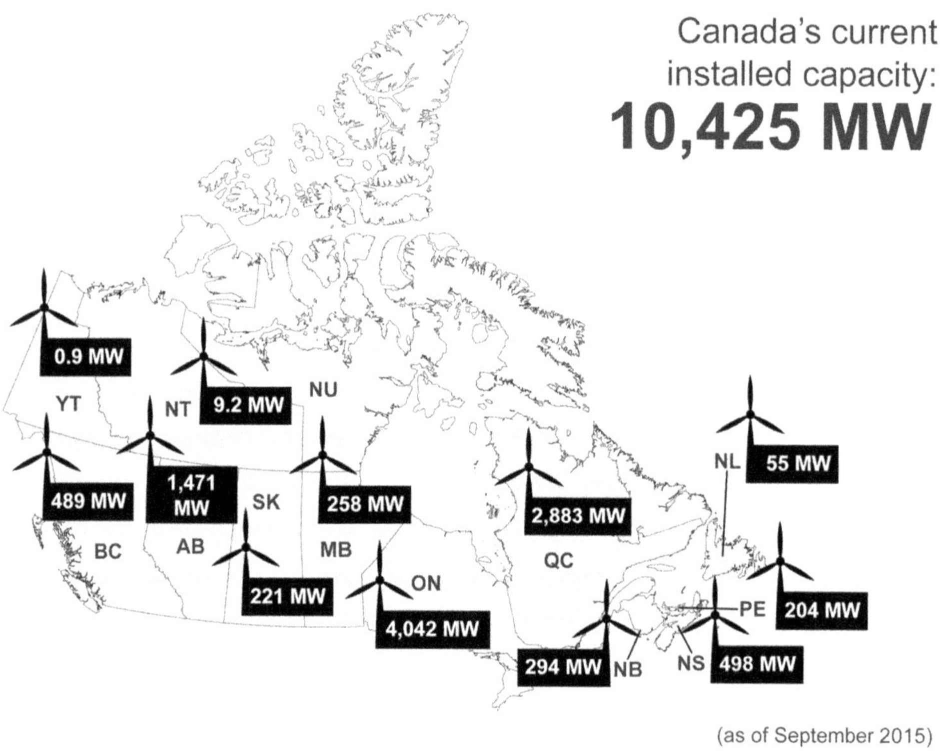

FIGURE 5.4 Installed wind capacity in Canada, as of September 2015.
(Source: After Canadian Wind Energy Association. Redrawn by Mark Warfel Jr.)

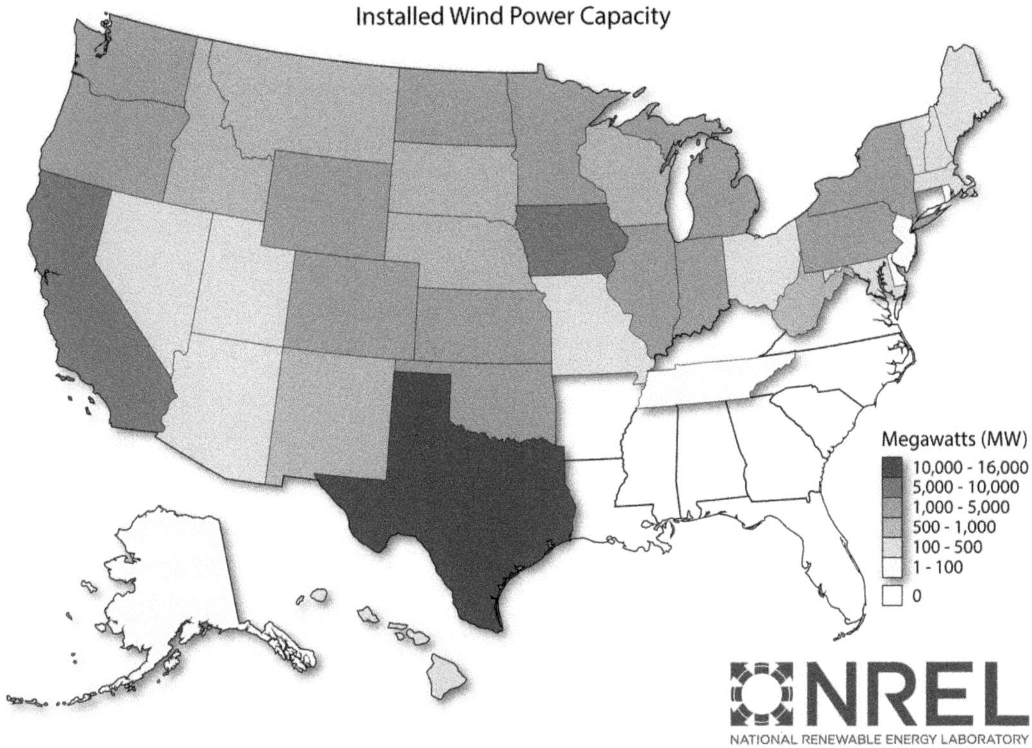

FIGURE 5.5 Installed wind power capacity in the United States, as of June 30, 2015.
(Source: National Renewable Energy Laboratory.)

In three counties in Indiana (1,700 MW), wind development was supported by the poorer cohort for its expected economic and social benefits (Mulvaney et al. 2013a, 2013b). Positive responses usually circled around the assumption that wind development would offer some protection of the farming lifestyle, although they also expressed resistance to change, concern over night lighting and changes in landscape, increased noise and possible negative impacts on health.

Swofford and Slattery (2010) conducted a survey in Cooke County in Northern Texas, the leading state in the US, with 14,000 MW. From the 200 surveys completed, about 60% were positive about wind energy development. Only 18% had a negative reaction. Again, older people were more disinclined toward wind power. In terms of perceived impacts, 30% reported disturbing noise, 47% judged turbines as unattractive, 90% reported seeing wind turbines while driving and those living closest had the greatest negative reaction, a common finding.

5.3.2 Europe

Recent years have witnessed rapid wind development in Europe, owing in large part to generous feed-in tariffs. This has been especially noticeable in Germany, Spain, and Italy, but several other countries have moved forward with speed as well, such as Denmark, Sweden, and the United Kingdom (Figure 5.6). Positive public opinion has been important.

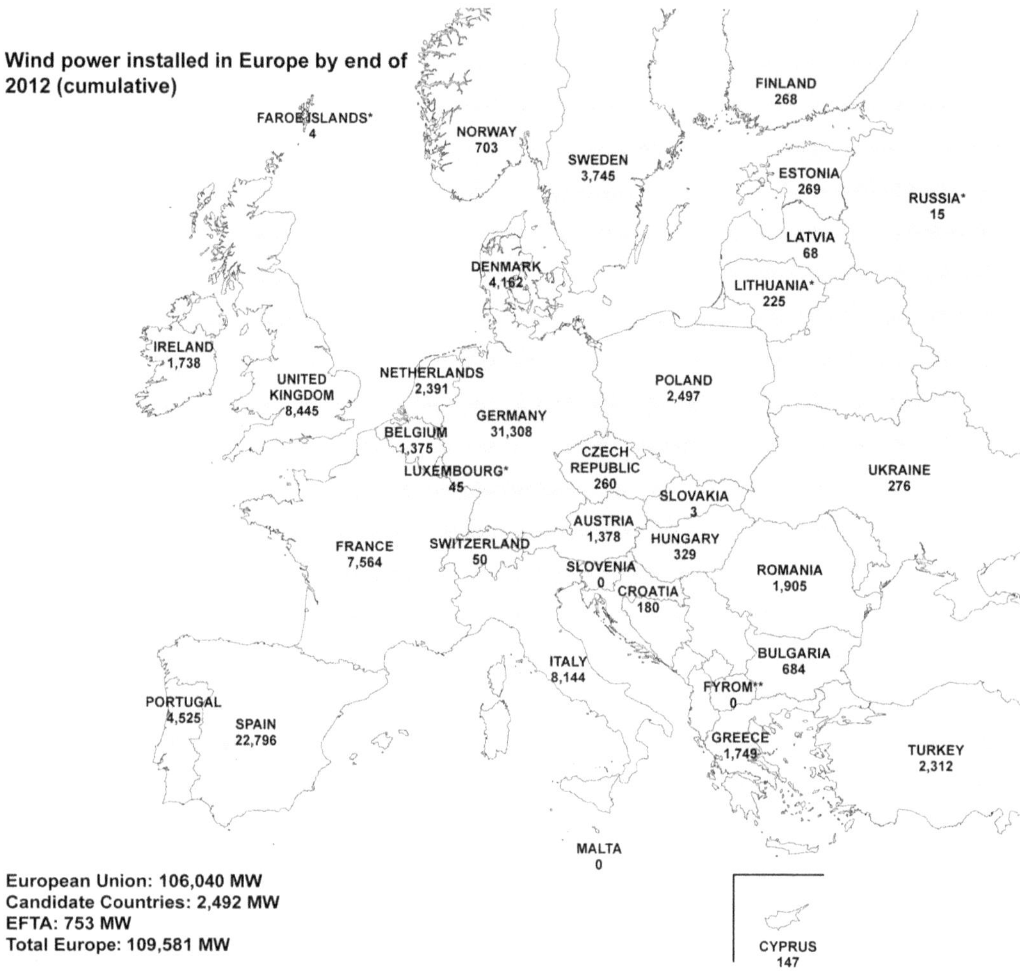

FIGURE 5.6 Installed wind power capacity in Europe through December 2014.
(Source: After European Wind Energy Association. Redrawn by Mark Warfel Jr.)

United Kingdom

After a rocky start, wind projects in the UK picked up to a rather rapid pace (Figure 5.7).

Onshore wind generated 7,001 GWh in the first quarter of 2015, followed by offshore wind at 4,322 GWh. Onshore wind has a cumulative capacity of 8,850 GW, a 12.1% increase on a year earlier. Offshore wind has a cumulative capacity of 4,749 GW. In total, 22.3% of total electricity is generated by renewables, including wind, solar, biomass, and hydro (Bell 2015). The entire country of the United Kingdom covers less land area than the State of Arizona – but with 10 times the people – so it is easy to accept that there are few places where wind installations do not attract attention.

While much of the UK is quite windy, west coast locations are the most attractive to developers. Almost completely rural, the county of Cornwall in the southwest has received more attention than most other areas of the country. For the 15 years between 1991 and 2006 – despite increasing wind

FIGURE 5.7 Bonus Anglesey Cemaes, wind development in Wales, with Wylfa nuclear power plant in the background.
(Photo: Paul Gipe. Used with permission.)

development there – there has been no significant change in opinion regarding residents' general acceptance of wind development (Elthan et al. 2008).

The central, more mountainous part of the UK is also attractive for wind development. In the area around Sheffield, the general attitudes were positive regarding wind development (Jones and Eiser 2009). No NIMBYism was evident, despite some concern over house depreciation, fear of change, unknown damage to landscape values, lack of trust in decision makers, and relative uncertainty over community support. In a later study in nearby Doncaster, Jones and Eiser (2010) discovered local development threshold predictors, including: general attitude, perceived knowledge, community attachment, environmental values, visual attractiveness, plus fairness and equity. There was expressed concern with cumulative effects on landscape change and visual impacts. Given the degree of centuries of industrial re-making of the landscapes in this area of the country, any public opposition to wind power does conjure up bemusement and irony.

Van der Horst and Toke (2009) conducted a GIS-based analysis of the response to wind development in rural England with a data set of 117 variables including education, health, demography, employment and housing. They uncovered a significant correlation between windfarm planning process outcomes and voter turnout and life expectancy. There are unequal outcomes in the windfarm site planning process related to coalitions of special interest groups with social capital that fare better because they are equipped to influence or shape outcomes. This suggests that wind proposals fare better when the public is engaged in the process and feel empowered about its results.

Concentrating in the northern regions of the country, Warren et al. (2005) surveyed two communities each in Scotland and Ireland. Their particular emphases included residents' reaction to windfarms before and after construction, as well as spatial effects such as proximity to windfarms. Objectors thought windfarms were noisy and visually intrusive. Significantly, these were held opinions regardless of actual noise or number of turbines visible. Aesthetic perceptions were the strongest influence on public attitudes, both negative and positive, an intuitive finding given the remote and open nature of these parts of the country. Respondents often noted the "seemingly instantaneous landscape transformation" (p. 888).

Denmark, the Netherlands and Germany

Wind development in these three countries has been especially notable. Taken together, they boast a generating capacity of about 41,000 MW in 2013 (EWEA 2014). They have also attracted almost continuous evaluation since the 1990s. No country has adopted wind power as enthusiastically as the Danes, suggesting these early findings have held true. In 2014, Denmark had 4,855 MW installed, and had the highest proportion of wind power in the world, with production in 2014 being 39% of total power consumption. For the month of January 2014, that share was over 61%. On November 3, 2013, wind power production exceeded the level of power consumption.

One benchmark for Denmark is the 1998 study by Krohn and Damborg on Sydthy, Denmark. They found the following:

- Younger residents were much more positive about windfarm development than older residents.
- People in the city zone were more negative than people in more rural areas.
- Four out of five are unbothered by noise and the larger visual presence.
- Men perceive louder noise from turbines than women.
- Public acceptance seems to increase in local areas after installation.
- About 61% indicated they would not mind more turbines.

Concentrating on nearby areas in the Netherlands and Germany in 2005, Wolsink compared several results from 1986 to 2002. In the Wadden Sea area, the primary concern was landscape impact, with other factors including equity and fairness of the siting process. A later paper (Wolsink 2007b) again found that the existing visual quality of existing landscape overshadows other attitudinal attributes such as design, number, and size of turbines.

In eastern and central Germany, wind development encounters little opposition over issues such as intrusion (Meyerhoff et al. 2009). Specifically, respondents said that current turbines do not influence the quality of the regional environment, that distance to residential areas is a key factor, and that smaller turbines are better. In southeastern Germany, Musall and Kuik (2011) found positive reactions to increased use of local wind energy, and the community ownership model was found to aid in the reconciliation of local opposition. Today, Germany has about 35,000 MW of installed wind generating capacity.

In total, in each of these three countries – countries that have seen some of the fastest and most important development of wind power, public acceptance to wind turbines has been notably favorable.

Sweden

Sweden, which has little fossil fuel, has a reputation for energy efficiency and a favorable disposition toward renewable energy development. In the past 25 years, the Swedish government has even proclaimed its desire to displace its nuclear power generation with these two alternatives. Although such a transition has not been possible, the country continues moving in that direction. It is with this background that we were not surprised to find strong support toward wind power (Ek 2005, Ek and Persson 2014, Johansson and Laike 2007). As with other studies noted earlier, however, support decreases with age and income, but increases with people who buy 'green' products. About 75% consider visual impact to the landscape as a major negative impact of windfarms, stressing the importance of how well wind turbines 'fit' within the landscape. Respondents preferred that wind development be located far offshore, and avoid mountainous areas, both of which have high recreational value (Figure 5.8). They found less concern for locations in or near residential landscapes.

Johansson and Laike (2007) conducted a survey of some 80 residents in agricultural landscapes in close proximity to urban residential areas. They were asked questions about the wind turbines' 'fit' within the landscape. Respondents felt that the most important consideration was whether the wind turbines fit within the landscape. Emotional states were not involved, except the perception of 'unity,' or how they 'fit' with the existing landscape.

In sum, responses to wind power are highly favorable, and developments are viewed positively as long as they avoid sensitive landscapes, and are designed to blend in as much as possible to other landscapes.

FIGURE 5.8 Avoiding the scenic coastal and mountainous areas of the country, wind development in Sweden is more common on farmland, here south of Gothenburg.

(Photo: M. Pasqualetti.)

Greece

Greece is a rather small country with a struggling economy and little in the way of fossil fuels. For these reasons, one might expect public opinions about renewable energy to be favorable, but it is a bit more complicated than that. Kaldellis et al. (2013) found that over 90% were familiar with wind energy and 78% perceived wind energy benefits. Almost two-thirds favored existing wind energy development, but only 35% were supportive of new wind energy installations. Two-thirds said they are not visually affected, 16% found windfarms attractive, whereas 21% found wind turbines annoying or aesthetically 'not right' (p. 203).

Kontogianni et al. (2014) conducted face-to-face interviews in picturesque Southern Eva. One principal finding was that noise and visual intrusion were problematic for 35–46%, but once they were operational, noise and visual intrusion was problematic for only 19–23% of respondents. Overall, respondents supported expansion of wind energy in other parts of Greece rather than their own region, a typical NIMBY response.

Dimitropoulos and Kontoleon (2009) developed and administered a choice experiment questionnaire to 212 respondents in 2007 for two island communities – Naxos and Skyros in southern Greece. The choice questionnaire included different energy development options. There was strong support for wind energy nationally for the islands. Naxos residents felt that wind energy development was unlikely to affect tourism whereas Skyros residents felt that it would have a significant negative impact. Being islands, their choices are limited and the price of their conventional sources of electricity is high.

Finally, in response to growing interest in solar power, Kaldellis et al. (2012) conducted a national survey to judge social acceptance of solar photovoltaic (PV) arrays. They found that 94% were in favor of PV parks, 53% were positive under certain conditions (such as incentives), and 30% needed proof of usefulness to gain their approval. Slightly over half perceived no visual impact from PV installations while 22% said they were visually annoying and not 'aesthetically right.' In general, public opinion in Greece for new PV installations was more positive than for new wind installations, though wind installations garnered mostly positive responses, whether on the mainland or on the islands.

5.3.3 Oceania

New Zealand and Australia

Wind developers are looking in just about every area of the world, and New Zealand and Australia have been receiving substantial attention. Graham et al. (2009) found no reliable relationship in New Zealand between proximity and attitudes toward windfarm development. That is, it made no difference whether the windfarm was closer or further from where respondents lived in terms of acceptance. The most frequently cited aspects of windfarm physical attributes were farm size and shape, which had some effect on acceptance.

Williams (2011) conducted surveys about rural landscape change from wind and biomass energy development in Tasmania and Western Australia. In general, renewable energy from both biomass and windfarms were viewed positively. Hall et al. (2013) examined seven ongoing wind development case studies. They found four common issues significant in forming public opinions about the wind development: trust in energy development companies or utilities, distribution justice, procedural justice, and place attachment – meaning visual changes to a place or landscape.

A third study (Lothian 2008), rated simulated windfarms in both inland and coastal landscape contexts. Respondent perception of scenic quality depended upon the existing scenic quality of the landscape, a recurring theme in every country. What is different, however, was specific evidence that

Lothian found that the addition of windfarms generally decreased visual quality for high quality landscapes but increased the visual quality of lower quality landscapes (p. 205). In this sense, wind can have a leveling impact on the landscape and, with all else being equal, should favor development in areas of lower visual quality. A common conclusion is: leave the nice places alone, but allow more renewable energy development in lower quality and already developed landscapes.

***

While the range of influences and reactions is wide, the one repeated more than any other among the 13 countries surveyed – and highlighted in italic and bold type – is opposition to changes imposed on the landscape (Table 5.2). In other words, the singularly most important concern about renewable energy is visual intrusion, a subjective matter of aesthetic values. It is essentially the same as NIMBYism. The public is saying, in effect, "We realize that renewable energy has many advantages, and we support its development. Just don't put it where I can see it."

Table 5.2 summarizes the principal themes noted in each country. These themes cut a wide swath, including economic benefits, place identity, public participation, equity, environmental justice, confidence, and trust in the development process, as well as differences in ownership, age, and education.

Many of the identified reactions emanate from wind power development more than from any other form of renewable energy, and we believe these reactions result from a combination of many factors, several of them obvious. For example, the turbines are as about as different from a stationary power plant as is possible, attracting attention in ways the public has little experience of absorbing. The blades spin, glint, and reflect sunlight, the entire apparatus rotates on its axis as the wind direction changes, aircraft warning lights blink day and night. Moreover, wind is a low-density energy resource, therefore resulting in a low ratio of kilowatts to land area. In short, the public is not accustomed to this form

TABLE 5.2 Social-cultural attitudes and renewable acceptance

Country	*Social-cultural attitudes*	*Supporting literature*
Australia	Trust in established procedures	Hall et al. 2013
	Place attachment	
	Democratic legitimacy, fairness	Hindmarsh 2010
	Place-based community interest	
	Areas of high scenic quality	Lothian 2008
Canada	Health risk perception	Baxter et al. 2013
	Community benefits	
	Community enhancement	
	Wind preference	
	Economic benefit and fairness	Corscadden et al. 2012
	Community consultation	
Czech Republic	***High landscape quality***	Molnarova et al. 2012
	Unattractive landscape acceptability	
	Visibility	
	General wind power attitude	
Denmark	Influence of prior experience	Ladenburg 2008, 2009b
	Local land usage patterns, location	
Germany North Sea	***Aesthetic seascape***	Gee 2010
	Sea as natural space	
	Local landscape identity	

(*Continued*)

TABLE 5.2 Continued

Country	*Social-cultural attitudes*	*Supporting literature*
	Renewable energy/climate change	
	Co-ownership	Musall and Kuik 2011
Germany mainland	Economic benefits	Zoellner et al. 2008
	Landscape evaluation	
	Procedural justice	
	Planning participation	
Greek islands	Conservation status	Dimitropoulos and Kontoleon 2009
	Governance characteristics	Kaldellis 2005
Greece mainland	***Visibility not-in-my-front-yard***	Kontogianni et al. 2014
	Physical landscape context	
	Socio-economic parameters	
Netherlands	Local planning context, participation	Breukers and Wolsink 2007
	Type of landscape	Wolsink 2007a, 2010
	Local involvement	De Vries et al. 2012
New Zealand	Positive general attitudes for renewable energy development	Graham et al. 2009
Scotland	***Landscape aesthetics***	Warren et al. 2005
	Community (dis)empowerment	
	Global/local factors	Warren and McFadyen 2010
	Intermittent production	
	Community ownership	
South Africa	Positive general attitude	Lombard & Ferreira 2014
	Place attachment	
Sweden	Local environmental factors	Söderholm et al. 2007
	Local community ownership	Ek and Persson 2014
	Local involvement process	
	Environmental intactness	Johansson and Laike 2007
	Landscape aesthetics and use	Waldo 2012
United Kingdom	Equity-impact distribution	Cowell 2009, van der Horst and Toke 2009, Mason and Milbourne 2014
	Project impact uncertainty	Jones and Eiser 2009
	Visual impact, place attachment	Haggett 2011
	Tangible benefits	Cass and Walker 2009
	Relationship with developers/others	Devine-Wright et al. 2007
	Visual attraction, energy security	Elthan et al. 2008
United States	Damage to the environment	Firestone et al. 2009
	Electricity rates, **aesthetics, noise**	Groth and Vogt 2014
	Impacts to specific activities	
	Proximity to the proposed project	Swofford and Slattery 2010
	Long-term impact uncertainty	
	Economic community benefits	Bidwell 2013
	Reduced energy bills	Mulvaney et al. 2013a, 2013b
	Landscape and rural lifestyle protection	
	Aesthetic intrusion	Pasqualetti 2011a

of development. As other forms of renewable energy development expand, we expect some of these concerns will abate.

For convenience and clarity, Table 5.3 summarizes the same country-by-country data sources and organizes the findings according to four major social cultural factors:

- physical environmental factors such as turbine acoustics, or odor from geothermal;
- contextual factors such as proximity to turbines and landscape context;
- political and institutional factors, such as energy policy support, self-efficacy, institutional capacity, public participation and consultation, and social and communication factors such as social influence processes (media, social networks);
- symbolic or ideological factors such as representations, place-identity processes, local community benefit and control, opposition to immediately local impacts (Not In My Back Yard, or NIMBY), and previous knowledge;
- socio-economic factors, such as shareholding, equity of impacts, and economic benefits.

TABLE 5.3 Socio-cultural factors affecting land-based wind development

Author/ Physical country environment	*Landscape context and proximity*	*Social, political, communication*	*Symbolic, ideological*	*Economic, equity, security*
United Kingdom				
Elthan et al. (2008)		✓		✓
Jones and Eiser (2009)		✓	✓	
Jones and Eiser (2010)		✓	✓	✓
van der Horst and Toke (2009)		✓		✓
Warren et al. (2005)	✓	✓		
Wadden Sea Area (Germany, Denmark and the Netherlands)				
Krohn and Damborg (1998)	✓	✓	✓	
Wolsink (2005)		✓		✓
Wolsink (2007b)	✓			
Meyerhoff et al. (2009)		✓	✓	
Musall and Kuik (2011)		✓		✓
Sweden				
Ek (2005)		✓	✓	✓
Ek and Persson (2014)		✓	✓	✓
Johansson and Laike (2007)		✓		
Greece				
Kaldellis (2005)		✓		
Kaldellis et al. (2013)		✓	✓	

(*Continued*)

TABLE 5.3 Continued

Author/ Physical country environment	*Landscape context and proximity*	*Social, political, communication*	*Symbolic, ideological*	*Economic, equity, security*
Kontogianni et al. (2014)	✓	✓		
Dimitropoulos and Kontoleon (2009)		✓		✓
New Zealand and Australia				
Graham et al. (2009)	✓	✓	✓	
Williams (2011)	✓			
Hall et al. (2013)	✓	✓	✓	✓
Lothian (2008)	✓			
United States				
Swofford and Slattery (2010)		✓		
Bidwell (2013)				✓
Mulvaney et al. (2013a, 2013b)		✓		✓

5.4 Offshore wind energy development social factors

Offshore wind development has become increasingly popular in certain areas, such as Denmark and the Netherlands, partly because it does not tend to trigger as much public opposition as developments on land in these more densely populated countries, and partly because some of the areas bordering the North Sea are relatively shallow, making offshore development more economical. The UK has about one-third of its wind development offshore, with several particularly large projects in the planning stages.

One of the major issues linked to offshore developments is the possible effect on tourism, especially ocean- and beach-related tourism. However, an early survey in Scotland (Braunholtz 2003) identified little concern from tourists. The Scottish Renewables survey of tourist willingness to revisit coastal areas found that 43% stated no difference, 28% were positive, 15% were completely positive and 7% were negative. Results indicate that only two out of five tourists were aware of windfarms in the vicinity, 43% said windfarms had a positive affect, 43% were neither positive nor negative, and 8% were negative to the tourism experience. When asked whether the presence of windfarms made a difference to likelihood of revisiting the area, about 91% stated it made no difference. In other words, despite reports of ongoing opposition to the idea of offshore wind developments, in Scotland at least, it has not been much of an issue for the intensity of development being considered at the time.

A few years later, Devine-Wright and Howes (2010) surveyed residents in two coastal towns in North Wales, UK, which were 15 kilometers (9 miles) from offshore wind sites. The investigators used in-depth interviews, focus groups and sampled 457 survey respondents. Because of historical differences and

TABLE 5.4 Socio-cultural factors affecting offshore wind development

Author/ Physical country environment	*Landscape context and proximity*	*Social, political, communication*	*Symbolic, ideological*	*Economic, equity, security*
United Kingdom				
Braunholtz (2003)	✓			
Devine-Wright and Howes (2010)	✓		✓	✓
Demark, the Netherlands and Germany				
Ladenburg (2009b)	✓			
Ladenburg (2010)	✓	✓		
Ladenburg and Möller (2011)	✓			
Gee (2010)	✓			
Wolsink (2010)	✓		✓	
United States				
Firestone and Kempton (2007)	✓	✓		✓
Lilley et al. (2010)	✓			

the perceived dependency on tourism, there were significant differences between residents of the two towns regarding anticipated negative impacts to tourism. Opposition to offshore windfarm development could be traced to place-based values such as scenic beauty and psychic restoration. Some respondents perceived such development as "'industrializing the area' and 'fencing in the bay'" (Devine-Wright and Howes 2010, p. 90). Other factors included the threat to identity or place attachment, and the level of trust in key actors.

A few studies have focused on Denmark, the Netherlands, and Germany, especially in the North and Wadden Sea areas. Ladenburg (2009b) conducted a study that included 700 respondents from a national survey plus two samples of wind-site communities of 350 respondents each. He found that having seen an offshore windfarm before influenced individual assessment by increasing acceptance of windfarms. Positive perceptions are influenced by location and distance to the coast from existing windfarms – meaning the further the distance, the better. In a second study, Ladenburg (2010) also analyzed a resident survey with 1,000 respondents and found positive attitudes related to male gender, higher income, higher education, lower frequency of beach visits, and type of beach use. Additional factors were whether there were views of land turbines from residences, as well as the types of beach use. In a third study, Ladenburg and Möller (2011) conducted a randomized national sample of some 1,050 respondents. Twenty-seven percent had a negative reaction to offshore wind. Travel time and attributes of the nearest windfarm were found to influence attitudes – meaning the shorter travel time would generate a more negative response. Counter intuitively, offshore windfarms with many turbines elicited more positive responses than farms with fewer turbines as seen from the shoreline.

Gee (2010) and Wolsink (2010) have also conducted studies of reactions to offshore wind in the North German Sea and Wadden seacoasts. Gee distributed a mail questionnaire in 2005 to some 15 communities in the North Frisian Islands. Aesthetic qualities of the shore and sea were found to be a significant driver affecting opinion, especially as the wind turbines were perceived as despoiling the open sea horizon. Wolsink (2010) surveyed the members of WaddenVereniging – the major Dutch Wadden Sea NGO for reactions to siting windfarms. Impacts on landscape and seascape are of major concern in this iconic landscape. Members of the NGO felt that there were other suitable windfarm sites – just not those chosen by the government or the project consultants.

In the United States, Firestone and Kempton (2007) conducted a survey of Cape Cod residents in the midst of the controversy over the proposed Cape Wind development between Martha's Vineyard and Cape Cod, Massachusetts. They had 504 survey respondents for a 39% response rate. More than 50% perceived negative impacts from windfarms based on changes to aesthetics, community harmony, local fishing industry and recreational boating. About 40% of respondents perceived negative impacts on property values, bird and marine life. In many cases, these perceived impacts were inconsistent with scientific studies and environmental impact statements. Younger, better educated respondents supported wind power development compared to older, higher income respondents who would see the wind development in their daily routines.

Lilley et al. (2010) surveyed more than 1,000 randomly sampled, out-of-state tourists at Delaware beaches in 2007.

> After providing respondents with wind turbine project photo-simulations at several distances, we inquired about the effect development would have on visitation. Approximately one-quarter stated that they would switch beaches if an offshore wind project was located 10 km from the coast, with avoidance diminishing with greater distance from shore.
>
> (Lilley et al. 2010, p. 1)

★★★

Jones and Eiser (2009) found little NIMBYism in people living close to proposed wind projects, but lots of uncertainty about effects from projects. Ladenburg (2008) found that people who had seen offshore windfarms located far from the coast had more positive attitudes than people who had seen windfarms close to the coast. Ek (2005) found that support of wind projects decreases with increasing age and income in Sweden, whereas Ladenburg (2009b) found that positive attitudes toward wind power increased for males, higher incomes, and higher education levels. Ladenburg (2008, 2009a, 2009b) also found that attitudes about offshore windfarms depended upon the type and frequency of outdoor activity (such as beach walking) and the amount of prior experience with windfarms. So specific coastal residents or coastal recreational user attitudes toward offshore windfarms are key to the acceptability of such development.

Ladenburg's 2010 Denmark study of beach-user attitudes towards offshore windfarms found that attitudes were different for frequent users compared to infrequent users, as well as differing with types of beach usage. Both Delaware and Long Island surveys asked respondents if they would support offshore windfarms being built off nearby coastal beaches. The Long Island survey response was 86% yes; in Delaware it was 78%. In the Delaware study, 11% of respondents indicated that they would not go back to the same beach but 83% said they would visit a beach not previously visited for seeing offshore wind turbines (Firestone et al. 2009). Clearly publicly accessible beaches and other coastal recreation areas and their users will be key ingredients influencing viewer sensitivity. Thus in terms of populations of 'sensitive receptors,' these would include residents who would see the wind turbines on a daily activity, plus recreational beach users.

5.5 Commercial solar energy and social acceptability factors

There are several solar energy technologies in commercial use, including PV modules, which convert sunlight into electricity directly, and solar concentrators such as parabolic trough concentrators and central tower receivers, which rely on a thermal cycle to generate electricity. Each form creates its own landscape signature, but collectively they differ visually in fundamental ways from wind energy equipment. For example, wind turbines spin while solar equipment does not, PV solar and trough concentrators have a relatively low vertical presence on the landscape compared to central receivers. In addition, solar technologies produce little noise, and seldom threaten birds or bats (except for central tower receivers), as wind turbines can in some locations. Solar installations require less land area for similar generation capacities than do wind turbines, which must be dispersed. Solar installations are generally the sole use of the land, while windfarms can share the land with farms, animals, and even other renewable energy development (Figure 5.9). Central receivers combine some of the characteristics of both, at least in their vertical dimension. For example, the Ivanpah central receiver towers, 40 miles southwest of Las Vegas, stand almost 460 feet (140 meters) off the desert floor, giving them a dominant impact on the landscape (Plate 9).

***

A study of community response to a proposed concentrating solar power (CSP) plant in the San Luis Valley (SLV) in Colorado was conducted by Farhar et al. (2010). The investigators did qualitative

FIGURE 5.9 Collocated solar and wind land use in San Gorgonio Pass, near Palm Springs, California. (Photo: M. Pasqualetti.)

in-depth interviews in 2008 and 2009 with stakeholders inside and outside the SLV regarding a hypothetical 100 MW CSP. Perceived benefits included: regional economic improvement, job creation, local tax revenue increase, clean energy, best land use, greenhouse gas reduction, improving energy security, providing dispatched energy, community pride and identity. Although there were no NIMBY-like responses, there were the usual array of concerns, including the need for water, land-use intensity, boom-bust economy, the need for transmission lines, efficacy of the technology, power intermittency, and the relationship with the utility developers/owners. Valley stakeholders emphasized environmental, social, and psychological benefits whereas outside the valley stakeholders emphasized technological benefits from the facility.

While Sovacool (2009) reports wide support for solar energy projects, Pasqualetti (2011a, 2011b, 2012) notes some of the barriers to renewable energy – especially in the desert southwest. He cites concerns about the amount of land in Arizona, Nevada, California, and New Mexico that may be utilized for large-scale solar energy development. Environmental groups are concerned with the proposed magnitude of solar development and its impact on the ecological landscape. This is also the focus of a $20 million grant from the National Science Foundation to the University of Nevada. Forty miles southwest of Las Vegas, the construction of the 392 MW Ivanpah central receivers was halted to allow for the relocation of protected desert tortoises. Once completed and energized, the Ivanpah receivers attracted a new level of attention from pilots alarmed by the glare produced from the heliostats. No public opinion surveys have yet been conducted, but it is just one of many solar installations planned or in operation in the desert southwest that will presumably be subject to such surveys in the near future.

Outside the US, national surveys are widely supportive of solar energy implementation including from Greece (Tsantopoulos et al. 2014, Kaldellis et al. 2013), Italy (Cicia et al. 2012), Spain (Heras-Saizarbitoria et al. 2011), and Germany (Zoellner et al. 2008). However, much of this support is for residential and commercial decentralized PV systems. There is little information or public surveys about the acceptability of utility scale solar power systems. There have been papers that describe the general environmental and social impacts of such systems. Turney and Fthenakis (2011) describe land use, human health, wildlife habitat, geohydrology, and climate/greenhouse gas impacts from large-scale solar development. Tsoutsos et al. (2005) also describe impacts such as the reduction of cultivatable land, visual intrusion, ecosystem, and health impacts.

TABLE 5.5 Socio-cultural factors affecting commercial solar development

Author/Physical country environment	*Landscape context and proximity*	*Social, political, communication*	*Symbolic, ideological*	*Economic, equity, security*
United States				
Farhar et al. (2010)	✓+	✓+		✓+
Pasqualetti (2011a, 2011b, 2012)	✓	✓		
Greece				
Tsantopoulos et al. (2014)		✓	✓	✓
Kaldellis et al. (2013)		✓		
Germany				
Zoellner et al. (2008)		✓		✓

Tsantopoulos et al. (2014) identified several reasons in support of PV development including financial, recognition/prestige, trust in relevant stakeholders, and environmental protection. Kaldellis et al. (2013) conducted a national Greek survey regarding social acceptance of solar PV arrays, including 24% in favor of PV parks, 53% in favor under certain conditions, and 30% in need of proof of usefulness. Fifty-three percent perceived no visual impact from PV installations, and "22% of the sample . . . stated that PV systems are either visually annoying or not aesthetically right" (Kaldellis et al. 2013, p. 41). In general, public opinion in Greece for new PV installations was more positive than for new wind installations.

In a survey of respondents in four different regions in Germany concerning implementation of ground-installed PV systems, Zoellner at al. (2008) found that support was between 75% and 85%. They said that there was "a strong connection between procedural justice criteria, such as transparency, early and accurate information as well as possibilities to participate during the planning and installation process and reported public acceptance became evident" (p. 40).

5.6 Social receptivity and geothermal energy development

Potential for geothermal development is widespread, and is currently in at least three dozen countries. Nonetheless, it has generated little published literature on social or landscape issues (see Table 5.6). Major impacts from geothermal development include land clearing for drill pads and the noise produced during drilling. Operational impacts include land subsidence with the loss of the geothermal reservoir, noise, odor and air quality impacts from hydrogen sulfide, ecological impacts from loss of habitat, and visual impacts through introduction of structures, pipes and escaping steam plumes.

In the low mountains near Clear Lake, California, Pasqualetti and Dellinger (1989) document that the key public issue related to geothermal development was conflict between the Clear Lake recreational landscape and the perceived industrial development and the hydrogen sulfide odor. The odor issues were addressed with scrubbers, but the scrubbers produced waste that needed to be trucked away, leading to accidents and other problems of disposal.

Far to the south, in the Imperial Valley of California, there were multiple concerns, including subsidence, induced seismicity, waste water disposal, and hydrogen sulfide emissions. In generation, the issue was whether geothermal development could co-exist with the valuable pre-existing agricultural land uses such as field crops and feed lots. Based on the simultaneous and substantial growth of revenues over the past 25 years, geothermal development and agriculture have proved largely compatible (Pasqualetti 2011b).

In the case of geothermal development in Hawaii there has been continuous local land use conflict, centered on the Puna geothermal development in the southeastern section of the big island of Hawaii. Issues have been odor, noise, and infringement with indigenous peoples' beliefs (Coe 1991, Edelstein and Kleese 1995).

Iceland has long hosted geothermal development for energy production, but there are still landscape conflicts as new proposals have emerged. Some of the proposed sites are "of distinctive and exceptional scenic landscapes. They include some of the most colorful scenery found anywhere in the world, many offer striking contrasts between 'fire and ice' with hot springs, lavas and glaciers" (Thórhallsdóttir 2007, p. 540). Sæþórsdóttir and Ólafsson (2010) also note the impact of geothermal development on nature tourism in Iceland, although one would assume that geothermal activity would be an attraction to tourists as well.

In New Zealand the geothermal energy and recreational spa development has exceeded the North Island's Spa Geyser basin's ability to replenish, causing both loss of energy production and tourism spa use. There have also been conflicts with traditional Maori religious beliefs (Barrick 2007, Kelly 2011).

TABLE 5.6 Socio-cultural factors affecting commercial geothermal development

Author/ Physical country environment	*Landscape context and proximity*	*Social, political, communication*	*Symbolic ideological*	*Economic, equity, security*
United States				
Pasqualetti and Dellinger (1989)	✓	✓		
Pasqualetti (2011b)		✓		
Coe (1991)	✓			✓
Edelstein and Kleese (1995)	✓			✓
Iceland				
Thórhallsdóttir (2007)		✓		
Sæþórsdóttir and Ólafsson (2010)		✓		✓
New Zealand				
Barrick (2007)			✓	✓
Kelly (2011)			✓	✓
Greece				
Polyzou and Stamataki (2010)	✓	✓		

The only social acceptance survey found anywhere for residents' reactions to geothermal development has been by Polyzou and Stamataki (2010) on the islands of Milos and Nisios, Greece, in 2004. Questions were asked about knowledge of geothermal development, annoyance, and environmental problems, plus negative and positive impacts. Almost 87% of Milos residents and 94% of Nisios residents knew about geothermal energy development. On Milios, 80% of residents perceived that geothermal energy was polluting. The most frequently cited impacts were air pollution (46% in Milos and 58% in Nisios) followed by soil and water pollution. Polyzou and Stamataki (2010) found that the surveyed residents did not consider noise or visual impact issues to be significant. Demographic factors were not significant except if respondent occupations were related to tourism.

Thus, like solar and wind, large-scale geothermal development has some perceived landscape issues for high quality landscapes and those of cultural significance, but also has some unique air quality and odor issues. In general, it has not produced significant public opposition, presumably for several reasons: higher energy density, and the relatively small degree of development that has taken place worldwide, especially compared to solar and wind.

5.7 Summary of acceptability by renewable energy type

In general, onshore wind developments find greater acceptability except for siting in mountainous and coastal landscapes that are highly dependent on recreational and tourism and/or have strong place-based values. Key demographic variables affecting acceptability are age (younger is more favorable), familiarity with the energy source (more familiar is better), and education (higher levels trend toward greater acceptability).

For offshore wind siting, acceptability increases with distance offshore except where visible from recreational- and tourism-dependent community landscapes. Age and educational level as well as familiarity with wind energy are key demographic variables.

For large-scale solar development, there is little survey data available yet, at least in the US. Large-scale PV seems to be more acceptable in most European countries regardless of demographic variables.

For geothermal development, we also have very little survey data, but from the few case studies there are we do learn that the quality of the existing landscape and indigenous place-based landscape values are key.

So, NIMBYism is not always an issue with renewable energy development when it comes to actual siting issues, but detailed place-based analysis of landscape values and uses are key in every case of renewable energy development. The next section will look in more detail at specific visual landscape impact issues and studies of the perception of these visual impacts.

5.8 Renewable wind energy facilities and visual perception

The literature generally states that people are more receptive to wind power if it is located far offshore or near visually non-sensitive landscapes without aesthetic, ecological or cultural value or features (Ladenburg 2010). However, Haggett (2011) states that there are equal problems with siting both onshore and offshore windfarms – but in slightly different ways. Firestone et al.'s (2009) survey indicated that there is higher acceptance (than onshore) of wind facilities in the eastern US, but as we have seen regarding the Cape Wind project, this is not always true, depending upon specific site location (Kempton et al. 2005). We should expect, therefore, that as the surveys become more site specific, so too will our understanding of public reactions.

5.8.1 Land-based windfarm visual impact perception

Bishop (2012) points out that distance from viewers is a major determinant of visual impact and that there are views from traveling roads and railroads that affect viewer response, meaning more daily visual exposure. Lothian (2008) states from his research that there is a positive effect of such facilities with landscapes of lower quality. There were similar findings in a survey in the Czech Republic (Molnarova et al. 2012). Wolsink (2007b) and de Vries et al. (2012) state that the landscape's character and quality overshadows the design of wind turbines, as well as their number and size. Lothian (2008) found that his sample favored white, blue or gray turbine colors.

The most recent work addressing visual thresholds of land-based windfarms is by Sullivan et al. (2012b). This work entailed a combination of field-based observations and geospatial analysis of three windfarms located in Colorado and Wyoming by experts, not public perception surveys. The objective of the study was to determine the visual characteristics of windfarms at varying distances, and the maximum limits of visibility under varying lighting and atmospheric conditions. Results include proposed impact thresholds for utility-scaled wind facilities in the western US landscape from 1 (not visible) to 6 (with high impact and visual preeminence) (Table 5.7).

The visibility rating scales in the Table 5.7 were then utilized by a number of professional raters to determine the thresholds of visibility (Figures 5.10 and 5.11).

As we can see, these visibility ratings and thresholds may hold for western US landscapes but have to be rescaled for other regional landscapes in the US.

TABLE 5.7 Visibility level rating descriptions

Visibility rating	*Description*
VISIBILITY LEVEL 1: *Visible only after extended, close viewing; otherwise invisible.*	An object/phenomenon that is near the extreme limit of visibility. It could not be seen by a person who was not aware of it in advance, and looking for it. Even under those circumstances, the object can only be seen after looking at it closely for an extended period of time.
VISIBILITY LEVEL 2: *Visible when scanning in general direction of study subject; otherwise likely to be missed by casual observer.*	An object/phenomenon that is very small and/or faint, but when the observer is scanning the horizon or looking more closely at an area, can be detected without extended viewing. It could sometimes be noticed by a casual observer; however, most people would not notice it without some active looking.
VISIBILITY LEVEL 3: *Visible after brief glance in general direction of study subject and unlikely to be missed by casual observer.*	An object/phenomenon that can be easily detected after a brief look and would be visible to most casual observers, but without sufficient size or contrast to compete with major landscape elements.
VISIBILITY LEVEL 4: *Plainly visible, could not be missed by casual observer, but does not strongly attract visual attention, or dominate view because of apparent size, for views in general direction of study subject.*	An object/phenomenon that is obvious and with sufficient size or contrast to compete with other landscape elements, but with insufficient visual contrast to strongly attract visual attention and insufficient size to occupy most of the observer's visual field.
VISIBILITY LEVEL 5: *Strongly attracts visual attention of views in general direction of study subject. Attention may be drawn by strong contrast in form, line, color, or texture, luminance, or motion.*	An object/phenomenon that is not of large size, but that contrasts with the surrounding landscape elements so strongly that it is a major focus of visual attention, drawing viewer attention immediately, and tending to hold viewer attention. In addition to strong contrasts in form, line, color, and texture, bright light sources (such as lighting and reflections) and moving objects associated with the study subject may contribute substantially to drawing viewer attention. The visual prominence of the study subject interferes noticeably with views of nearby landscape elements.
VISIBILITY LEVEL 6: *Dominates view because study subject fills most of visual field for views in its general direction. Strong contrasts in form, line, color, texture, luminance, or motion may contribute to view dominance.*	An object/phenomenon with strong visual contrasts that is of such large size that it occupies most of the visual field, and views of it cannot be avoided except by turning the head more than 45 degrees from a direct view of the object. The object/phenomenon is the major focus of visual attention, and its large apparent size is a major factor in its view dominance. In addition to size, contrasts in form, line, color, and texture, bright light sources and moving objects associated with the study subject may contribute substantially to drawing viewer attention. The visual prominence of the study subject detracts noticeably from views of other landscape elements.

(*Source:* Sullivan et al. 2012a, p. 17.)

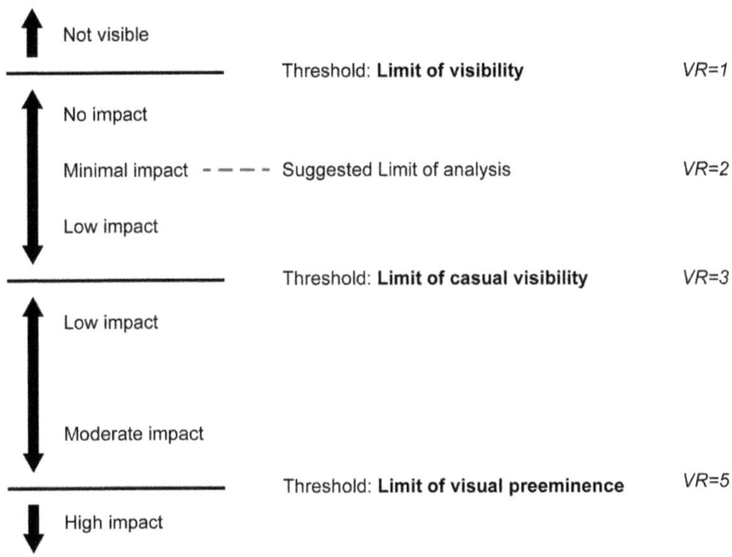

FIGURE 5.10 Visibility thresholds linked to ratings.
(Source: Redrawn by Mark Warfel Jr. from Sullivan et al. 2012a, p. 36.)

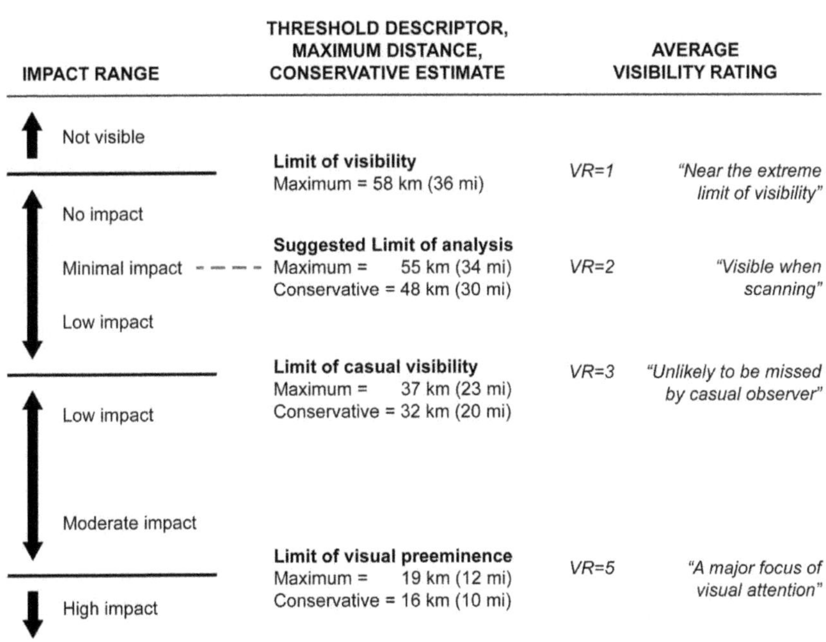

FIGURE 5.11 Visibility impact range, thresholds and visibility ratings.
(Source: Redrawn by Mark Warfel Jr. from Sullivan et al. 2012a, p. 40.)

The Sullivan et al. (2012b) study also addressed night lighting on the three study sites where red flashing hazard navigation lighting was officially recorded as visible at 36.2 miles at the Cedar Creek windfarm in Colorado. For all three windfarms, average visibility varied widely for a given distance but decreased as to distance from the observation point to the facility. Also the higher the contrast between the turbines and backdrop, the higher the average visibility ratings.

5.8.2 Offshore windfarms impact perception

Studies of visual thresholds and causal factors of visual impact related to offshore windfarms have been done in Australia (Bishop 2002, Bishop and Miller 2007, Lothian 2008, Shang and Bishop 2000), Denmark (Ladenburg 2009c, Ladenburg and Dahlgaard 2012, Möller 2006) and Sweden (Johansson and Laike 2007). Visual perception studies of offshore wind turbines in the US are just beginning. The only older related work of interest is Nassauer's (1983) study of oil and gas development in the coastal landscape of Louisiana, and Thayer's (1987, 1994) work on the early California inland windfarms.

The Bishop and Miller (2007) study assesses wind turbine visibility and impact at different distances from the viewer, in different lighting and atmospheric conditions, and with moving or stationary blades. From this online survey, distance and visual contrast were found to be very good predictions of perceived impact. Lothian (2008) maintains that visual impacts rise in areas of higher scenic quality, and that negative reaction to offshore facilities does not decrease with distance. There was significant difference in perceived impact between simulations with moving versus stationary blades – moving was viewed more positively.

Another significant visual perception study was done by Ladenburg (2009a) in Demark. He found that visual perception is influenced by prior experience. People with experience of seeing windfarms located far offshore have a significantly more positive perception of the visual impact than people with experience of seeing windfarms closer to the coast. It is noted that future acceptance of offshore windfarms is interdependent on the location of existing and new offshore windfarms.

Sullivan et al. (2013) also did a study of offshore wind turbine visibility and visual impact thresholds. This study involved professionals observing eleven wind turbines facilities from 29 onshore locations in the UK including six nighttime observations. The four professional raters filled out 98 visibility-rating forms for these observation points and agreement was stated to be "very high between the raters" (p. 10).

Results of the visibility ratings "indicates a gradual drop of the ratings with distance" but the "change is nonlinear because of the variability in lighting, contrast of the wind turbines with the background, facility size and layout, blade orientation and rotation rate, and various other factors that affect visibility" (p. 11). They found that "moderately sized offshore wind facilities may be visible at distance exceeding 35km (22mi.)" and "they were visible at a maximum distance of 44km (27mi.)" (p. 11). Offshore windfarms were judged a major foci of visual attention at distances of 16 kilometers (10 miles) or less suggesting potentially high levels of impact for sensitive viewers (Sullivan et al. 2013, p. 12).

Turbine blade movement

> was visible at distance as great as 42km (26mi.) in 42 of the 49 daytime observations and was observed routinely at 34km (21mi.) or less. Contrary to expectation, lighting conditions, sun angle, and contrast between turbines and sky backdrop did not substantially affect the likelihood of observing blade motion, blade motion was visible at distances beyond 30km (19mi.).
>
> (Sullivan et al 2013, p. 12)

Thus from these studies, we can draw that distance and visual contrast are two key variables needed to assess the visual impact of land and offshore windfarms, and second, familiarity and experience of such development will be key when assessing cumulative visual impacts. Another interpretation of familiarity is that people are influenced by the loud stories coming from the opposition, and do not trust the corporate developers. They are more afraid of what they have not experienced, but calm down when they have actually seen what it will be like. This does not mean that opposition goes away, just that it is not as great as their stirred-up fears.

5.8.3 Large-scale solar power plant visual perception

There is very little literature on large-scale solar installations but glare, visual contrast and landscape scenic quality are key elements to be considered (Torres-Sibille et al. 2009). Sullivan et al. (2012a) describe the potential landscape and visual impacts of utility scale solar energy facilities – especially on National Park units, wilderness areas, national historic sites, and scenic trail corridors in the US Southwestern region. Such facilities tend to be large in size, have regular, strong geometry, highly reflective surfaces and are visible for long distances. Sullivan et al. (2012a) is working with BLM and the NPS to analyze the visual impacts of parabolic trough, PV and power tower facilities.

As part of this work at the Argonne National Laboratory, Sullivan et al. (2012a) have cited the visual contrast from parabolic trough facilities, including glare from heat transfer fluid tubes, geometric patterns of reflected light creating strong scintillations, plumes from cooling towers, plus reflections from support facilities. Glare was bright enough to cause visual discomfort at distances greater than 6 kilometers (4 miles).

> A 1.6 sq km (400 acre) parabolic trough facility was found to be visible at distances greater than 23 km (14 mi.) A thin-film PV facility was visible at a distance of 35 km (22 mi.). Other effects include dramatic and rapid color changes and reflectivity of solar collector arrays from parabolic trough and thin film PV facilities.
>
> (Sullivan et al. 2012a)

A study of the visual impact of the Ivanpah Solar Electric Generating System power tower facility conducted by Argonne National Laboratory (Sullivan and Abplanalp 2015) indicated that "reflected sunlight from the receivers was the primary source of visual contrast regardless of viewing distance or viewing geometry" (p. 1). Glare from heliostats at Ivanpah was observed to cause visual discomfort for some viewers at a distance of 32 kilometers (20 miles), and the facility was easily visible from a mountaintop 56 kilometers (35 miles) away (Sullivan and Abplanalp 2015, p. ES-2). Glare from heliostats at the Ivanpah facility as seen from a commercial airliner is shown in Figure 5.12.

In Europe, there are landscape impact and method development studies for PV solar systems in Italy (Chiabrando et al. 2009, Fabrizio and Garnero 2012), Greece (Kapetanakis et al. 2014) and Spain (Torres-Sibille et al. 2009). Chiabrando et al. (2009) have developed a visual assessment tool that assesses the reduction of cultivatable land and the amount of visual intrusion on components of the landscape. Kapetanakis et al. (2014) have proposed that visual impact be addressed by looking at visibility, color, and fractal dimension. Visibility is the total area covered by the PV modules that is visible, color is the contrast between PV modal panels and the surrounding landscape, and fractal dimension measures the degree of artificial geometry between the landscape and the PV plant.

Only Torres-Sibille et al. (2009) have used landscape perception studies to test a visual impact assessment method. They utilized semantic differential to obtain data from individuals' reactions to before

FIGURE 5.12 Glare from the Ivanpah solar installation. View is from the southeast, 30 miles south of Las Vegas, looking northwest from 30,000 feet in a commercial airliner.
(Photo: M. Pasqualetti.)

and after landscape scenes with and without solar PV installations. They utilized 40 college age students to examine and compare before and after photo simulations of solar plant installations. Plant size and color contrast were the two most often mentioned criteria. The investigators also compared these subjective perception results with more objective measures of plant visibility, color and fractal dimension plus atmospheric conditions.

Thus for utility scale solar plant installations, glare, visibility and landscape contrast are key variables affecting visual impact.

5.8.4 Visual perception summary

For land based windfarms, visibility affects the degree of impact with decreasing distance as well as contrast or fit in the landscape. Also high-visual, quality recreational landscapes and users are more sensitive to visual impacts from on-land wind. For offshore wind, increasing distance from viewers has less impact. Again, frequent shoreline recreational users may be more sensitive to visual impacts. For PV solar arrays, fit within the landscape is key if near high-visible, high-quality landscapes. Reflection and glare are also issues causing visual impacts depending on lighting conditions and direction.

Concentrated solar power plants have their own issues related to introducing large scale vertical forms in the landscape. The social acceptance of renewable energy installations is an evolving concern with many variables. Clearly, there are places where they will substantially interfere with preferred lifestyles, as well as other places where the impact will be less dramatic. Care must be taken in all cases to avoid the most egregious landscape conflicts. At the same time, strategies for sensitive deployment and enhanced mitigation strategies should continue to aim for further refinement.

Notes

1 See, for example, a rather comprehensive Internet site dedicated to the opposition of wind power, European Platform Against Windfarms, at www.epaw.org/index.php?lang=en
2 See www.w-wind.com.au/downloads/CFS4Tourism.pdf
3 See www.gallup.com/poll/161519/americans-emphasis-solar-wind-natural-gas.aspx and www.ewea.org/file admin/ewea_documents/documents/publications/WD/WD22vi_public.pdf
4 See www.ewea.org/fileadmin/ewea_documents/documents/publications/WD/WD22vi_public.pdf
5 See www.ewea.org/fileadmin/ewea_documents/documents/publications/WD/WD22vi_public.pdf
6 See http://mthink.com/article/public-attitudes-wind-energy/
7 See www.oddzialywaniawiatrakow.pl/upload/file/502.pdf and www.ecodyfi.org.uk/tourism/Windfarms_research_eng.pdf
8 See www.ewea.org/fileadmin/ewea_documents/documents/publications/WD/WD22vi_public.pdf
9 See www.appa.es/ (in Spanish).
10 See www.ewea.org/fileadmin/ewea_documents/documents/publications/WD/WD22vi_public.pdf
11 See www.ewea.org/fileadmin/ewea_documents/documents/publications/WD/WD22vi_public.pdf
12 See www.renewableuk.com/ and www.gov.uk/government/uploads/system/uploads/attachment_data/file/65520/6410-decc-public-att-track-surv-wave2-summary.pdf

References

Barrick, K.A. (2007). Geyser decline and extinction in New Zealand: Energy development impacts and implications for environmental management. *Environmental Management 39*(6), 783–805. doi: 10.1007/s00267-005-0195-1.

Baxter, J., R. Morzaria, and R. Hirsch. (2013). A case-control study of support/opposition to wind turbines: Perceptions of health risk, economic benefits, and community conflict. *Energy Policy 61*, 931–43. doi: 10.1016/j.enpol.2013.06.050.

Bell, D., T. Gray, and C. Haggett. (2005). The 'social gap' in windfarm siting decisions: Explanations and policy responses. *Environmental Politics 14*(4), 460–477. doi: 10.1080/09644010500175833.

Bell, J. (2015). Renewables share of UK electricity generation hits record in Q1'15. *Clean Technica*, August 17. http://cleantechnica.com/2015/08/17/renewables-share-uk-electricity-generation-hits-record-q115/

Bidwell, D. (2013). The Role of Values in Public Beliefs and Attitudes towards Commercial Wind Energy. *Energy Policy 58*, 189–99. doi: 10.1016/j.enpol.2013.03.010.

Bishop, I.D. (2002). Determination of thresholds of visual impact: The case of wind turbines. *Environment and Planning B: Planning and Design 29*(5), 707–718. doi: 10.1068/b12854.

Bishop, I.D. (2012). Understanding new landscapes: Support for renewable energy planning. *International Journal of E-Planning Research 1*(14), 1–16. doi: 10.4018/ijepr.2012100101.

Bishop, I.D. and D.R. Miller. (2007). Visual assessment of offshore wind turbines: The influence of distance, contrast, movement and social variables. *Renewable Energy 32*(5), 814–31. doi: 10.1016/j.renene.2006.03.009.

Bohn, C. and C. Lant. (2009). Welcoming the wind? Determinants of wind power development among the US States. *The Professional Geographer 6*(1), 87–100.

Braunholtz, S. (2003). *Public Attitudes to Windfarms: A Survey of Local Residents in Scotland.* Scottish Executive Social Research, Edinburgh.

Breukers, S. and M. Wolsink. (2007). Wind power implementation in changing institutional landscapes: An international comparison. *Energy Policy 35*(5), 2737–2750. doi: 10.1016/j.enpol.2006.12.004.

Cass, N., and G. Walker. (2009). Emotion and rationality: The characterization and evaluation of opposition to renewable energy projects. *Emotion, Space and Society 2*(1), 62–69. doi: 10.1016/j.emospa.2009.05.006.

Chiabrando, R., E. Fabrizio, and G. Garnero. (2009). The territorial and landscape impacts of photovoltaic systems: Definition of impacts and assessment of the glare risk. *Renewable and Sustainable Energy Reviews 13*(9), 2441–2451. doi: 10.1016/j.rser.2009.06.008.

Cicia, G., L. Cembalo, T. Del Giudice, and A. Palladino. (2012). Fossil energy versus nuclear, wind, solar and agricultural biomass: Insights from an Italian National Survey. *Energy Policy 42,* 59–66. doi: 10.1016/j.enpol.2011.11.030.

Coe, B.A. (1991). The state versus Pele: Structure and leadership in geothermal energy development. *Environ Impact Assessment Review 11*(4), 333–352.

Corscadden, K., A. Wile, and E. Yiridoe. (2012). Social license and consultation criteria for community wind projects. *Renewable Energy 44*, 392–397. doi: 10.1016/j.renene.2012.02.009.

Cowell, R. (2009). Wind power, landscape and strategic, spatial planning: The construction of 'acceptable locations' in Wales. *Land Use Policy 27*(2), 222–232. doi: 10.101016/j.landusepol.2009.01.006.

Devine-Wright, P. (2011). Enhancing local distinctiveness fosters public acceptance of tidal energy: A UK case study. *Energy Policy 39*(1), 83–93. doi:10.1016/j.enpol.2010.09.012.

Devine-Wright, P. and Y. Howes. (2010). Disruption to place attachment and the protection of restorative environments: A wind energy case study. *Journal of Environmental Psychology 30*(3), 1–10. doi 10.1016/j.jenvp.2010.01.008.

Devine-Wright, P., Walker, G., Hunter, S., High, H., and Evans, B. (2007). *An Empirical Study of Public Beliefs about Community Renewable Energy Projects in England and Wales.* Manchester: Manchester Architecture Research Centre, School of Environment and Development.

de Vries, S., M. de Groot, and J. Boers. (2012). Eyesores in sight: Quantifying the impact of man-made elements on the scenic beauty of Dutch landscapes. *Landscape and Urban Planning 105*(1–2), 118–127. doi: 10.1016/j.landurbplan.2011.12.005.

Dimitropoulos, A. and A. Kontoleon. (2009). Assessing the determinants of local acceptability of wind-farm investment: A choice experiment in the Greek Aegean Islands. *Energy Policy 37*(5), 1842–1854. doi: 10.1016/j.enpol.2009.01.002.

Edelstein, M.R. and D.A. Kleese. (1995). Cultural relativity of impact assessment: Native Hawaiian opposition to geothermal energy development. *Society and Natural Resources: An International Journal 8*(1), 19–31. doi: 10.1080/08941929509380896.

Ek, K. (2005). Public and private attitudes towards 'green' electricity: The case of Swedish wind power. *Energy Policy 33*(13), 1677–1689. doi: 10.1016/j.enpol.2004.02.005.

Ek, K. and L. Persson. (2014). Windfarms – where and how to place them? A choice experiment approach to measure consumer preferences for characteristics of windfarm establishments in Sweden. *Ecological Economics 105*, 193–203. doi: 10.10.1016/j.ecolecon.2014.06.001.

Elthan, D.C., G.P. Harrison, and S.J. Allen. (2008). Change in public attitudes towards a Cornish windfarm: Implications for planning. *Energy Policy 36*(1), 23–33. doi: 10.1016/j.enpol.2007.09.010.

Erbil, A.Ö. (2011). Social acceptance of the clean energy concept: Exploring the clean energy understanding of Istanbul residents. *Renewable and Sustainable Energy Reviews 15*(9), 4498–4506. doi: 10.1016/j.rser.2011.07.101.

Ertör-Akyazı, P., F. Adaman, B. Özkaynak, and Ü. Zenginobuz. (2012). Citizens' preferences on nuclear and renewable energy sources: Evidence from Turkey. *Energy Policy 47*, 309–20. doi: 10.1016/j.enpol.2012.04.072.

European Wind Energy Association (EWEA). (2014). Wind in power 2013: European statistics. www.ewea.org/fileadmin/files/library/publications/statistics/EWEA_Annual_Statistics_2013.pdf

Fabrizio, E. and G. Garnero. (2012). Visual impact, landscape and renewable energy plants: The case of PV. TSO6L – Implementation of Plans and Infrastructure Paper No. 5946. In *FIG Working Week 2012: Knowing to Manage Territory, Protect the Environment, Evaluate the Cultural Heritage.* Rome, Italy, May 6–10, 2102.

Farhar, B.C., L.M. Hunter, T.M. Kirkland and K.J. Tierney. (2010). Community Response to Concentrated Solar Power in the San Luis Valley. Sub contract Report NREL/SR-550-48041. US National Renewable Energy Laboratory.

Firestone, J. and W. Kempton. (2007). Public opinion about large offshore wind power: Underlying factors. *Energy Policy 35*(3), 1584–1598. doi: 10.1016/j.enpol.2006.04.010.

Firestone, J., W. Kempton, and A. Krueger. (2009). Public acceptance of offshore wind power projects in the USA. *Wind Energy 12*(2), 183–202. doi: 10.1002/we.316.

Gee, K. (2010). Offshore wind power development as affected by seascape values on the German North Sea coast. *Land Use Policy 27*(2), 185–194. doi: 10.1016/j.landusepol.2009.05.003.

Graham, J.B., J.R. Stephensen, I.J. Smith. (2009). Public perceptions of wind energy developments: Case studies from New Zealand. *Energy Policy 37*(9), 3348–3357. doi: 10:1016/j.enpol.2008.12.035.

Groth, T.M., and C. Vogt. (2014). Residents' perceptions of wind turbines: An analysis of two townships in Michigan. *Energy Policy 65*, 251–260. doi: 10.1016/j.enpol.2013.10.055.

Haggett, C. (2011). Understanding public responses to offshore wind power. *Energy Policy 39*(1), 503–510. doi 10.1016/j.enpol.2010.10.014.

Hall, N., P. Ashworth, and P. Devine-Wright. (2013). Societal acceptance of windfarms: Analysis of four common themes across Australian case studies. *Energy Policy 58,* 200–208. doi: 10.1016/j.enpol.2013.03.009.

Heras-Saizarbitoria, I., E. Cilleruelo, and I. Zamanillo. (2011). Public acceptance of renewables and the media: An analysis of the Spanish PV solar experience. *Renewable and Sustainable Energy Reviews 15*(9), 4685–4696. doi: 10.1016/j.rser.2011.07.083.

Hindmarsh, R. (2010). Windfarms and community engagement in Australia: A critical analysis for policy learning. *East Asian Science, Technology and Society: An International Journal 4*(4), 541–563. doi 10.1007/s12280-010-9155-9.

Jobert, A., P. Laborgne, and S. Mimler. (2007). Local acceptance of wind energy: Factors of success identified in French and German case studies. *Energy Policy 35*(5), 2751–2760. doi: 10.1016/j.enpol.2006.12.005.

Johansson, M. and T. Laike. (2007). Intention to respond to local wind turbines: The role of attitudes and visual perception. *Wind Energy 10*(5), 430–451.

Jones, C.R. and J. Eiser. (2009). Identifying predictors of attitudes towards local onshore wind development with reference to an English case study. *Energy Policy 37*(11), 4604–4614 doi: 10.1016/j.enpol.2009.06.015.

Jones, C.R. and J. Eiser. (2010). Understanding 'local' opposition to wind development in the UK: How big is a backyard? *Energy Policy 38*(6), 3106–3117 doi: 10:1016/j.enpol.2010.01.051.

Kaldellis, J.K. (2005). Social attitude towards wind energy applications in Greece. *Energy Policy 33*(5), 595–602. doi: 10.1016/j.enpol.2003.09.003.

Kaldellis, J.K., M. Kapsali, El. Kaldelli, and Ev. Katsanou. (2013). Comparing recent views of public attitude on wind energy, photovoltaic and small hydro applications. *Renewable Energy* 52, 197–208. doi: 10.1016/j.renene.2012.10.045.

Kaldellis, J.K., M. Kapsali, and Ev. Katsanou. (2012). Renewable energy applications in Greece – What is the public attitude? *Energy Policy 42,* 37–48. doi: 10.1016/j.enpol.2011.11.017.

Kapetanakis, L. A, D. Kolokotsa, E.A. Maria. (2014). Parametric analysis and assessment of the photovoltaics landscape integration: technical and legal aspects. *Renewable Energy 67,* 207–214. doi: 10.101016/j.renene.2013.11.043.

Kelly, G. (2011). History and potential of renewable energy development in New Zealand. *Renewable and Sustainable Energy Review 15*(5), 2501–2509. doi 10.1016/j.rser2011.01.021.

Kempton, W., J. Firestone, J. Lilley, T. Rouleau, and P. Whitaker. (2005). The offshore wind power debate: Views from Cape Cod. *Coastal Management 33*(2), 119–149. doi: 10.1080/08920750590917530.

Klick, H. and E. Smith. (2010). Public understanding of and support for wind power in the United States. *Renewable Energy 35*(7), 1585–1591. doi: 10.1016/j.renene.2009.11.028.

Kontogianni, A., Ch. Tourkolias, M. Skourtos, and D. Damigos. (2014). Planning globally, protesting locally: Patterns in community perceptions towards the installation of windfarms. *Renewable Energy 66*, 170–177. doi:10.1016/j.renene.2013.11.074.

Krohn, S. and S. Damborg. (1998). On public attitudes towards wind power. *Renewable Energy 16*(1–4), 954–960. doi: 10.1016/S0960-1481(98)00339-5.

Ladenburg, J. (2008). Attitudes towards on-land and offshore wind power development in Denmark: Choice of development strategy. *Renewable Energy 33*(1), 111–118. doi: 10.1016/j.renene.2007.01.011.

Ladenburg, J. (2009a). Visual impact assessment of offshore windfarms and prior experience. *Applied Energy 86* (3), 380–387. doi: 10.1016/j.apenergy.2008.05.005.

Ladenburg, J. (2009b). Stated public preferences for on-land and offshore wind power generation – a review. *Wind Energy 12*(2), 171–181 doi: 10.1002/we.308.
Ladenburg, J. (2009c). Preferences of coastal zone user groups regarding the siting of offshore windfarms. *Ocean & Coastal Management 52*(5): 233–242. doi: 10.1016/j.ocecoaman.2009.02.002.
Ladenburg, J. (2010). Attitudes towards offshore windfarms: The role of beach visits on attitude and demographic and attitude relations. *Energy Policy 38*(3), 1297–1304. doi: 10.1016/j.enpol.2009.11.005.
Ladenburg, J. and J.-O. Dahlgaard. (2012). Attitudes, threshold levels and cumulative effects of the daily wind-turbine encounters. *Applied Energy 98*, 40–46. doi: 10.1016/j.apenergy.2012.02.070.
Ladenburg, J. and B. Möller. (2011). Attitude and acceptance of offshore windfarms: The influence of travel time and windfarm attributes. *Renewable and Sustainable Energy Reviews 15*(9), 4223–4235. doi: 10.1016/j.rser.2011.07.130.
Lilley, M.B., J. Firestone, and W. Kempton. (2010). The effect of wind power installations on coastal tourism. *Energies 3*(1), 1–22. doi:10.3390/en3010001.
Lombard, A. and S. Ferreira. (2014). Residents' attitudes to proposed windfarms in the West Coast region of South Africa: A social perspective from the south. *Energy Policy 66*, 390–399. doi: 10.1016/j.enpol.2013.11.005.
Lothian, A. (2008). Scenic perceptions of the visual effects of windfarms on South Australian landscapes. *Geographical Research 46*(2), 196–207 doi: 10.1111/j.1745-5871.2008.00510.x.
Mason, K. and P. Milbourne. (2014). Constructing a 'landscape justice' for windfarm development: The case of Nant Y Moch, Wales. *Geoforum 53*, 104–115. doi: 10.1016/j.geoforum.2014.02.012.
Meyerhoff, J., C. Ohl, and V. Hartje. (2009). Landscape externalities from onshore wind power. *Energy Policy 38*(1), 82–92. doi: 10.10116/j.enpol.2009.08.055.
Möller, B. (2006). Changing wind-power landscapes: Regional assessment of visual impact on land use and population in Northern Jutland, Denmark. *Applied Energy* 83(5), 477–494. doi: 10.1016/j.apenergy.2005.04.004.
Molnarova, K., P. Sklenicka, J. Stiborek, K. Svobodova, M. Salek, and E. Brabec. (2012). Visual preferences for wind turbines: Location, numbers and respondent characteristics. *Applied Energy 92*, 269–278. doi: 10.1016/j.apenergy.2011.11.001.
Moula, M.E., J. Maula, M. Hamdy, T. Fang, N. Jung and R. Lahdelma. (2013). Researching social acceptability of renewable energy technologies in Finland. *International Journal of Sustainable Built Environment 2*(1), 89–98. doi: 10.1016/j.ijsbe.2013.10.001.
Mulvaney, K.K., P. Woodson, and L.S. Prokopy. (2013a). Different shades of green: A case study of support for windfarms in the rural Midwest. *Environmental Management 51*(5), 1012–1024. doi: 10.1007/s00267-013-0026-8.
Mulvaney, K.K., P. Woodson, and L.S. Prokopy. (2013b). A tale of three counties: Understanding wind development in the rural Midwestern United States. *Energy Policy 56*, 322–330. doi: 10.1016/j.enpol.2012.12.064.
Musall, F.D. and O. Kuik. (2011). Local acceptance of renewable energy: A case study from Southeast Germany. *Energy Policy 39*(6), 3252–3260. doi: 10.1016/j.enpol.2011.03.017.
Nassauer, J.I. (1983). Oil development in a coastal landscape: Visual preferences and management implications, *Coastal Management 11*(3), 199–217.
Oikonomou, E.K., V. Kilias, A. Goumas, A. Rigopoulos, E. Karakatsani, M. Damasiotis, D. Papastefanakis, and N. Marini. (2009). Renewable energy sources (RES) projects and their barriers on a regional scale: The case of wind parks in the Doecanese islands, Greece. *Energy Policy 37*(11), 4874–4883. doi: 10.1016/j.enpol.2009.06.050.
Pasqualetti, M.J. (2011a). Opposing wind energy landscapes: A search for common cause. *Annals of Association of American Geographers 101*(4), 907–917.
Pasqualetti, M.J. (2011b). Social barriers to renewable energy landscapes. *Geographical Review 101*(2), 201–223.
Pasqualetti, M.J. (2012). The misdirected opposition to wind power. In *Learning from Wind Power: Governance, Societal and Policy Perspectives on Sustainable Energy*. J. Szarka, R. Cowell, G. Ellis, P. Strachan and C. Warren, Eds. Palgrave, pp. 133–152.
Pasqualetti, M.J. and M. Dellinger. (1989). Hazardous waste from geothermal energy: A case study. *The Journal of Energy and Development, 13*(2), 275–295.
Pasqualetti, M.J., P. Gipe, and R. Righter. (2002). *Wind Power in View: Energy Landscapes in a Crowded World*, Academic Press.
Polyzou, O. and Stamataki, S. (2010). Geothermal energy and local societies – a NIMBY syndrome contradiction? Presentation at the World Geothermal Congress, Bali, Indonesia, April 25–29, 2010.

Ribero, F., P. Ferrerira, M. Araujo, and A.C. Braga. (2013). *Public Opinion on Renewable Energy Technologies: The Portuguese Case.* Sustainable Electricity Power Planning and Department of Production and Systems, University of Minho, Guimaraes, Portugal.

Sæpórsdóttir, A.D. and R. Ólafsson. (2010). Nature tourism assessment in the Icelandic Master Plan for geothermal and hydropower development, Part II: Assessing the impact of proposed power plants on tourism and recreation. *Journal of Heritage Tourism 5*(4), 333–349. doi: 10.1080/1743873X.2010.517840.

Shang, H.-D. and I.D. Bishop. (2000). Visual thresholds for detection, recognition and visual impact in landscape settings. *Journal of Environmental Psychology 20*(2), 125–140. doi: 10.1006/jevp.1999.0153.

Söderholm, P., K. Ek, and M. Pettersson. (2007). Wind power development in Sweden: Global policies and local obstacles. *Renewable and Sustainable Energy Reviews 11*(3), 365–400. doi: 10.1016/j.rser.2005.03.001.

Sovacool, B.K. (2009). The cultural barriers to renewable energy and energy efficiency in the United States. *Technology in Society 31*(4), 365–373. doi: 10.1016/j.techsoc.2009.10.009.

Sovacool, B.K., and P.L. Ratan. (2012). Conceptualizing the acceptance of wind and solar electricity. *Renewable and Sustainable Energy Reviews 16*(7), 5268–5279. doi: 10.1016/j.rser.2012.04.048.

Sullivan, R.G. and J. Abplanalp. (2015). Visibility and Visual Characteristics of the Ivanpah Solar Electric Generating Power Tower Facility Project Report, Argonne National Laboratory.

Sullivan, R.G., L.B. Kirchler, C. McCoy, J. McCarty, K. Beckman, and P.D. Richmond. (2012a). Visual impacts of utility-scale solar energy facilities on southwestern desert landscapes. Paper presented at the National Association of Environmental Professionals, 37th Annual Conference. Portland, OR, May 21–24, 2012.

Sullivan, R.G., L.B. Kirchler, T. Lahti, S. Roche, K. Beckman, B. Cantwell, and P. Richmond. (2012b). Wind turbine visibility and visual impact distances in western landscapes. In *Science, Politics and Policy: Environmental Nexus*: Conference proceedings, National Association of Environmental Professionals, 37th Annual Conference. Portland, OR, May 21–24, 2012, pp. 897–942.

Sullivan, R.G., L.B. Kirchler, J. Cothren, and S.L. Winters. (2013). Offshore wind turbines visibility and visual impact threshold distances. *Environmental Practice 15*(1), 33–49. doi 10.1017/s1466046612000464.

Swofford, J. and M. Slattery. (2010). Public attitudes of wind energy in Texas: Local communities in close proximity to windfarms and their effect on decision-making. *Energy Policy 38*(5), 2508–2519. doi: 10.1016/j.enpol.2009.12.046.

Thayer, R.L. (1994). *Gray World, Green Heart: Technology, Nature and the Sustainable Landscape.* John Wiley & Sons.

Thayer, R.L. and C. Freeman. (1987). Altamont: Public perceptions of wind energy landscape. *Landscape and Urban Planning, 14*, 379–398. doi: 10.1016/0169-2046(87)90051-X.

Thórhallsdóttir, T.E. (2007). Environment and energy in Iceland: A comparative analysis of values and Impacts. *Environmental Impact Assessment Review 27*(6), 522–544. doi 10.1016/j.eiar.2006.12.004.

Toke, D., S. Breukers, and M. Wolsink. (2008). Wind power deployment outcomes: How can we account for the differences? *Renewable and Sustainable Energy Reviews 12*(4), 1129–1147. doi: 10.1016/j.rser.2006.10.021.

Torres-Sibille, A.C., V.-A. Cloquell-Ballester, V.-A. Cloquell-Ballester, M.A.A. Ramirez. (2009). Aesthetic impact assessment of solar plants: an objective and a subjective approach. *Renewable and Sustainable Energy Reviews 13*(5), 986–999. doi: 10.1016/j.rser.2008.03.012.

Tsantopoulos, G., G. Arabatzis, and S. Tampakis. (2014). Public attitudes towards photovoltaic developments: Case study from Greece. *Energy Policy 71*, 94–106. doi: 10.1016/j.enpol.2014.03.025.

Tsoutsos, T., N. Franzeskaki, and V. Gekas. (2005). Environmental impacts from solar energy technologies. *Energy Policy 33*(3), 289–296. doi: 10.1016/S0301-4215(03)00241-6.

Turney, D. and V. Fthenakis. (2011). Environmental impacts from the installation and operation of large-scale solar power plants. *Renewable and Sustainable Energy Reviews 15*(6), 3261–3270. doi: 10.1016/j.rser.2011.04.023.

van der Horst, D. and D. Toke. (2009). Exploring the landscape of windfarm developments: local area characteristics and planning process outcomes in rural England. *Land Use Policy 27*(2), 214–221. doi: 10.1016/j.landusepol.2009.05.006.

Waldo, Å. (2012). Offshore wind power in Sweden: A qualitative analysis of attitudes with particular focus on opponents. *Energy Policy 41*, 692–702. doi: 10.1016/j.enpol.2011.11.033.

Warren, C.R., C. Lumsden, S. O'Dowd, and R.V. Birnie. (2005). Green on green: Public perceptions of wind power in Scotland and Ireland. *Journal of Environmental Planning and Management 48*(6), 853–875. doi: 10.1080/09640560500294376.

Warren, C.R., and M. McFadyen. (2010). Does community ownership affect public attitudes to wind energy? A case study from South-West Scotland. *Land Use Policy 27*(2), 204–213. doi: 10.1016/j.landusepol.2008.12.010.
Williams, K. (2011). Relative acceptance of traditional and non-traditional rural land uses: Views of residents in two regions, Southern Australia. *Landscape and Urban Planning 103*(1), 55–63. doi:10.1016/j.landurbplan.2011.05.012.
Wolsink, M. (2005). Wind power and the NIMBY-myth: Institutional capacity and the limited significance of public support. *Energy 21*(1), 49–64.
Wolsink, M. (2007a). Planning of renewables schemes: Deliberative and fair decision-making on landscape issues instead of reproachful accusations of non-cooperation. *Energy Policy 35*(5), 2692–2704. doi: 10.1016/j.enpol.2006.12.002.
Wolsink, M. (2007b). Wind power implementation: The nature of public attitudes: Equity and fairness instead of 'backyard motives'. *Renewable and Sustainable Energy Reviews 11*(6), 1188–1207. doi: 10.1016/j.rser.2005.10.005.
Wolsink, M. (2010). Near-shore wind power: Protected seascapes, environmentalists' attitudes, and the technocratic planning perspective. *Land Use Policy 27*(2), 195–203. doi: 10.1016/j.landusepol.2009.04.004.
Zoellner, J., P. Schweizer-Ries, and C. Wemheuer. (2008). Public acceptance of renewable energies: Results from case studies in Germany. *Energy Policy 36*(11), 4136–4141. doi: 10.1016/j.enpol.2008.06.026.

PART II

6

THE VISUAL SIGNATURES OF RENEWABLE ENERGY PROJECTS[1]

Robert Sullivan

Highlights

- Renewable energy and associated electric transmission facilities have unique visual characteristics that increase their visibility in the landscape.
- Renewable energy facilities can be visible for very long distances in both day- and night-time views.
- The visual experience of renewable energy facilities is dynamic; their appearance can change dramatically as the lighting changes over the course of the day.
- Wind turbine blade movement increases wind facility visual contrast. Glare increases solar facility visual contrast.

6.1 Introduction

One's first sight of a vast symmetrical array of tens of thousands of black solar panels across a desert valley or the sweeping white blades of huge wind turbines stretching across a prairie landscape is a visual experience not soon forgotten. The large size, ordered angular geometry, highly reflective surfaces, and unique but distinctly "manufactured" appearance of renewable energy facilities often contrast strongly with the natural or rural settings in which they are located. The facilities may be visible for long distances, and in many situations they are simply too big, too bright, or too unusual to be overlooked. Understanding the unique visual characteristics of renewable energy facilities and how they create visual contrast is important to predicting associated visual impacts and identifying effective impact mitigation strategies. Stronger visual contrast generally results in larger visual impacts. This means that altering the visual characteristics of a project so that the contrasts with its landscape setting are reduced can lessen visual impacts. This chapter examines the unique visual characteristics of renewable energy facilities and the types of visual contrasts they create.

6.2 Visual contrast

In simple terms, *visual contrast* is the difference in color and brightness between an object and its surroundings. Visual contrast is what allows a viewer to recognize change in the landscape. For example, if wind turbines are built in a natural-appearing prairie landscape (as in Figure 6.1), the tall shapes of the towers, their white color, smooth textures, and moving blades will differ noticeably with the flat line of the horizon and the green or brown colors and varying textures of vegetation that dominate the natural-appearing landscape. The differences between the turbines' visual properties and those of the existing landscape are visual contrasts. In fact, it is these contrasts that make it possible for us to see the turbines as distinct from their background; without visual contrast, an object cannot be noticed by viewers, and therefore cannot trigger the emotional response that causes the viewers to regard the object as a positive or negative visual impact.

FIGURE 6.1 Wind turbines create strong vertical line contrasts in flat landscapes with strong horizon lines. Dunlap Ranch Wind Energy Project near Medicine Bow, Wyoming.

(Credit: Robert Sullivan, Argonne National Laboratory.)

Visual contrast is typically described as the differences in the four basic visual design elements of form, line, color, and texture between a proposed project and the surrounding landscape. These design elements are summarized below, as described in the U.S. Department of the Interior Bureau of Land Management's *Manual 8431 Visual Resource Contrast Rating* (BLM 1986):

Form: The mass or shape of an object or of objects that appear unified. Forms can appear to be flat (two-dimensional forms) or to have volume (three-dimensional forms).

Examples of natural forms commonly encountered in landscape views are masses of mountains, valley floors, plains, and lakes. Large masses of similar-appearing vegetation, such as expanses of shrubs or clumps of forest viewed against a backdrop of grasses, can also create strong forms in the landscape. Forms can also be artificial, such as buildings, paved areas, or the large rectangular block of a solar collector array at a solar energy facility.

Geometry is a sub-element of form, and some forms can appear as standard geometrical figures of two or three dimensions (e.g., square, circle, triangle, cube, sphere, cone). This is especially true of artificial forms; for example, solar collector arrays often appear as rectangles, parallelograms, or ellipses as viewed from elevated viewpoints. Naturally occurring forms tend to have more irregular geometry (e.g., mountains, lakes, or vegetation patches).

Line: The path, real or imagined, that the eye follows when perceiving abrupt differences in form, color, or texture or when objects are aligned in a one-dimensional sequence. Line is often evident as the edge of shapes or masses in the landscape, such as the silhouette of a mountain against the sky, or the shoreline of a lake.

Boldness, complexity, and orientation are sub-elements of line. Boldness refers to the visual strength of the line; complexity refers to the degree of intricacy of a line; and orientation refers to the overall relationship of the line to the (horizontal) axis of the landscape.

Examples of lines commonly encountered in natural-appearing landscapes are the horizon line; lines of stratified layers of topography (e.g., successive ridges – see Figure 6.2); the silhouette lines of mountains or ridges against the sky; strata in rock formations; shorelines, streams; and the edges of vegetation masses. Like forms, lines in the landscape can be human-made, including roads, fences, transmission towers and conductors, wind turbine towers, the edges of solar arrays, and the pipelines of geothermal plants.

Because wind, solar, and geothermal facilities typically have many straight-line, straight-edged, or curved components (e.g., turbine towers, steam pipes, solar panels, mirrors, heliostats, or electricity conductors), line contrast from these facilities can be very strong if the lines are bold, especially when the orientation of the facility lines are perpendicular to the predominant natural line, e.g. a tall wind turbine contrasting with the strong horizon line in a flat landscape.

Color: The property of reflecting light of a particular intensity and wavelength (or mixture of wavelengths) to which the eye is sensitive. Color is the major visual property of surfaces.

Sub-elements of color include hue, value, and chroma:

- *Hue* – the aspect of color we know by particular names (e.g., red, blue, orange), and that is determined by the particular wavelengths of light emitted or reflected from an object.
- *Value* – the lightness or darkness of the color, ranging from black to white.
- *Chroma* – the degree of color saturation or brilliance, ranging from pure or highly saturated (high chroma) to dull or washed out (low chroma).

Natural-appearing landscapes are usually dominated by the hues of vegetation, rock, and soil, which tend toward muted (low chroma) greens, browns, and grays. Artificial elements, of course, can be any

FIGURE 6.2 The massive forms of mountain ridges are an important element in this landscape. (Credit: Robert Sullivan, Argonne National Laboratory.)

color, may have high chroma values and can sometimes be highly reflective. For example, depending on the technology, solar facilities use thousands of mirrored surfaces that in some instances are sources of glinting (a brief flash of light) or glare (light bright enough to cause annoyance or discomfort). When glinting and glare are absent, the mirrors or heliostats may reflect the sky, clouds, or, at certain angles, even the ground or surrounding vegetation (see Figure 6.3). Modern wind turbines in the U.S. are always white, which often contrasts strongly with the surrounding natural colors. Geothermal facilities can be any color, but if their sometimes extensive pipeline networks are not painted or coated to match the backdrop, their surfaces may be highly reflective.

Texture: The aggregation of small forms or color mixtures into a continuous surface pattern; the aggregated parts are small enough that they do not appear as discrete objects in the visible landscape. Two types of texture are recognized:

- *Color Mixture (mottling)* – intrinsic surface color contrasts of very small scale, which may be due to hue, chroma, or value, alone or in combination.
- *Light and Shade* – the color contrast, particularly in value, created by differences in lighting on a varied surface or repeated forms. It consists of the repetition of a lit side, a shaded side, and the shadow cast.

Perception of texture is highly dependent on distance; a texture that appears coarse at short distances may appear as a fine texture at longer distances and will appear to be smooth at even longer distances.

FIGURE 6.3 Reflections of the sky in this solar parabolic trough facility's mirrors cause strong color contrast with the surrounding vegetation. Nevada Solar One near Boulder City, Nevada.
(Credit: Robert Sullivan, Argonne National Laboratory.)

Naturally occurring textures include those of vegetation, soils, and rocks. Artificial structures including renewable energy facilities often have monotone, smooth surfaces that lack texture even at very close viewing distances; however, light and shade textures may be important contrast sources from longer distances. They may be seen as the interplay of shadows and lit surfaces from complex piping and other elements of a power block at a geothermal plant, or from thousands of visually overlapping sunlit solar collectors and the shadows they cast on the ground.

6.3 Visibility factors

Wind, solar, and geothermal facilities cover very large areas, and the structures involved can be very tall or highly reflective. These structures have distinctly artificial geometry that can contrast strongly with natural-appearing backgrounds when lighting conditions and viewing angles are favorable. However, at other times, even when viewed from the same location, facilities may be invisible or hard to distinguish from the background. Or, they may be plainly visible but appear substantially different than they did at other times of the day. The visibility of an object and its apparent visual characteristics in any given view result from a complex interplay between the observer, the observed object, and various factors that affect perception. These are referred to as *visibility factors*.

At a general level, visibility factors can be thought of as factors that determine how easy it is to see an object in a landscape and include such things as the object's distance away from the observer, its size and color, but also the lighting falling on it. Because some visibility factors are highly variable over time, they largely determine the hour-to-hour, day-to-day, and seasonal variation in visual contrasts that are an important part of the visual experience of the landscape. Visibility factors determine the distances at which facilities become visible, and the nature and magnitude of the visual contrasts they create.

There are eight major types of visibility factors that affect perception of large objects in the landscape:

- Viewshed limiting factors,
- Viewer characteristics,
- Lighting factors,
- Atmospheric conditions,
- Distance,
- Viewing geometry,
- Backdrop, and
- Object visual characteristics.

The visibility factors and their spatial relationships in the landscape are depicted conceptually in Figure 6.4.

6.3.1 Viewshed limiting factors

Viewshed limiting factors include variables associated with line-of-sight visibility between the viewer and elements in the viewed landscape, including curvature of the Earth, atmospheric refraction, and screening by topography, vegetation, or structures.

6.3.2 Viewer characteristics

Characteristics of the viewer affect the perception of contrast and the ability to discern objects in the landscape.

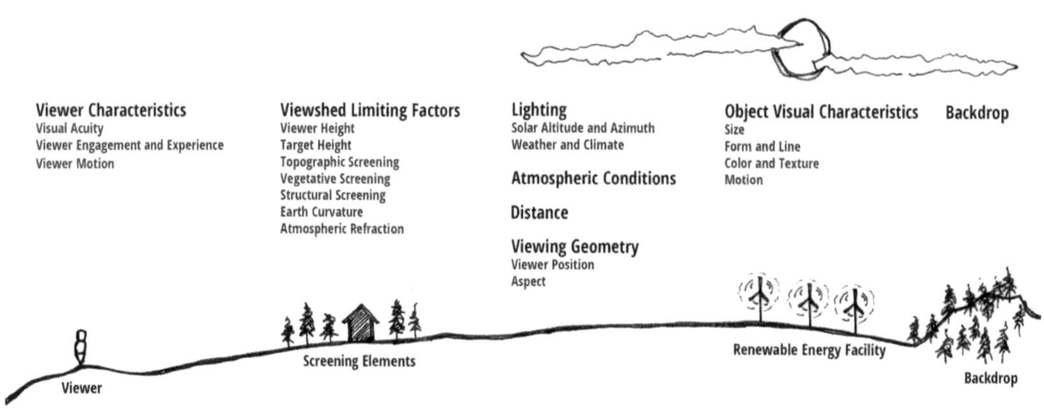

FIGURE 6.4 Schematic diagram of visibility factors (elements are not shown to scale).
(Credit: Robert Sullivan, Argonne National Laboratory.)

- *Visual acuity* is the sharpness or clarity of vision. Visual acuity varies from person to person, though major deficiencies are usually corrected by appropriate eyewear. Nonetheless, some viewers are more or less able to distinguish fine details and slight contrasts in the visual field.
- *Visual engagement and experience* refer to how closely the viewer is looking at the landscape, whether he/she is looking for a particular object or type of object, and his/her familiarity with the type of object. Looking more closely at the landscape will reveal details that go unnoticed in a casual glance, and viewers familiar with particular types of objects (e.g., wind turbines) may spot them more quickly than viewers unfamiliar with them.
- *Viewer motion* may change the aspect of the viewed facility, and it can also limit the duration of views, and the portion of the view that is in sharp focus.

6.3.3 Lighting

The intensity and distribution of lighting has a profound effect on the apparent color of objects and their backgrounds. The angle of sunlight falling on an object may result in shadows that greatly increase its apparent contrast with the background. The sun angle is expressed in two ways:

- Solar altitude (the angle of the sun above the horizon), and
- Solar azimuth (the horizontal angle of the sun, i.e., its compass direction).

Solar altitude and azimuth determine the direction and intensity of lighting on the facilities and the length and direction of shadows cast by facility components, both of which affect facility visibility.

Lighting angle and intensity changes dramatically in the course of the day, and lighting intensity can also change rapidly with the passage of clouds in front of the sun. These changes can have profound effects on the visual contrasts created by facilities, and their appearance can change dramatically as a result. Figure 6.5 shows how different wind turbines may appear in different lighting.

FIGURE 6.5 A passing cloud has shaded the two foreground turbines, causing a dramatic change in apparent color. (Credit: Robert Sullivan, Argonne National Laboratory.)

6.3.4 Atmospheric conditions

Water vapor (humidity) and particulate matter (dust, air pollution, and other particles) within the air can affect visibility by scattering and absorbing light coming from an object, which diminishes contrast and subdues colors. Atmospheric conditions affect the sharpness, brightness, and color of objects, and affect the visibility of objects more as the view distance increases.

6.3.5 Distance

Distance affects both the apparent size of an object and the perceived degree of contrast between the object and its surroundings. In general, visual contrasts are greater when objects are seen at close range. If other visibility factors are held constant, the greater the distance, the less detail is observable and the more difficult it will be for an observer to distinguish individual features.

As noted above, visibility is affected by humidity and air quality, and thus varies widely by region. In some landscapes with low humidity and good air quality such as the western United States, onshore wind and solar facilities may be visible beyond 35 mi (56 km) in the daytime (Sullivan et al. 2012a; Sullivan & Abplanalp 2015), and the aviation obstruction lighting on wind turbines and solar power towers may be visible at similar distances at night (Sullivan et al. 2012a). Offshore wind facilities in the United Kingdom have been observed beyond 25 mi (40 km) (Sullivan et al. 2013).

6.3.6 Viewing geometry

Viewing geometry refers to the spatial relationship of the viewer to the viewed object (e.g., a renewable energy facility), including the *observer position* and the *bearing* of the view.

- *Observer position* refers to the viewer's elevation with respect to the viewed object: whether the viewer is elevated with respect to the facility and therefore looking downward at it, lower in elevation than the facility and therefore looking upward at it, or level with the facility and looking across the landscape at it.
- *Bearing* refers to the compass direction of the view from the viewer to the object.

Both the observer position and the bearing have important effects on facility visibility and contrast levels. An elevated view often shows more of the facility and makes it appear larger. The bearing of the view determines which side of the facility is in view and the angle of surfaces with respect to the viewer. Figure 6.6 shows how an elevated viewpoint causes a large increase in the visual contrast associated with a low-profile solar facility. Plate 10 shows how changing the bearing of the view interacts with the lighting on a solar facility to cause dramatic changes in the apparent color of the facility.

The viewer position may have especially large effects on the visual contrasts created by solar facilities, which generally have relatively low-height collector arrays and cover large areas. Views from ground level may show the solar collector array as a thin line on the horizon, while views from elevated viewpoints often include the top surfaces of the structures in the facility, causing it to occupy more of the field of view and making the full areal extent and the angular geometry of the facility more apparent. Elevated views also tend to show more of the often highly reflective solar arrays, which can greatly increase visual contrast, especially if glare or glinting occurs.

6.3.7 Backdrop

Objects that stand out against the visual backdrop (the background behind the facility) typically command a viewer's attention. As contrast between an object and its background is reduced, the ability

FIGURE 6.6 Two views of the same solar facility from ground-level and elevated viewpoints show increased visual contrast for the elevated view.

(Credit: Robert Sullivan, Argonne National Laboratory.)

to distinguish the object from the background diminishes. When the contrast becomes too small, the object will no longer be visible as separate from its background.

The visual backdrop of the facility is a key factor in determining the visual contrasts it creates as seen from a given viewpoint. As shown in Figure 6.7 the color and complexity of the background can have a large effect on the visibility of lattice transmission towers. When transmission towers are located on ridges, such that they are silhouetted against a uniform bright sky backdrop (often referred to as skylining), the towers typically are much more visible than they would be against a darker and more varied ground backdrop, and skylined towers are usually visible at longer distances than towers viewed against ground backdrops. On the other hand, sunlit, white wind turbines may be much more visible against dark ground backdrops than they are against bright sky backdrops (see Figure 6.8).

FIGURE 6.7 The background can affect the visibility of lattice transmission towers.
(Credit: Robert Sullivan, Argonne National Laboratory.)

FIGURE 6.8 White wind turbines visible against a dark ground backdrop.
(Credit: Robert Sullivan, Argonne National Laboratory.)

6.3.8 Object visual characteristics

The inherent visual characteristics of an object (e.g., a renewable energy facility) will obviously affect its visibility and the level of visual contrast it creates. The facility and structure size; the scale relative to other objects in view; the form, line, surface colors, and textures of the facility components; and any visible motion of the facility components will all affect the facility's apparent visual contrast.

At certain distances, the movement of wind turbine blades may strongly attract visual attention. Although solar collector arrays may change their orientation during the day, the movement is usually very slow, and not apparent in short-duration views. Geothermal and transmission facilities generally do not have moving parts, and, as a result, the viewing experience is less dynamic than wind or solar facilities; however, for both geothermal and certain types of solar facilities, the movement of visible steam or water vapor plumes can attract visual attention.

Reflective surfaces can cause glinting (brief flashes of reflected light) and glare (light bright enough to be annoying or distracting). All types of renewable energy facilities and transmission facilities are capable of causing glinting and glare, but solar facilities are most subject to glinting and glare because the solar collectors/reflectors are highly polished surfaces. Transmission line conductors and insulators may also be glare sources, but the use of low-reflectance materials in these components can substantially reduce associated glinting and glare (BLM 2013).

6.4 Visual contrasts of onshore and offshore wind, solar, geothermal, and electric transmission facilities

The visual characteristics and major sources of visual contrast associated with utility-scale onshore and offshore wind, solar, geothermal, and electrical transmission facilities are described in the following sections. The discussion is limited to operating facilities; however, the site exploration and evaluation, construction, and decommissioning activities also have particular visual characteristics and visual contrasts associated with them, and while most (but not all) impacts from these activities are temporary in nature, the visual contrasts may in some cases exceed the contrasts associated with facility operations. This is particularly true of construction activities.

6.4.1 Onshore wind energy facilities

Land-based utility-scale wind energy facilities vary greatly in size, from hundreds of acres for very small facilities to hundreds of thousands of acres for the very largest facilities which may contain 1,000 or more turbines. The primary visible components of wind energy facilities are of course the wind turbines themselves, but other above-ground ancillary structures include permanent meteorological towers, control buildings, electrical power conditioning facilities, and substations to connect the wind energy facility to the electrical grid.

The primary sources of visual contrast associated with operating onshore wind energy facilities include:

- Vertical line contrasts associated with the wind turbine towers;
- Color contrast from the white tower and blade structures, seen against a sky or ground backdrop (the Federal Aviation Administration [FAA] requires that utility-scale wind turbines exhibit a color contrast with their surroundings when viewed from the air as an aide to aerial navigation safety, which usually means they are painted white);
- Form and scale contrast from the height of individual wind turbines and the large expanse of the wind turbine array as a whole;

- Motion of the wind turbine blades;
- Shadow flicker;
- Line, color, and texture contrasts from roads and other cleared areas; and
- Color contrast from aviation obstruction lighting at night.

Other sources of visual contrast include blade glinting (momentary flashes of reflected light from turning turbine blades); and form, line, color, and texture contrasts from substations, meteorological towers, and from ancillary structures, such as administration or maintenance buildings. Various sources of visual contrasts associated with wind facilities are shown in Figures 6.9 through 6.11.

FIGURE 6.9 Ancillary structures at a wind energy facility.
(Credit: Robert Sullivan, Argonne National Laboratory.)

FIGURE 6.10 A wind facility substation.
(Credit: Robert Sullivan, Argonne National Laboratory.)

FIGURE 6.11 Wind turbines on a mountain ridge in Maine.
(Credit: James F. Palmer, Burlington, Vermont.)

The dark vertical lines of wind turbines silhouetted against the sky can be visible for long distances (beyond 35 mi [56 km] in excellent viewing conditions), and, similarly, sunlit turbines may be conspicuous against darker vegetation, rock, or dark cloud backgrounds (Sullivan et al. 2012a). Although the individual turbines may appear to be very small, because wind facilities typically have many turbines spread out over a wide area (typical wind facilities cover several thousand acres, and the largest wind facilities may cover several hundred thousand acres), even a distant facility may occupy a significant portion of the *horizontal field of view* (the horizontal extent of the human view) and be easily noticeable, though not causing strong visual contrast. On flat ground such as plains or mesas, the mass of wind turbines may appear as a banded but not solid form at long distances.

At shorter distances, the blade motion becomes visible, which may add substantially to visual contrasts. At even shorter distances, although the individual wind turbines may still not appear particularly large, large wind facilities may stretch across much of the visible horizon and the sweeping blades may strongly attract and hold attention.

Although utility-scale wind turbines are very tall (currently they may be the size of 35- to 60-story buildings e.g., 350 to 600 ft), they are also very narrow and lack "visual mass." Their visibility depends greatly on whether they are sunlit or shaded, and the interaction of turbine lighting with the lightness, color, and complexity of the background, as shown in Plate 11. The orientation of the blades to the viewer and the wind speed affect the viewers' ability to see the spinning motion of the wind turbine blades, which also affects visibility.

Wind turbine contrasts are also highly dependent on the spatial relationship of the viewpoint, the wind facility, and the apparent path of the sun across the sky in the course of the day. For example, wind facilities to the east of the viewpoint with a sky backdrop will be silhouetted by the rising sun in the morning, but if viewed against a ground backdrop they may be difficult to see. In the afternoon, the turbines will be lit by the setting sun, and the white towers may be very conspicuous against a darker vegetation or rock backdrop. Facilities west of the viewpoint will be sunlit in the morning and shaded in the afternoon. Because the sun is in the southern sky most of the year, wind turbines south of a viewpoint will seldom be sunlit but will often be silhouetted by the sun behind them. Facilities north of a viewpoint will almost always have at least some sun on them, and this may greatly affect their overall visibility.

To avoid collisions with aircraft, at least some of the turbines in utility-scale wind energy facilities must have flashing red warning lights operating at night. Night-sky contrasts of utility-scale wind facilities can be substantial, particularly in rural or undeveloped areas, where there are few other light sources and there is a uniform and generally featureless dark background (see Plate 12). The effect can be particularly striking when seen reflected from a body of water. In recently built facilities, not every turbine has an aviation obstruction light; however, in a large facility, dozens or even hundreds of turbines may have synchronized red flashing lights, so that all of the lights flash on and off at the same time. For viewers near to the wind facility, the flashing effect can be very conspicuous and is a unique visual experience. The lights can be visible for very long distances (beyond 35 mi [56 km]) (Sullivan et al. 2012a). While they may not be as bright as other visible lights at long distances, the synchronized flashing makes them more conspicuous and instantly recognizable as wind facilities.

6.4.2 Offshore wind energy facilities

Similarly to land-based wind energy facilities, offshore wind energy facilities vary greatly in size, and currently, the largest facilities may cover tens of thousands of acres. Typically, the only offshore components of wind energy facilities that are visible from shore are the wind turbines and electrical service platforms, which are essentially offshore substations.

The primary sources of visual contrast associated with operating offshore wind energy facilities include:

- Line contrasts between the vertical wind turbine towers and the flat sea horizon;
- Color contrast from the white or light gray tower and blade structures, generally seen against a sky backdrop when viewed from shore but occasionally viewed against a water backdrop from elevated onshore locations;
- Form and scale contrast from individual turbines and the array as a whole;
- Motion of the wind turbine blades;
- Color contrast from aviation obstruction lighting at night; and
- Color contrast from marine navigation lighting at night.

Other sources of visual contrast include blade glinting (momentary flashes of reflected light from turning turbine blades); color contrasts from marine paint (typically yellow) at the base of the towers for marine hazard navigation; and form, line, and color contrasts from electrical service platforms. Figure 6.12 shows a small offshore wind facility in the United Kingdom.

FIGURE 6.12 Burbo Bank Offshore Wind Facility in the United Kingdom.
(Credit: Robert Sullivan, Argonne National Laboratory.)

Research conducted in the U.K. has shown that modern offshore wind turbines can be visible for very long distances (beyond 26 mi [42 km] under favorable lighting and atmospheric conditions in daytime views, and beyond 24 mi [39 km] in night-time views) (Sullivan et al. 2013).

In the absence of islands or other land backdrops, most sea views are visually very simple, and views are dominated by the sea/sky horizon line. The often stark white, light gray, or very dark (when silhouetted or shaded) vertical lines of offshore wind turbines are thus particularly noticeable as the eye scans along the horizon line, even though the turbines may not appear to be particularly large. The silhouettes of wind turbines may be particularly striking during the sunrise or sun set – which are favored times to view the ocean.

The visually simple seascape also makes blade movement more noticeable compared with a more complex and varied land or vegetation setting. From certain onshore locations, where offshore wind facilities are laid out in regularly spaced grids, viewers may be looking down parallel rows of turbines that form a striking symmetrical arrangement that tends to command and hold visual attention (see Figure 6.13).

It is important to note that for technical reasons, offshore wind turbines may be larger than onshore wind turbines, and there are currently 6-MW wind turbines in production that are almost 650 ft tall from the water surface to blade tip. Larger turbines are likely to cause greater visual contrasts at a given distance and to be visible at longer distances. The trend is toward developing ever larger turbines for offshore use.

6.4.3 Solar energy facilities

On average, utility-scale solar energy facilities are much smaller than wind facilities, and depending on the technology employed, may range from tens of acres to several thousand acres. Three main classes

FIGURE 6.13 Parallel rows of turbines in an offshore wind facility. An electrical service platform is visible between the two leftmost rows of turbines.
(Credit: Robert Sullivan, Argonne National Laboratory.)

of utility-scale solar energy facilities are in use in the United States, and within each group there are variants that employ slightly different components that give rise to differing visual contrasts, but within each main type, the contrasts are generally similar.

Solar power tower facilities

Power towers typically use tens of thousands of large flat or nearly flat mirrors (heliostats) to reflect sunlight onto the top of a very tall tower (typically 400 to 800 ft [120 to 240 m] tall) (see Figures 6.14 and 6.15). The heliostats track the sun during the course of the day to maximize the amount of sunlight concentrated onto the tower. The top portion of the tower is made of a special material that absorbs the sunlight and uses the heat from the sunlight to heat a fluid inside the tower to extremely high temperatures. The fluid is circulated through the tower to a steam turbine generator (STG), where the heat from the fluid is used to create steam that drives a conventional steam turbine to generate the electricity, which is fed to an electrical substation that connects to the electrical grid.

Solar parabolic trough

Parabolic trough facilities also use mirrors to reflect sunlight, but the mirrors are curved and are arranged in long parallel rows (troughs) generally about 20 to 30 ft above the ground (see Figures 6.16 and 6.17). The curved mirrors focus the reflected sunlight onto tubes that run parallel to the mirrors and are located just above the mirrors. The reflected sunlight heats a fluid in the tubes to very high temperatures, and, similarly to the power towers, the heat from the fluid is pumped to an STG to drive a steam turbine to generate electricity. The mirrors track the sun from east to west over the course of the day to maximize the amount of sunlight falling on the tubes.

FIGURE 6.14 Receiver tower and heliostat array of the Ivanpah Solar Electric Generation Facility. The heliostat array is approximately 1.3 mi in diameter.
(Credit: Robert Sullivan, Argonne National Laboratory.)

FIGURE 6.15 Illuminated receiver tower of a 20-MW power tower facility in Spain.
(Credit: Robert Sullivan, Argonne National Laboratory.)

FIGURE 6.16 Mirrors of a parabolic trough facility.
(Credit: Robert Sullivan, Argonne National Laboratory.)

FIGURE 6.17 A parabolic trough facility as seen from an elevated viewpoint 4 miles away.
(Credit: Robert Sullivan, Argonne National Laboratory.)

Photovoltaic facilities

Conventional photovoltaic (PV) facilities are fundamentally different from power tower and parabolic trough facilities in that they do not involve the generation of heat to drive a steam turbine to generate electricity. Instead, thousands of sunlight-absorbing solar panels convert the sunlight that falls on them directly into electricity (see Figures 6.18 and 6.19). The electricity generated by the panels is fed into power conversion units which are housed in small structures scattered throughout the collector array. Underground cables feed the electricity into a substation for connection to the electrical grid. Conventional PV facilities do not use mirrors to reflect or concentrate sunlight, although there is a small subset of PV facilities called *concentrating PV facilities* that do use mirrors or lenses to concentrate the sunlight onto the solar panels. There are several types of PV technologies, and, depending on the facility, the panels may or may not track the sun during the course of the day, and this affects the appearance of the facility and how its components reflect light.

FIGURE 6.18 A thin-film PV facility seen from a slightly elevated viewpoint about 2 miles away.
(Credit: Robert Sullivan, Argonne National Laboratory.)

FIGURE 6.19 PV solar panels convert sunlight directly into electricity.
(Credit: Robert Sullivan, Argonne National Laboratory.)

For economic reasons, utility-scale solar facilities are concentrated in the southwestern United States. In fact, power tower and parabolic trough facilities are effectively limited to the southwestern states because they require very strong direct sunlight to reach acceptable levels of generation efficiency. PV systems do not have the same limits on direct solar radiation, and thus utility-scale PV facilities are found in some states outside the southwestern United States. However, there are far fewer facilities outside the southwestern states and they tend to be much smaller.

The following list presents the primary sources of visual contrast associated with operating solar energy facilities. Because there are several types of solar technologies, the types of visual contrasts associated with the facilities vary, and not all of the contrasts listed below would be associated with every type of facility. The primary sources of visual contrast associated with operating solar energy facilities include:

- Form and line contrasts from changes to landform (not all facilities require landform changes such as site grading);
- Color and texture contrasts from vegetation clearing or management;
- Form, line, color, and texture contrasts from the solar collector/reflector array (see Plates 10 and 14 and Figures 6.6 and 6.20);
- Form, line, color, and texture contrasts from STGs (power towers and parabolic trough facilities only), and cooling towers (power towers and parabolic trough facilities using wet or hybrid cooling only);
- Scale contrasts because of the large extent of the collector/reflector arrays for all types of solar facilities and the height of receiver towers (power towers only) (see Figures 6.14 and 6.15);
- Line, color, and texture contrasts from roads;
- Movement and color contrast from water vapor plumes (for some power towers and parabolic trough facilities only);
- Glare and glinting from solar collectors/reflectors (see Plate 14 and Figures 6.21 and 6.22);
- Other light reflections from solar collectors/reflectors and ancillary components such as wind fences and site boundary fences; and
- Color contrast from facility lighting at night (see Plate 13).

Other sources of visual contrast include color contrast from aviation obstruction lighting (for power towers only) both during the day (typically white strobe lights) and at night (typically slowly flashing red lights); and form, line, color, and texture contrasts from transmission lines and substations, and from ancillary structures, such as administration or maintenance buildings.

Most of the land occupied by a solar facility is devoted to the solar collector/reflector array. Typically, the arrays are arranged in circles, rectangles, or other straight-sided polygons, which are densely packed with rows of collectors (PV panels) or reflectors (heliostats for power towers or parabolic trough mirrors). The arrays are generally low in height, and thus may be screened from view by tall vegetation, structures, or terrain, unless the viewer is elevated with respect to the facility. However, power towers and parabolic trough facilities have other infrastructure elements, such as the building that houses the steam turbine generator and steam cooling structures, that are much taller than the reflector arrays and are much harder to screen.

The most conspicuous element of power tower facilities is the power tower itself, often referred to as the central receiving tower. Receiver tower height and design vary, but they typically are poured concrete or steel structures ranging in height from about 400 to 800 ft (122 to 244 m), and so they are usually much taller than any nearby structures in the desert areas where they are typically found (see Figures 6.14 and 6.15). When the facility is in operation and the sun is shining, the top portion of the tower structure

FIGURE 6.20 Small (<550 acres) PV arrays (black) and a parabolic trough facility (white) viewed from a mountaintop 10 miles away.
(Credit: Robert Sullivan, Argonne National Laboratory.)

FIGURE 6.21 Glare spot on parabolic trough facility.
(Credit: Robert Sullivan, Argonne National Laboratory.)

is very brightly illuminated by the reflected light from the tens of thousands of heliostats in the reflector array, to the extent that the top of the tower appears to glow a brilliant white (see Figure 6.15). If there is sufficient particulate matter in the surrounding air, faint streamers of reflected sunlight may radiate both diagonally upwards and downwards from the receiver, as shown in Figure 6.15. The overall visual effect of the streamers is unique and dramatic and may be visible beyond 10 mi (16 km) (Sullivan & Abplanalp 2015). BLM- and U.S. Department of Energy (DOE)-sponsored research on the visual

FIGURE 6.22 Glare from solar power tower, Ivanpah Solar Electric Generating System. Power block visible at base of tower.

(Credit: Robert Sullivan, Argonne National Laboratory.)

characteristics of solar energy facilities has shown that the receiver tower structures may be visible for very long distances (beyond 35 mi [56 km]) (Sullivan & Abplanalp 2013; Sullivan & Abplanalp 2015).

The research cited above has also shown that both power towers and parabolic trough facilities are capable of creating strong glare; that is, reflected light bright enough to cause annoyance, and, in some cases, for distances of at least 10 mi (see Plate 14 and Figures 6.21 and 6.22) (Sullivan & Abplanalp 2013; Sullivan & Abplanalp 2015). Depending on the facility, glare may be observed at various times throughout the day at different times of the year. It may be painfully bright in some circumstances (Sullivan et al. 2012b).

PV facilities do not have steam turbine generators and the associated infrastructure, such as cooling towers, cooling ponds, water vapor plumes, and piping that both power towers and parabolic trough facilities have. PV facilities use lower profile structures (typically 6 to 20 ft [2 to 6 m]), and they also utilize black panels rather than the silvered mirrors that power towers and parabolic troughs use. Thus they are less subject to glinting and glare than power towers and parabolic trough facilities, although PV facilities can and do cause glare from sunlight reflected off the surface of the panels. PV facilities require very few employees to run the facilities, so there are fewer vehicles and less worker activity than at trough and power tower facilities. Fewer workers and structures at PV facilities mean less lighting is required at night than power tower and parabolic trough facilities. They do not need the site grading that parabolic trough facilities require to ensure efficient transport of the heat transfer fluid (although the sites are sometimes graded nonetheless). For these reasons, in general, PV facilities have substantially lower visual contrasts than either parabolic trough or power tower facilities. However, some PV facilities may be subject to dramatic changes in color because of reflected sunlight/sky off the reflective glassy surfaces of the panels, despite the black color of the underlying PV material, as shown in Plate 10.

Concentrating PV designs vary but sometimes involve the placement of a sheet of Fresnel lenses in front of the PV panels, which are mounted on pedestals much taller than normal PV panels; the assemblies may be as high as 40 ft (12 m) and resemble very large tilted tables on pylons (Figure 6.23). They generally track the sun precisely to maximize the sunlight concentrated on the panels. In some instances, they may cause bright light reflections visible for very long distances (Sullivan & Abplanalp 2013).

Limited observations of solar facilities at night suggest that ground-based lighting at the facilities may be visible for long distances (Sullivan et al. 2012b). Power tower central receiving towers are tall enough that they would require aviation obstruction lighting, which may be visible for very long distances. Lighting at a parabolic trough facility is shown in Plate 13.

FIGURE 6.23 Concentrating PV facility.
(Credit: Robert Sullivan, Argonne National Laboratory.)

CASE STUDY 6.1 COMPARING VISIBILITY OF SOLAR FACILITIES

The Silver State North PV Facility and the Ivanpah Solar Electric Generating System (ISEGS) power tower facility are both located on BLM-administered lands in the Ivanpah Valley close to the California and Nevada border along Interstate 15, southwest of Las Vegas. Comparing the visual characteristics of the two facilities illustrates how visibility factors interact to determine the nature and magnitude of visual contrasts created by utility-scale energy facilities.

The 50-MW Silver State North Thin-film PV Facility has two rectangular solar panel arrays covering approximately 600 acres (240 hectares) (see Figure 6.24). The collector arrays consist of black PV panels mounted side-by-side on sloped galvanized steel fixed-tilt mounting structures, with a total height of about 7 ft (2 m). The panels are arrayed in long east-west oriented rows, which form modular blocks. The power conversion units (PCUs), which house electrical equipment, are interspersed at regular intervals amongst the collector arrays. The PCUs protrude several feet above the solar module array and are painted shadow gray to reduce visual contrast. Other on-site facilities consist of a substation, collection and transmission lines, an operation and maintenance area, a small operations and maintenance building, and access roads.

The 392-MW ISEGS power tower facility consists of three 460-ft (140 m) "power towers" surrounded by 173,000 flat, garage-door sized mirrors called heliostats about 11 ft (3.3 m) tall (see Figure 6.25). These mirrors reflect sunlight onto the top portions of the power towers, which are referred to as "receivers." The heliostats cover more than 3,100 acres (1,300 ha) and are arranged in concentric circles around each of the towers. Each tower is accompanied by a steam turbine generator set, air-cooled condensers, and other auxiliary systems that are located at the base of the tower. These structures project well above the heliostat arrays. Other ISEGS facility components include an administration building, an operation and maintenance building, a substation, a transmission line, and access roads.

Both facilities are within 1.5 mi (2.4 km) of Interstate 15 (I-15), close enough for viewers on I-15 to see details of the facilities' structure, but not far enough that haze and atmospheric effects normally affect visibility. Because both facilities are slightly elevated with respect to viewpoints on I-15, the viewing angle is low. The small size and the relatively low height of the Silver State North array make it much less noticeable than the ISEGS facility. Because of their low height and dark color, the Silver State North PV panels are inconspicuous from many viewpoints, but as motorists on I-15 pass the facility, they will see lighter bare soil between the rows. Depending on their relative positions and the angle of the sun, viewers may perceive the panels as blue or white as the glassy surfaces reflect the sky. The panels quickly fade back to black as viewers drive past the facility. From most viewing angles, the shadow gray PCUs blend well with the olive green of the desert vegetation surrounding the facility. The administration and maintenance building is also shadow gray, and its low profile does not attract attention.

In contrast, when the sun is shining and the ISEGS plant is operating, the reflected light from the heliostats makes the receivers atop the power towers shine with a brilliant white color that is easily visible at distances exceeding 35 mi (Sullivan and Abplanalp 2015). The very tall towers and complex, large blocky structures at the tower bases are conspicuous because they are strong vertical elements in a largely horizontal landscape. Because the heliostats are arranged in concentric circles, some part of the heliostat array is always facing the viewer, causing that portion of the heliostats array to appear as glittering silver-white, standing out from the olive green vegetation behind them. Occasionally, direct reflections from the heliostats cause small spots of glare bright enough to cause physical discomfort to viewers as far as 20 mi (32 km) from the facility.

In general, the height, complex geometry, and high reflectivity of power tower facilities cause significantly higher contrasts than PV facilities. As a result, power towers are more likely to attract attention and, potentially, opposition.

PLATE 1 Visual simulation of Cape Wind from the Hyannis Port Golf Club on Cape Cod, Massachusetts.
(Source: Save Our Sound.)

PLATE 2 Landscape contrast between growing wind development and the declining oil fields of Wolf Ridge, Texas.
(Photo: M. Pasqualetti.)

PLATE 3 In the early days for wind development in Southern California, substantial public opposition arose from the intrusion of the turbines on the natural landscape. Mt. San Jacinto is the backdrop to the City of Palm Springs, California.

(Photo: M. Pasqualetti.)

PLATE 4 Glare from Ivanpah solar power tower in Ivanpah Valley. View is from the SE at a distance of 40 miles at 35,000 feet in a commercial airliner. Ivanpah Dry Lake is crossed by the straight line of I-15 on the left side (south) of the glare. The airliner was 25 miles SSW of Las Vegas, Nevada.

(Photo: M. Pasqualetti.)

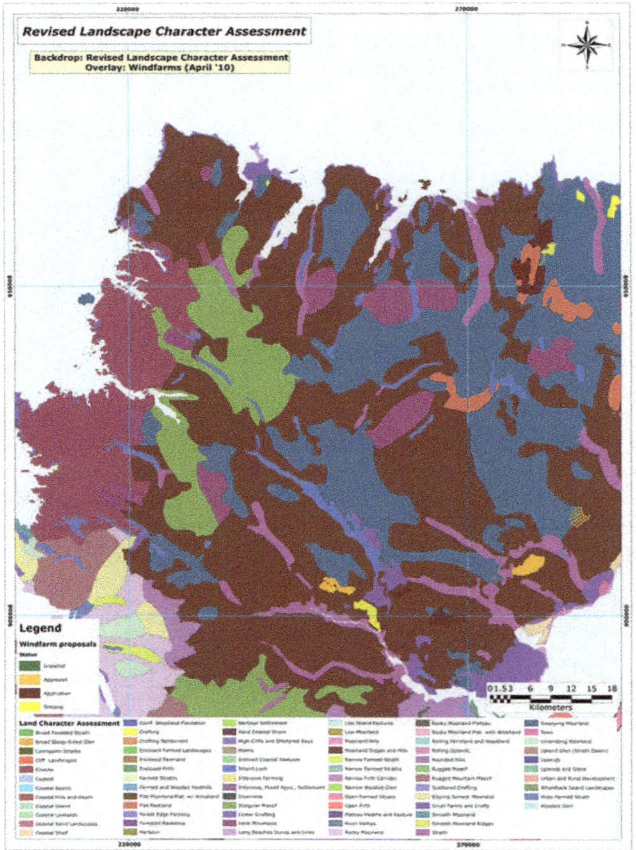

(A)

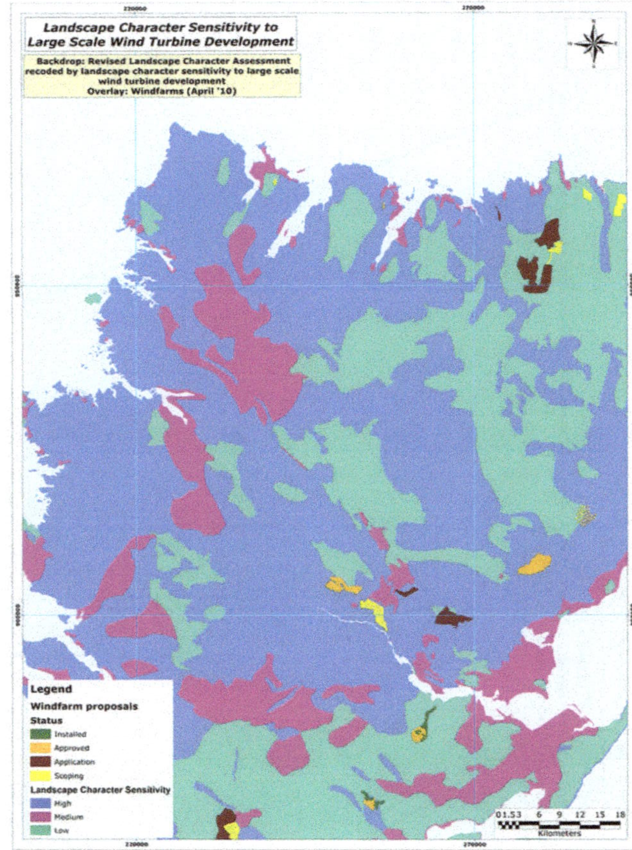

(B)

PLATE 5(A) AND (B) Two maps from the Scottish Highland windfarm landscape supplementary planning guidance project: (a) the landscape character map for the study area (upper left); and (b) the map of landscape character sensitivity for the study area (upper right). Also on the map are the locations of the then constructed, approved or planned windfarm developments in order to see how well they related to the sensitivity assessment. This is reasonably satisfactory, showing that the project-based assessments were sound.

(Reprinted by permission from James Hutton Institute.)

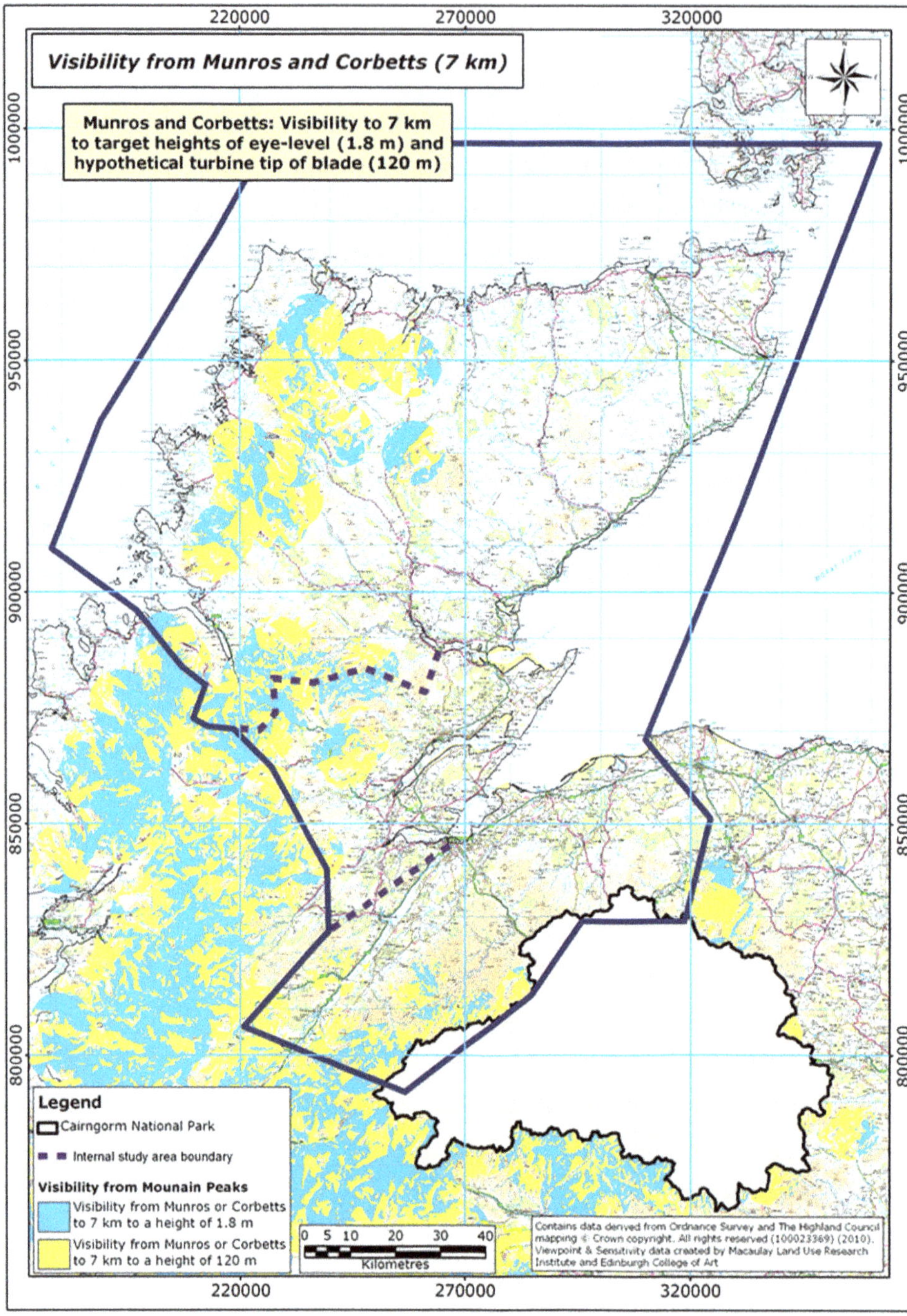

PLATE 6 A map showing an example of the extent of the viewsheds from one set of selected viewpoints – those from the summits of Munros and Corbetts – out to 7 kilometers (short-medium distance visibility). (Reprinted by permission from James Hutton Institute.)

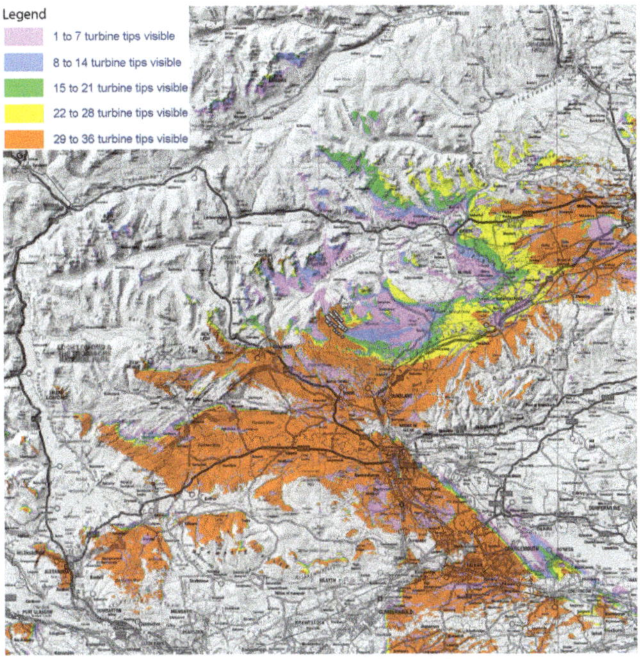

PLATE 7 An example of a computer-generated zone of theoretical visibility for a project in Scotland, based on landform, to 35 kilometers radius from a perimeter around the extent of the proposed development. The colors show the number of turbine blade tips visible from different places, reflecting the variability in visibility across the ZTV. The cluster of small circles represents the sites of the proposed turbines.

(Reprinted by permission from Envision 3D Ltd.)

PLATE 8 An example of a typical visualization as presented for an LVIA. The original photograph (top) is matched first to a wireframe perspective (middle) and then used to create a photomontage of a development from a single viewpoint (bottom), using the photography and visualisation standards recommended by SNH.

(Reprinted by permission from Envision 3D Ltd.)

PLATE 9 Glare from the Ivanpah Solar electric generating system power towers, 40 miles southwest of Las Vegas, Nevada.

(Photo: Robert Sullivan, Argonne National Laboratory.)

PLATE 10 Different views of a photovoltaic solar facility at the same time of day along different bearings show dramatically different apparent color. Copper Mountain Solar Facility near Boulder City, Nevada.

(Photo: Robert Sullivan, Argonne National Laboratory.)

PLATE 11 Three views of the Dunlap Wind Energy Project in Wyoming at distances between 6 and 10 miles illustrate how the complex interaction of viewing geometry, lighting, and visual backdrop affect visibility.
(Photo: Robert Sullivan, Argonne National Laboratory.)

PLATE 12 Hazard navigation lighting atop wind turbines.
(Photo: Terrence J. DeWan & Associates.)

PLATE 13 Lighting at the Nevada Solar One parabolic trough facility near Boulder City, Nevada. (Photo: USDI Bureau of Land Management.)

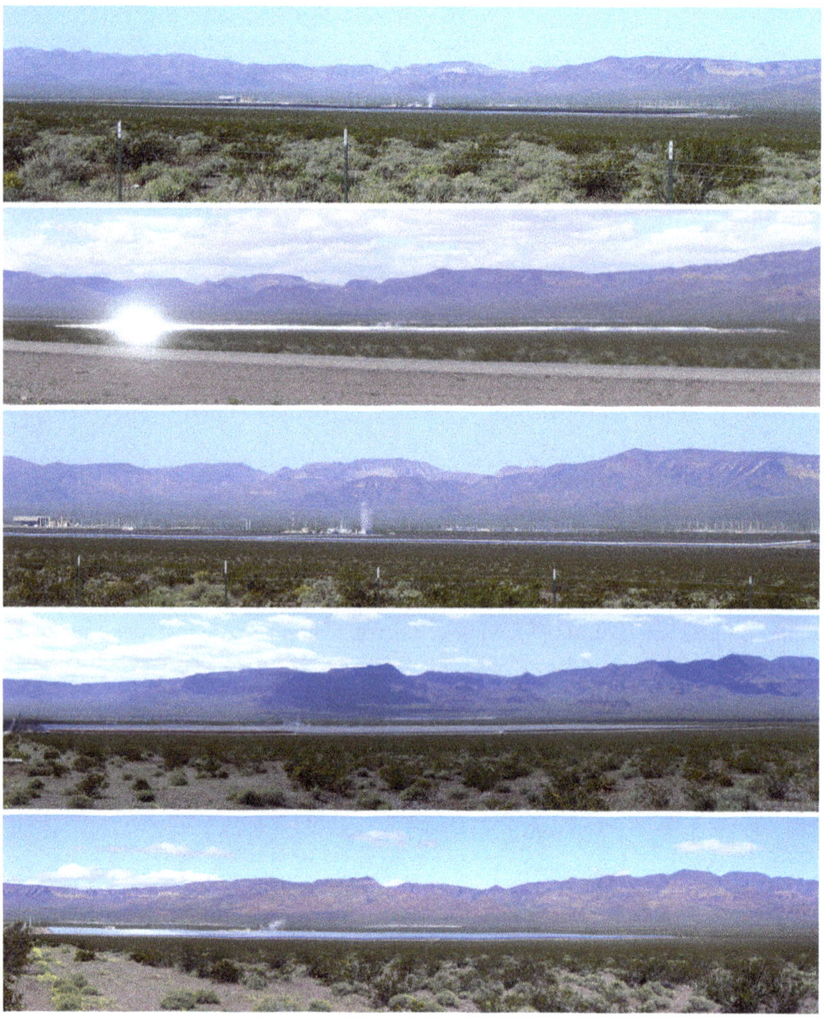

PLATE 14 Several views of the Nevada Solar One parabolic trough facility from different angles and under different lighting conditions show a wide range of color contrasts, including glare (second image from top). (Photo: Robert Sullivan, Argonne National Laboratory.)

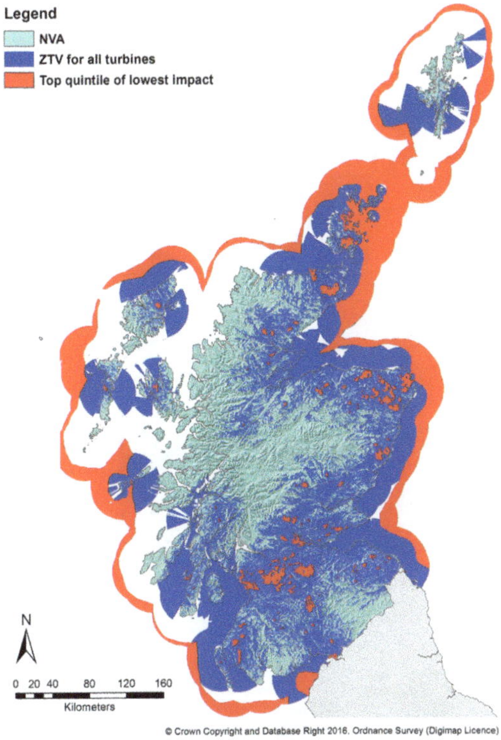

PLATE 15 Map showing the zones of theoretical visibility (ZTV) for existing wind turbines in Scotland and areas where new turbines could be installed with minimum impact on the remaining areas without a view of one or more turbines (non-visible areas or NVA).

(Map and data supplied by Wildland Research Institute, 2016.)

PLATE 16 Color treatment of geothermal plant (near Mammoth, California) to reduce contrast.

(Photo: M. Pasqualetti.)

PLATE 17 Visual clutter from mixing turbine and tower types. San Gorgonio Pass, CA.
(Photo: M. Pasqualetti.)

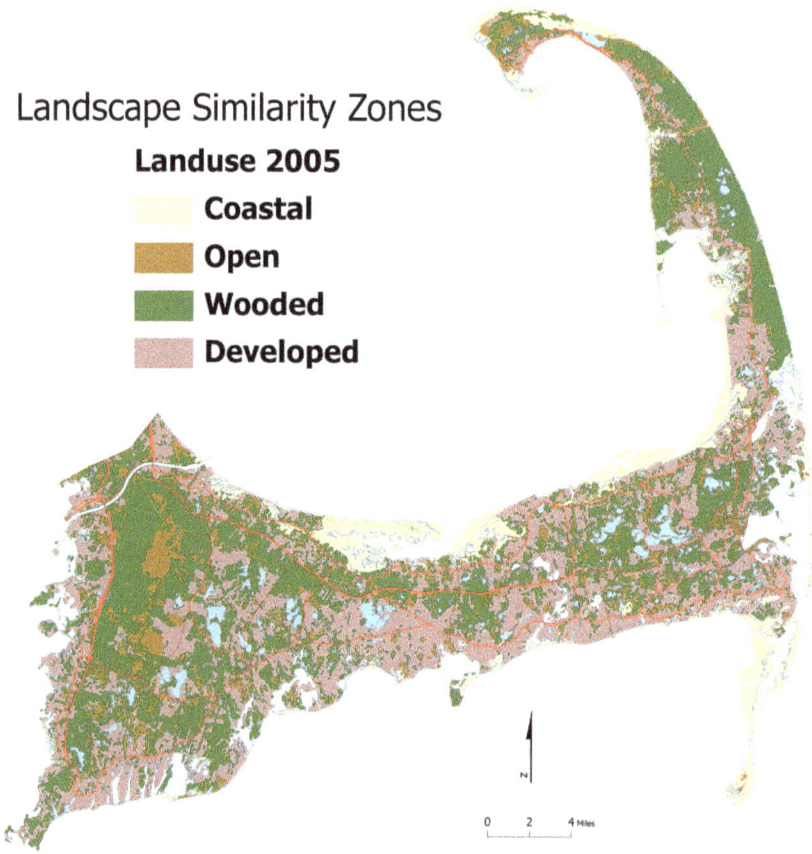

PLATE 18 Cape Cod landscape similarity zone map.
(Reproduced by permission from the Cape Cod Commission.)

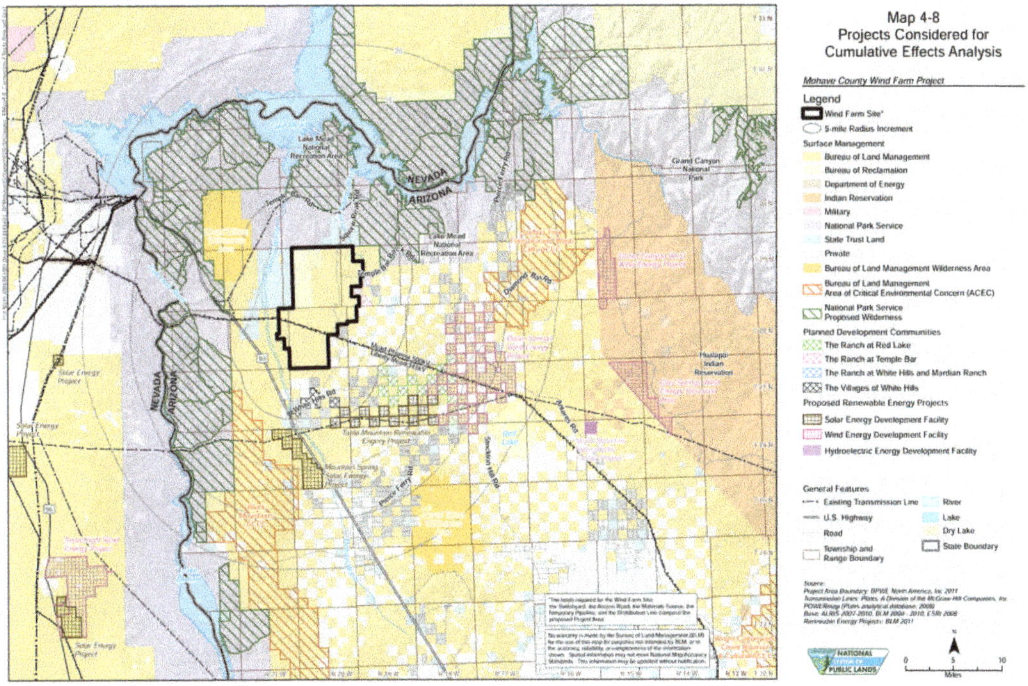

PLATE 19 Map of the Mohave County Wind Farm cumulative effects analysis, shows spatial relationship with sensitive areas such as wilderness and recreation areas.

(Source: USDI Bureau of Land Management, Mohave County Wind Farm Project Final Environmental Impact Statement (2013).)

PLATE 20 A photograph of the project area for an offshore wind facility used as the base photograph for a photomontage.

(Credit: Bureau of Ocean Energy Management.)

PLATE 21 A 3D wireframe model of wind turbines overlaid onto a base photograph during the photomontage production. The wire frame turbines are visible just above the horizon.

(Credit: Bureau of Ocean Energy Management.)

PLATE 22 A spatially accurate and realistic simulation of an offshore wind facility. The simulated wind turbines are visible just above the horizon. (Note that the original simulation is 11 inches by 17 inches, at a much higher resolution, and with better color reproduction.)

(Credit: Bureau of Ocean Energy Management.)

Simulation

010 Long Point Camps

Early Morning

Siemens SWT-3.6-107

10 nm

CONTEXT MAP

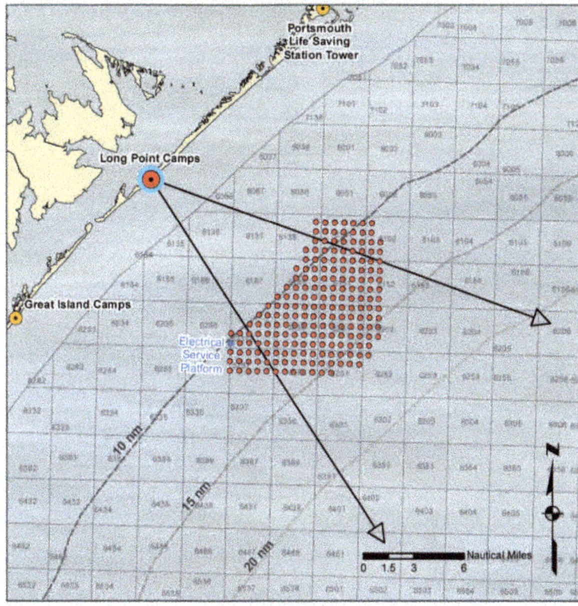

GENERAL INFORMATION

Base Photograph

Photo Name: LPM_0170-UV2
Date: June 3, 2012
Time: 6:17 AM
GPS Coordinates[1]: lat 34.898961°, long -76.255103°
Viewpoint Elevation: 11'

Sun and Weather

Sun Angle/Azimuth: 74°
Sun Elevation: 16°
Lighting Angle: Back lit
Weather Conditions: Sunny
Visibility[2]: 10 mi
Wave Height: 1 - 2'
Period: 7 - 8 sec.

Camera

Camera Make/Model: Nikon D7000
Sensor Dimensions: 23.6 mm X 15.6 mm
Lens Make/Model: Nikkor DX AF-S 35 mm
Lens Focal Length: 35 mm
35 mm Equivalent Focal Length: 52.5 mm
Horizontal and Vertical Angles of View: 37.3° wide and 25.3° high
Camera Height: 1.5 m (5')
Camera Azimuth[3]: 117°

Wind Turbine Information

Number: 200
Make and Model: Siemens SWT-3.6-107
Height/Dimensions:
Support Structure/Monopile Ht.: 13 m (43')
Hub Ht. (above Monopile): 80 m (262')
Rotor Diameter: 107 m (351')
Total Height to Tip of Blade: 147 m (481')
Service Platform: A bldg. 50'H X 100'W X 200' L elevated 50' above the water

VIEWING INSTRUCTIONS

The simulation is properly printed on an 11" X 17" sheet at actual size. If viewed on a computer monitor, use the highest screen resolution. The simulated image is at the proper perspective when viewed at 23.5" from the eye, or at a distance of approximately twice the image height.

NOTES

- The image was taken with a UV filter.
- Refraction Coefficient[4] (k) = .075

PANORAMA

Simulation location within the panorama view (190° X 60°) from the Long Point Camps site

T. J. Boyle Associates
landscape architects • planning consultants

PLATE 23 Appropriate supplementary documentation for a simulation provided separately from the simulation. (Credit: Bureau of Ocean Energy Management.)

Simulation A. Wind turbines as close as 0.3 nautical miles (0.6 km).

Simulation B. Wind turbines as close as 1.5 nautical miles (2.8 km).

Simulation C. Wind turbines as close as 3 nautical miles (5.6 km).

Simulation D. Wind turbines as close as 5 nautical miles (9.3 km).

PLATE 24 Simulated wind turbines at four distances from a Cape Cod beach used as part of a public preference survey.
(Reproduced by permission from the Cape Cod Commission.)

PLATE 25 Turbines in San Gorgonio Pass. Because the project is next to a busy highway, it is unavoidably visible to thousands of people every day, starting from at least 35 miles away as in this photo.

(Photo: M. Pasqualetti.)

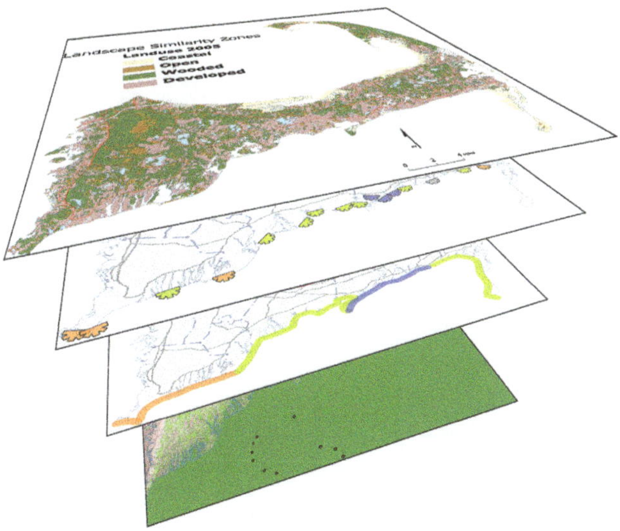

PLATE 26 Landscape and seascape maps used by the Cape Cod Commission identified 52 critical viewpoints, thus reducing ambiguity about scenic resources.

(Reproduced by permission from the Cape Cod Commission.)

PLATE 27 Solar sharing. Panels make electricity while providing gathering space at Arizona State University.

(Photo: M. Northum.)

PLATE 28 Parcels of farmland near the crippled Fukushima nuclear plant are returning to life and being covered with solar panels, reflecting government incentives to invest in renewable power. Solar sharing is being tested by Iitate Denryoku, a power company set up by residents.

(Source: The Japan Times, Sept. 16, 2015.)

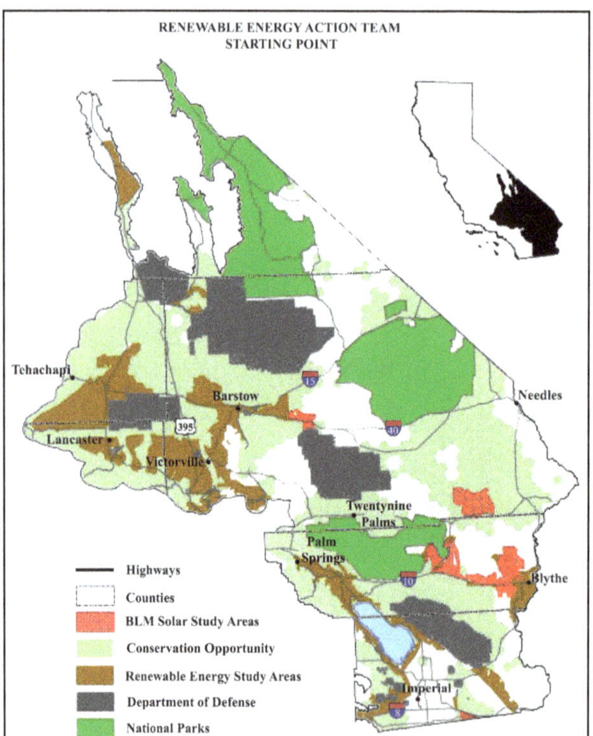

PLATE 29 Map of the Draft California Desert Renewable Energy Plan.

(Source: USDI Bureau of Land Management.)

FIGURE 6.24 Silver State North Thin-film PV Facility from near I-15.
(Credit: Robert Sullivan, Argonne National Laboratory.)

FIGURE 6.25 ISEGS facility from I-15.
(Credit: Robert Sullivan, Argonne National Laboratory.)

6.4.4 Geothermal facilities

Utility-scale geothermal facilities use steam, or the heat from steam or very hot water, to boil water or other substances to drive a turbine to generate electricity. Utility-scale geothermal energy facilities are typically much smaller in size than wind or solar facilities. A 50-MW plant might use an area of up to 20 to 25 acres; a smaller 20-MW plant would require about five to 10 acres. However, the land area encompassing all on-site and off-site facility components (well pads, pipelines, power plants, access roads, and transmission lines) could be as much as 370 acres for a 50-MW plant.

The three basic types of utility-scale geothermal power plants are flash steam, binary cycle, and dry steam. The type of power plant used depends on the temperature, depth, and quality of the water and steam in the geothermal reservoir. These power plants can also be hybridized by including elements of the different power plant types at a single location.

The primary sources of visual contrast associated with operating geothermal facilities include:

- Form, line, and color contrast from building structures, cooling towers, and extensive piping;
- Color and texture contrasts from vegetation clearing or management on the right-of-way (ROW) where clearing and vegetation management are required;
- Color contrast and movement of steam or water vapor plumes;
- Line, color, and texture contrasts from roads; and
- Form, line, and color contrast associated with substations and associated roadways, fencing, and landform alterations.

Other sources of visual contrast during facility operations include further development of the well field, which includes drilling, development, and maintenance of new wells, and pipeline, road, and transmission line construction and maintenance. These activities would be periodic and temporary in nature, but could occur throughout the life of the facility.

Visible geothermal energy facility components include multiple production and injection wells. The number of wells is dependent on the characteristics of the geothermal reservoir and its power generation capacity. For example, a 50-MW plant could require up to 25 production wells and 10 injection wells. Wells would be installed on a well pad, typically with an area of about one to five acres.

Geothermal power plants are typically supported by a more or less extensive pipeline system in the plant's vicinity. A pipeline system includes a gathering system for produced geothermal fluids and an injection system for the reinjection of geothermal fluids after heat extraction takes place. Pipelines are usually 24 to 36 inches in diameter and typically are constructed on supports above the ground surface. The pipelines may have a few feet of clearance underneath them. Every 100 to 200 ft or so, the pipeline may have an expansion loop, or U-shaped bends, to allow for expansion due to heating and cooling. Pipelines transporting hot fluids to the plant are wrapped in insulation, but injection pipelines generally are not.

Some geothermal facilities will at times produce visible steam or water vapor plumes. In some instances, plumes may substantially increase the overall visibility of the geothermal facility (see Figure 6.26). Binary and flash/binary power plants normally do not emit visible steam or water vapor plumes, whereas flash and steam plants do (Lund 2007). Two other types of plumes may be visible. These consist of bore head plumes and natural vents. Bore head plumes are pressurized steam, and are quite dense in comparison to those emitted from cooling towers. The bore head plume has greater vertical scale and generally is visible during all weather conditions. Steam plumes from natural vents can vary in size and density.

Geothermal energy facilities require a number of ancillary facilities and components. These include buildings (other than the power plant) such as an administration or control building, fire control, and a

FIGURE 6.26 Vapor rising from cooling towers in a dry steam plant at the Geysers, 50 miles north of Napa, California.
(Credit: U.S. Geological Survey.)

maintenance building. Other visible components would include fencing, drilling rigs or derricks, water use, storage, and discharge facilities, and off-site components such as access roads and a transmission line. On-site tanks would be used to store water. On-site evaporation ponds may be required to contain circulating water, blowdown, or other discharges.

During the operations phase, the well field would be developed, roads would be built or improved, well field equipment, including piping, would be installed, transmission lines would be built, and well work-overs, repairs, and maintenance would occur.

Drilling new wells in the well field is an important activity during production. Drill rigs used in the production phase may be larger than those used during the exploration phase. The drill rig derrick could be as much as 180 ft in height. These rigs typically are equipped with diesel engines, fuel and drilling mud storage tanks, mud pumps, and other ancillary equipment. Full-size tractors may be used to get this equipment to a drilling site.

Drilling and flow-testing activities associated with geothermal energy production are often conducted 24 hours per day, and would require night lighting during operations. Nighttime light sources during drilling and flow testing would be confined to the drill rigs and other operational areas as necessary for safety. Additional lighting may be needed in areas used for the movement of personnel.

6.4.5 Electric transmission facilities

Although there are many transmission tower types that vary widely in design, size, and materials, three major types are lattice towers, monopoles, and H-frames (see Figure 6.27). While there is overlap in the size range by voltage, in general, the majority of operating higher voltage towers in the United States (500 kV and above) are lattice towers, and the majority of lower voltage towers are H-frames or monopoles.

Most transmission towers are made of steel, but some smaller towers (generally H-frames) are constructed of wood. Surface treatments include unpainted galvanized steel; the rusted brown surface of Corten steel; paint (often white); or the dark brown to gray of wooden poles. Heights vary widely but rarely exceed 200 ft, the height at which aviation obstruction lighting would be required. Consequently, transmission lines generally do not have lighting; however, important exceptions include towers near airports, towers used to cross rivers, and other areas where they might pose a special hazard to air navigation.

The primary sources of visual contrast associated with operating transmission facilities include:

- Form, line, and color contrast from transmission tower structures;
- Color and texture contrasts from vegetation clearing or management on the ROW where clearing and vegetation management are required;
- Line, color, and texture contrasts from roads;
- Line contrast and sometimes reflections from the conductors (the transmission cables);
- Refracted light from glass insulators where they are used; and
- Form, line, and color contrast associated with substations and associated roadways, fencing, and landform alterations.

FIGURE 6.27 Primary transmission tower types. Note vehicles and person at base of towers for scale purposes. (Credit: Robert Sullivan, Argonne National Laboratory.)

Other sources of visual contrast include bird deflectors and aerial marker balls that are attached to the conductors; guy wires used to support certain types of lattice towers; and aviation obstruction lighting on towers in certain situations, as noted above.

The limited research available on high-voltage transmission facility visibility suggests that transmission tower visibility is highly dependent on the color and visual complexity of the background (Jones and Jones 1976; Sullivan et al. 2014). A dark and more visually complex background, such as rough textured shrubs, is better able to "mask" the towers than more uniform, finely grained, and lighter backgrounds such as dried grasses, sand, or the sky. A recent study found that lattice towers viewed against sky backdrops are visible at longer distances and are generally more visible at a given distance than towers viewed against ground backdrops (Sullivan et al. 2014). Figure 6.28 shows skylined transmission towers. The same study found that 500 kV lattice transmission towers in western landscapes are sometimes plainly visible at distances of up to 10 mi, although the individual towers appear very small at that distance. In addition, the study found that the 500 kV lattice towers became a major focus of visual attention at distances of up to 3 mi; smaller H-frame towers are substantially less visible than larger lattice or monopole towers.

Access roads and cleared ROWs in forested landscapes create strong line and color contrasts that can greatly increase visibility of transmission facilities; at long distances, the cleared ROW may be visible even though the towers are not. Figure 6.29 shows a cleared transmission ROW. Cleared ROWs have been observed at 40 mi (64 km), and individual towers at distances of 23 mi (37 km), but in these observations, the objects were at the limit of visibility; that is, barely discernible (Jones and Jones 1976).

Substations are an important component of the visual contrasts from transmission facilities. Substations vary in size and configuration but may cover several acres; they are cleared of vegetation and typically surfaced with gravel. They are fenced and graded and are reached by an associated access road. In general, substations include a variety of structures, conductors, fencing, lighting, and other features that result in a visually complex and "industrial" appearance.

FIGURE 6.28 Skylining makes these 500-kV lattice towers visible at 8 mi.
(Credit: Robert Sullivan, Argonne National Laboratory.)

FIGURE 6.29 Cleared ROWs in forested areas can be visible for very long distances.
(Credit: BLM.)

6.5 Summary and conclusions

Renewable energy facilities have unique visual properties combined with large size, ordered angular geometry, and highly reflective surfaces that add strongly contrasting and distinctly artificial elements to the landscape. Their large sizes and reflective surfaces make them visible for long distances, but their appearance and visibility ultimately depend on the complex interplay of visibility factors. Understanding the visual characteristics of the facilities and how they interact with visibility factors to create visual contrast is essential to accurate visual impact assessment and effective visual impact mitigation.

Unusual and important visual properties of renewable energy facilities include the motion of wind turbine blades and the propensity of solar facilities to cause glare under certain conditions. These phenomena contribute substantially to the visual contrasts created by wind and solar facilities. Interestingly, the phenomena are not captured well in conventional visual simulations that are commonly used to assess potential impacts of proposed wind and solar facilities, as discussed in Chapters 7 and 9. The motion of wind turbine blades, glare from solar facilities, and the changes in facility visibility resulting from changing visibility factors over time demonstrate that the visual experience of renewable energy facilities is dynamic in nature, and differs markedly from the unchanging static view of facilities depicted in visual simulations used in many visual impact assessments.

Note

1 This chapter is based in part on Chapter 2 of BLM 2013, prepared by the author and J. Abplanalp for the U.S. Department of the Interior, Bureau of Land Management and publicly released in March 2013.

References

Bureau of Land Management (BLM). (1986). *Visual Resource Contrast Rating: BLM Manual Handbook 8431, Release 8–30.* Washington, DC: U.S. Department of the Interior.

Bureau of Land Management (BLM). (2013). *Best Management Practices for Reducing Visual Impacts of Renewable Energy Facilities on BLM-Administered Lands.* Cheyenne, WY: U.S. Department of the Interior.

Jones and Jones. (1976). *Measuring the Visibility of High Voltage Transmission Facilities in the Pacific Northwest: Final Report to the Bonneville Power Administration.* Seattle, WA: U.S. Department of the Interior.

Lund, J. (2007). Characteristics, development, and utilization of geothermal resources. *GHC Bulletin*, June 2007.

Sullivan, R. G., and Abplanalp, J. M. (2013). *Utility-Scale Solar Energy Facility Visual Impact Characterization and Mitigation: Project Final Report.* Argonne, IL: Argonne National Laboratory.

Sullivan, R. G., and Abplanalp, J. M. (2015). *Visibility and Visual Characteristics of the Ivanpah Solar Electric Generating System Power Tower Facility: Project Final Report.* Argonne, IL: Argonne National Laboratory.

Sullivan, R. G., Abplanalp, J. M., Lahti, S., Beckman, K. J., Cantwell, B. L., and Richmond, P. D. (2014). *Electric Transmission Visibility and Visual Contrast Threshold Distances in Western Landscapes.* Paper presented at the National Association of Environmental Professionals, 39th Annual Conference, April 7–10, St. Petersburg, FL.

Sullivan, R. G., Kirchler, L. B., Cothren, J., and Winters, S. L. (2013). Offshore wind turbine visibility and visual impact threshold distances. *Environment Practice, 15*(1), 33–49.

Sullivan, R. G., Kirchler, L. B., Lahti, T., Roché, S., Beckman, K., Cantwell, B., and Richmond, P. D. (2012a). *Wind Turbine Visibility and Visual Impact Threshold Distances in Western Landscapes.* In *Science, Politics and Policy: Environmental Nexus.* Proceedings of the 37th Annual National Association of Environmental Professionals Conference, May 21–24, 2012, Portland, OR, pp. 897–942.

Sullivan, R. G., Kirchler, L. B., McCoy, C., McCarty, J., Beckman, K., and Richmond, P. D. (2012b). *Visual Impacts of Utility-scale Solar Energy Facilities on Southwestern Desert Landscapes.* Paper presented at the National Association of Environmental Professionals, 37th Annual Conference, May 21–24, 2012, Portland, OR.

7

IMPROVING THE VISUAL FIT OF RENEWABLE ENERGY PROJECTS

Dean Apostol, John McCarty, and Robert Sullivan

Highlights

- Planning is the backdrop for design and mitigation.
- Landscape aesthetic principles are the foundation for good design.
- Multiple techniques are available to renewable energy planners to avoid or reduce scenic impacts, which can be thought of as "Best Practices".
- Some techniques apply to any project type, while others apply to specific renewable energy facility types.

7.1 Introduction

No matter the extent that regulators or communities may want to shield the public from the view of renewable energy development, there will be times and places where that goal is not feasible. As was pointed out in Chapter 6, renewable energy facilities are very large, can extend over huge areas of land, often have high reflectivity or contrasting colors, and can be seen for very long distances, making full screening extremely hard to achieve. Paying attention to design and mitigation is thus crucial in areas where aesthetic resources are a concern.

Design and mitigation are often used interchangeably in environmental planning, but their meanings are distinct and it is useful to separate the two. In this book we use the word *design* to mean a conscious arrangement of elements to create a desirable appearance. Since renewable energy facilities are designed from an optimal engineering perspective, i.e., to generate the maximum power output, we do not advocate redesigning them, but rather giving attention to how they are configured within land or seascapes to minimize negative visual impacts. We use *mitigation* to mean avoiding or reducing impacts that cannot be addressed through design. *Compensatory mitigation* means off-setting unavoidable impacts through various means. This approach is discussed in Chapter Eight.

The authors accept that renewable energy is necessary and desirable, and that to have enough of it to be effective (using existing technologies), a lot of land and sea surface area will need to be populated by turbines, towers, collectors, and other industrial facilities, inevitably causing visual impacts. However, there are land and seascapes that may be too valuable as scenery to commit to renewable energy. These special places ought to be protected, but to what extent and how? There are additional areas that have some scenic value, but perhaps not so much as to warrant a ban on energy development. A key question this chapter seeks to answer is: How should regulators and developers modify their plans to conserve these areas while still allowing some energy development?

7.1.1 Background

The regulatory and site selection phases of a renewable energy project are where design and mitigation should begin. An optimal design starts with the selection of a site appropriate to the proposed use. In the case of renewable energy projects, this means the energy resource has to be present, but also means adequate site size, proximity to transmission lines, roads, appropriate soils and landforms, compatible adjacent land uses, and the relative absence of natural and cultural resources that would be unacceptably harmed by developing energy resources at that location. These resources include plants, wildlife, cultural sites, and highly valued scenic landscapes.

Local land use plans can provide a good accounting of resources and the potential compatibility or appropriateness of a proposed use. Chapter 3 discusses the range of land use regulations that apply to renewable energy in North America and Australia, while Chapter 4 provides background on this issue in Europe. For many if not most jurisdictions, including federal lands in the US, a "pre-application" process allows identification of issues before submitting a formal proposal.

Too often, sites for renewable energy projects are chosen in advance of thoroughly researching land use allowances and planning objectives identified in relevant land use plans. In the absence of conducting a thorough due diligence process, development proposals can deviate from desirable or permissible uses, and became stalled or rejected on either rules or public resistance. Deviating from the vision and terms of the land use plan often leads to supplemental processes, and added time and expense with no guarantee of approval. It may also galvanize public resistance to the project and establish legal standing for opponents. Significant problems can arise if the proposed project does not integrate desired land use outcomes described within the land use plans, including the requirements for protecting scenic values. This can prove to be a critical and expensive error in project planning and design.

While it is rare for local land use plans to explicitly protect scenic landscapes from energy development, scenic values are often mentioned or described in planning vision statements, if not embedded in the zoning and land use code. These statements should serve as a heads up to developers that scenery may be an issue even where it is not clearly protected, because the land use plans are usually developed with extensive community involvement, and thus reflect local values.

7.2 Assessing and incorporating landscape aesthetic characteristics

Renewable energy projects are inherently engineering intensive efforts with little or no application of the *environmental design arts* aimed at approving appearance and fit. Typically, projects are designed and sited to optimize the generation of energy and its efficient distribution. A key factor in assessing the visual impact of a given development is the extent to which it visually dominates the natural or valued cultural characteristics of the landscape.

Where conservation of scenic resources is desired, energy facilities should be designed to fit within the land or seascape as much as possible, avoiding wholesale changes that obliterate or overwhelm desired attributes. Fitting a project into its context and avoiding or reducing visual dominance requires an *assessment of landscape character* to identify places and patterns that a development should respond to visually.

Four sets of visual information are often included to help incorporate visual design into a project. These include: *natural visual elements*, *landscape character types*, *visual landscape traits*, and *visual variables*. Careful examination of these can help the planners fit a project into its surroundings and help reduce or minimize the degree of negative visual change.

7.2.1 Natural visual elements

Form, line, color, and texture are often used to describe the visual character of landscapes. Identifying and if possible repeating these in the design of a renewable energy project can reduce visual dominance and impacts. Although very large-scale projects may still be visually dominant even when repeating these four visual elements, the degree of dominance can be reduced.

Form, line, color, and texture are present in all landscapes, but one or more of these elements may have a greater degree of visual importance over the others. Understanding which is most important can inform those designing a project on how to make it more visually subordinate to the valued visual character (USDA 1973).

7.2.2 Landscape character

Landscape character refers to attributes that give a place its particular aesthetic qualities, including landform, vegetation, water, and special features like rock outcrops. These attributes can be natural or cultural, including historic farmsteads or field patterns, views, and viewpoints.

Landscape character is often heavily influenced by features that relate to scale. There are five basic scale types: panoramic, enclosed, feature, focal, and canopied (USDA 1973). Considering the scale of a proposed energy development within the scale type can give early insight on the potential level of visual dominance to be expected.

Panoramic land and seascapes are visually open and offer wide, expansive views, such as those commonly associated with the Great Plains of the Western US and Canada, prairies and steppes, large water bodies like the Great Lakes and coastal areas, and large open valleys within otherwise hilly or mountainous regions. Panoramic landscapes may also be experienced from high points above tree line, such as the High Sierras or Rockies in North America, and the mostly open hills and low mountains of Great Britain. Viewing distance can range from 10 to 40 mi (60 km) or more in optimal conditions, and are sometimes limited only by the curvature of the Earth and the clarity of the air. Small-scale objects located in the middle or background may pass notice, especially if built features repeat form, line, color, and texture. Even large-scale projects that have a horizontal form, like a solar photovoltaic (PV) facility, can avoid being visually dominant if care is taken to reduce color contrast, and if the outer edge of the project is sympathetic to natural forms and patterns.

Very tall objects, such as wind turbines or solar towers, can be visually dominant in panoramic landscapes because of the contrast between the vertical towers and the strong horizontal lines of the land. Dominance can be reduced through design by carefully positioning facilities, reducing the amount of the visual field they occupy, increasing distance from the observer, and/or repeating form, line, color, and texture where possible. When facilities are large and expansive, distance is the most effective tool, and the further away a large-scale development is from key viewpoints, the less it will visually dominate.

Enclosed landscapes are those with well-defined visual boundaries where the limit of visibility is set by a backdrop of hillsides, cliffs, and/or tall vegetation (USDA 1973). Canyons and gorges are classic enclosed landscapes. They may have a large vertical dimension, but typically have a restricted horizontal one. Enclosed landscapes often have a high degree of visual variety that includes a composite of landform and mixed heights of vegetation. Large-scale renewable energy development will tend to be visually dominant in such settings unless limited to well-concealed locations.

Feature landscapes have a distinct element, such as a tall mountain, butte, or a cultural landmark that dominates the view. They may be found in combination with a panoramic or enclosed setting. Devils Tower in Wyoming, Shiprock in New Mexico, many of the Cascade Mountain volcanoes, and the classic hill town of San Gimignano in Tuscany are examples of feature landscapes. Where dominant features exist, care should be taken to avoid placement of facilities in close proximity to the dominant feature, which will attract attention.

Focal landscapes typically have converging lines from landforms, vegetation, and/or water bodies, or a progression of aligned features that direct attention to a point of convergence (USDA 1973). Placing development near the focal points should be avoided to minimize visual attention to the facilities.

Canopied landscapes have an overhead plane that restricts the view. Forests and woodlands are classic examples of canopied landscapes. Large scale facilities can be concealed from viewers within canopied landscapes, especially if a forest blocks views close to the viewpoint.

7.2.3 Visual landscape traits

The interplay between form, line, color, and texture create distinct visual traits within a landscape setting. Depending on where a facility is placed relative to these, visual dominance can either be accentuated or reduced. Reading the landscape for visual cues and eye-catching visual traits can help the project designer avoid placing facilities where visual dominance will be accentuated. There are six visual landscape traits to consider: contrast, sequence, axis, convergence, co-dominance, and enframement (Simonds 1961).

Visual contrast can have a harmonious visual effect or a discordant one. Generally, lower contrast means less visual distinction, desirable when one seeks to make something less noticeable. However, in some situations, contrast can add variety and interest that is pleasing. Consider the case of a lighthouse on a coastal bluff, and how its contrast creates a desirable visual effect. In some landscapes, a few well-placed wind turbines can have a similar, enhancing effect due to their contrast. Similarly, a solar facility located in a generally featureless desert landscape may add visual interest.

Sequence has a dual meaning in visual design. One involves features being experienced one after the other, in "visual rhythm." Disrupting this sequence by placing facilities between the features may be a disharmonious change. The second meaning is the sequence of human experience while moving through a landscape. Landscape architects have long used planned sequencing that creates an interplay between built and natural appearing elements that highlights visual variety and can create a sense of intrigue and relief. Project designers can use this to plan when and how people will be visually introduced to a project and include varying views, vistas, and features, and convey a sense of considerate placement as opposed to unconsidered and sprawling visual pollution.

A *visual axis* is a prominent line that guides the eye (USDA 1973, Simonds 1961). Placing a facility along or at the terminus of a strong visual axis attracts attention and increases visual dominance. Roads are the most common example of an axis in many landscapes. Placing a wind turbine along the axis of a road emphasizes it to a greater extent than if it was placed well to one side. A transmission corridor built straight up a forested hillside can create a visually discordant axis.

Convergence occurs when parallel lines appear to meet due to perspective. They can be created by linear transitions in vegetation, landforms, and textures. The eye tends to follow converging lines to a focal point at the apex. Placing facilities at or near focal points can attract attention to the facility and increase visual dominance.

Co-dominance occurs where two or more features have equal visual "weight," i.e., they attract visual attention roughly equally. Landscapes with naturally occurring co-dominant features can be unique and have a striking sense of appeal, like the Three Sisters volcanoes in Central Oregon. Multi-dominance, or many dominant features, is uncommon in natural-appearing landscapes. It is more commonly associated with cultural changes, such as a series of evenly spaced wind turbines along a ridgeline, or multiple large buildings in an urban setting creating many roughly equal features.

Enframement occurs when features frame a view and direct attention, emphasizing a particular feature; for example, trees bordering both sides of the roadway, or a gap between landforms that border the view of a mountain (USDA 1973). Enframement and axis can work together to create a vista that emphasizes the visual dominance of the feature. Placing energy facilities within an enframed landscape can compete with the valued feature.

7.2.4 Visibility factors

Nine variables influence the visibility of renewable energy facilities (see also Chapter 6):

1. *Observer's position:* The horizontal and vertical angle of proximity between the observer and the facility can significantly influence visibility. Observers can be above, below, or at the same general level with the object(s) being viewed.
2. *Distance:* Visibility is reduced as distance increases.
3. *Duration:* Brief glimpses or fleeting glances reduce visibility compared with durations of several seconds to minutes or more.
4. *Seasonality of use:* Some landscapes have a defined and limited seasonal use, while they otherwise might have very low use.
5. *Light conditions:* Direction and angle of light can intensify visibility (front lit), or downplay it when it is in shadow (backlit). In the northern hemisphere, objects on south facing landforms are front lit for much of the day, while those on north facing landforms are backlit.
6. *Atmospheric conditions:* Natural haze, humidity, and air pollution can limit visibility of facilities.
7. *Motion:* Moving objects attract and hold attention, such as the rotating blades of a wind turbine.
8. *Scale:* The size relationship between a facility and its surroundings affects the degree of visibility.
9. *Spatial relationship:* Spatial relationship of facilities to landscape features influences their visibility (USDI 1986).

These variables are often overlooked until late in the visual impact analysis phase of a project, when making changes is difficult. It is much better to consider them early in a project in order to influence site selection and design, especially location. Paying attention to them allows developers and regulators to integrate the degree of visibility directly into the project.

On the other hand, if these variables are effectively integrated into project plans, then visual impact analysis can have more positive findings and result in only minor refinements or mitigation.

7.3 Recommended best practices

With the previous discussion in mind, the next section addresses the application of planning and design techniques to the problem of renewable energy impacts to scenery. Development of renewable energy is still a relatively new phenomenon, and examples of good, proactive planning and design to avoid or reduce visual impacts are few and far between. The authors consulted multiple sources to identify design and mitigation techniques commonly suggested or used in renewable energy projects. In addition we drew upon our own experience as designers and planners to supplement these approaches, and applied standard visual resource practices where appropriate.

7.3.1 Best practices applied to all renewable energy types

Avoidance of impacts

It is standard visual resource practice that the first and surest way to protect scenic resources from impacts of energy (or other) development is to identify and avoid areas of high aesthetic value, including in some cases entire *viewsheds* (the area seen from a viewpoint or corridor). Avoidance means preventing the impact altogether by placing a higher value on landscape aesthetics than on the benefits of developing renewable energy in a given location, and often results from a regulatory decision or public concerns.

Some lands are legally protected from energy development. Wilderness, national parks, national historic sites, UNESCO World Heritage sites, designated scenic areas (like the Columbia River Gorge in the United States), Wild and Scenic Rivers, State or Provincial Parks are all examples of protected landscapes, at least within their administrative boundaries.

However, lands just outside of protected areas, and perhaps visible from them, may not be protected from visual impacts, or may be protected at a much lower level. Standing at a high viewpoint in a National Park or wilderness for example, one may be able to see many miles into the distance, well outside the boundaries. If these views are protected from development by jurisdictions outside the park, then avoidance of impact is achieved. Avoidance may also be de facto where other resources, habitat for sensitive species for example, precludes energy development. Developers normally will not attempt to site a project within a legislatively protected area, regardless of how tempting the energy resource is. But in some cases viewsheds are only identified after a project has been proposed, initiating additional protection measures extending outside of the protected area.

Plate 15 is a map that shows zones of theoretical visibility (ZTV) for existing wind turbines in Scotland. It identifies areas where new turbines could be installed with minimum visual impact. The tool ("Viewshed Explorer") uses highly efficient ray-casting techniques borrowed from 3D gaming software with voxel terrain models that enable real-time calculation of viewsheds and batch processing of cumulative viewsheds across large landscapes as an aid in regional planning (see Carver and Washtell 2012). This tool has been used by Scottish Natural Heritage and the US National Park Service to assess landscape impacts on wildness and wilderness character (Carver et al. 2012, Carver et al. 2013). In Plate 15 it calculated the impact of existing windfarms on the Scottish landscape and, using a reverse viewshed calculation, identified locations where new turbines could be built without impacting areas that are presently free from views of existing turbines (Carver and Markieta 2012). The speed and functionality of the Viewshed Explorer tool makes this kind of analysis possible allowing cumulative "viewshed transforms" to be calculated across large terrain datasets within practical timeframes.

Analyze landscape character before designing project

As noted, a landscape character assessment includes mapping and describing features that give an area its particular aesthetic qualities, including landform, vegetation, water, and special features like rock outcrops. Features can also be natural or cultural, including historic farmsteads or field patterns, views and viewpoints. Landscape character assessment done early enough in the planning process can help identify places that should be protected within the project area, and begin to suggest how a project might best conform to a site or locale. A landscape character analysis is standard visual resource practice and a critical tool. Once a site has been selected for a development project, but prior to beginning the design, an analysis of landscape character should be done alongside other site and environmental analyses. Too often, landscape character is simply described in the visual impact assessment for the proposed project rather than actually used to help shape the project from the start.

Minimize and repair ground disturbance

A proven method for reducing visual impacts is to limit the area of ground disturbance and quickly repair those areas that are disturbed, including construction staging areas as well as the development site. Ground disturbance introduces a visual contrast, since recently disturbed areas are typically much lighter in color than undisturbed areas. Cleared areas that are barren of vegetation in an otherwise vegetated landscape often show strong color and texture contrasts. Minimizing and repairing disturbed ground often benefits other resources as well as aesthetics, for example by reducing erosion and limiting opportunities for invasive species.

Identify key observation points

Key observation points (KOPs) are locations from which people are most likely to see a project. These could be a single viewpoint, such as a park or use area, or multiple points along a corridor, like a trail, road, or river. If design measures are to be used to lessen impacts, it is the design as seen from these viewpoints that most matters aesthetically. Project planners can often identify KOPs from existing plans for an area, or failing that, through public consultation (Sullivan and Meyer 2014).

Site facilities away from most prominent land features

As described earlier in the discussion about landscape features, a project should be designed so that prominent facilities are not placed on, in front of, or too near important local or more distant aesthetic features. For example, the high point on a ridgeline may be a good place for a wind turbine, but it also may be a prominent feature people are most interested in viewing. Keeping turbines on the ridge, but off the high point can reduce their prominence. A scenic lake at the bottom of a valley can have appropriate setbacks so that visitors can continue to enjoy the setting without facilities looming over them. A line of sight to a distant mountain ideally should not be interrupted with a transmission tower or turbine. (USDI 2013, Vissering 2011, Scottish Natural Heritage 2012.)

Avoid night sky impacts

Renewable energy facilities, especially in the US, are often located in sparsely populated desert or grassland regions where the ability to see the stars is highly valued, both by residents and perhaps as a revenue-generating tourist attraction. Hundreds of synchronized flashing warning lights for aircraft can detract from enjoyment of night sky views, while sky glow and glare from facility lighting can make it difficult to see night sky objects, such as stars and nebulae. Field studies by the authors show that in dry, open landscapes hazard navigation lighting on wind turbines can be seen for 35 mi (59 km) or more, and may be a more serious visual intrusion than the towers in daylight hours. Fortunately, technologies are now available (referred to as audiovisual warning systems or AVWSs) that reduce or eliminate the need for continuously blinking lights on tall towers. These activate lights only when a plane is in the vicinity. AVWSs can greatly reduce night sky impacts of wind energy facilities, and they should be used wherever they are permitted. In addition, night lighting on other facilities can be designed to be directed downward and not "leak" into the starlit night (Sullivan and Meyer 2014). (See Plate 12 and Figure 7.1.)

The following practices can help reduce night sky impacts:

- The minimum number of lights and the lowest luminosity consistent with safe and secure operation of the facility should be used;
- Retro-reflective or luminescent markers should be used in lieu of permanent lighting where feasible;
- Lights of the proper design should be used. They should be fully shielded to eliminate uplight, placed and directed to eliminate light spill and trespass to offsite locations;
- Lights of the proper color should be used;
- Steps should be taken to reduce unnecessary lighting use through alternatives to permanent lighting, such as restricting lighting usage to certain time periods;
- Where feasible, activities should be restricted to avoid night-sky impacts; and
- A process for promptly addressing and mitigating complaints about potential lighting impacts should be developed (Sullivan and Meyer 2014).

Site new facilities in already disturbed landscapes or clearings

Sites that are already visually disturbed generally have less scenic integrity than undisturbed sites. Therefore concentrating new development in these areas may reduce the loss of scenic quality. This technique has been used in regional energy facility siting studies, such as the Desert Renewable Energy Conservation Plan in southern California. (USDI 2013.)

Increase distance to reduce visual dominance

As previously noted, distance is an important factor in visual impact. Generally, the farther away a discordant object or set of objects is, the lower the impact. In the case of commercial wind turbines, given their huge size, spinning blades, and white color, it takes a lot of distance to significantly reduce visual impacts. As was pointed out in Chapter 6, large commercial wind turbines are major visual features even at what are considered background distances of 10 mi (16 km) or more (Sullivan et al. 2012, Sullivan et al. 2013). Solar power towers may produce strong visual contrasts from more than 20 mi (32 km) away (Sullivan and Abplanalp 2015). Ground mounted PV panels seen against a dark ground plane on the other hand, may recede into insignificance at relatively short distances. (USDI 2013, Vissering 2011, Scottish Natural Heritage 2012.)

Locate facilities in less prominent locations and away from focal points

As previously noted, locating turbines or other facilities away from the most visually prominent features and views is a good method for reducing visual impacts. (USDI 2013.)

FIGURE 7.1 The Daggett area along Historic Route 66 in California has been identified in Scenic Byway plans as a place where renewable energy development should be permitted because it has already disturbed land. Other much more pristine areas to the east will receive a higher level of protection from development.
(Photo: Dean Apostol.)

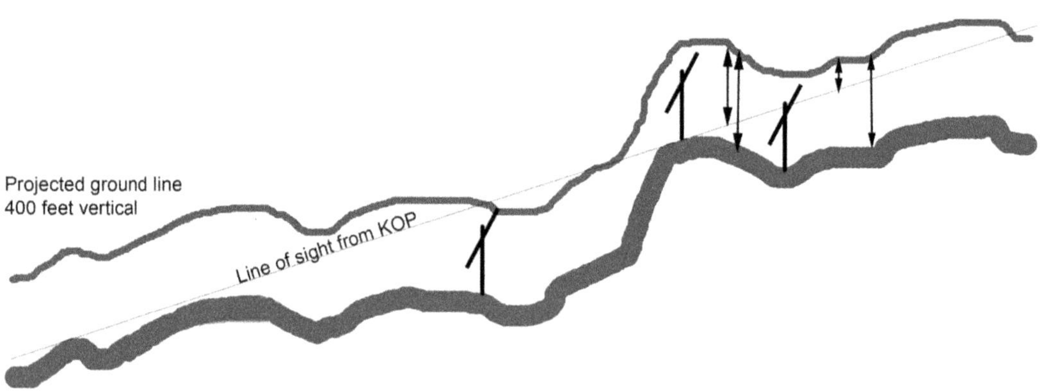

FIGURE 7.2 Using local topography to screen wind turbines from a KOP.
(Image courtesy of BLM.)

Use site-specific location and topographic features to reduce visibility

The diagram above shows how local topography can be used to screen energy facilities, in this case wind turbines, from a viewpoint through micro-siting. The ground plane is projected up to the height of turbines to illustrate where they would be partially or fully hidden from a KOP. (McCarty 2013a.)

Use color to reduce contrast

Some energy facilities, such as geothermal plants and outbuildings, can be painted to reduce visual contrast with the natural colors of an area. For turbines this is difficult due to aviation safety requirements. But at the least, lower height support structures can be painted to reduce contrast (see Plate 16).

7.3.2 Design tools specific to land-based wind turbines

The following techniques are useful ways to reduce aesthetic impacts from land-based wind energy development.

Ensure that all turbines are balanced in color, shape, size, and style

This is an issue primarily where new turbines are added to landscapes that have older designs in them. Having different designs in the same visual field creates "business" or visual chaos that adds to aesthetic problems. (Pasqualetti and Gipe 2002.)

Provide visual order and avoid chaos, clutter, or disarray

Turbines are inherently visually dominant features that draw the eye. Commercial wind projects in the US average 50 turbines, and in the western US can have up to 1,000 turbines. It is very important to ensure that the visual composition of turbines as seen from KOPs is orderly and "readable" as a pattern. What will be successful visually varies from place to place. And it is not always technically feasible to simply place turbines where they might look best. Achieving a good composition may mean starting with an optimum technical spacing and then removing, or perhaps slightly moving some number of turbines to increase visual order from key viewpoints. Energy developers may oppose this because every turbine taken away or moved to a suboptimal location means lost revenues, but it is part of balancing development against valued resources. (See Plate 17.)

Visual order can be provided in a number of ways, including:

- Maintaining order and unity among turbine strings;
- Keeping a visual flow by avoiding conspicuous gaps in turbine strings;
- Treating turbine clusters as distinct visual units and using the space between turbine clusters or strings as a compositional element;
- Keeping heights generally uniform, or tapering them with the landforms;
- Avoiding excessive visual density within clusters (too much visual overlap of turbines and/or blades); and
- Creating a pattern of turbines that is responsive to the surroundings (terrain, vegetation, lines, cultural patterns). For example, non-linear, clustered turbines make visual sense in rolling or topographically variable landscapes, while linear or grid arrangements fit the pattern of flat, agricultural landscapes with strong field patterns.

(A)

(B)

FIGURE 7.3(A)–(B) Examples of visual order in turbine layout: (a) visual order, but still interfering with agreeable mountain scenery; (b) visual order by clustering turbines.

(Photos: Martin Pasqualetti and Dean Apostol.)

Break long lines of arrays with open, undeveloped spaces

This is an especially useful technique in landscapes where turbines occupy prominent ridgelines.

Avoid siting turbine clusters where they are framed by landforms

As noted earlier, people tend to focus on prominent landforms. If landforms frame a view viewers will focus on the framed area. Placing turbines in that frame will maximize turbine visibility.

It is generally better to have turbines off to one side rather than in the center of a view

Most views have a center, or focal area, as viewed from a KOP. If a view center is identified, visual impacts can be reduced by placing turbines to one or the other side, away from the view center.

Viewing objects against the sky

Viewing objects against the sky, known as "skylining," tends to increase their visual prominence. Visual contrast may be lower if turbines are viewed against the ground or a landform as opposed to being seen against the sky. The possibility of avoiding skylining is dependent on the nature of the topography and the elevation of the viewpoint. Given the great height of modern turbines, in flatter landscapes turbines will inevitably be seen against the sky. But there are cases where visual analysis can identify places in a landscape that are more likely to be viewed against ground than sky.

FIGURE 7.4 Turbines framed by landforms at Tehachapi Pass, California.
(Photo: John McCarty.)

7.3.3 Offshore wind turbines

The sea is a flat expanse that does not generally offer anything in the way of features to design with or around. Near-shore environments may have features that can be responded to much in the same way as landscapes, including cliffs, coves, vegetation, and more. But near-shore seascapes are highly valued aesthetically, and any commercial-scale wind project sited along a popular public shoreline may face a battle over aesthetics.

The most reliable tool in the design and mitigation toolbox for offshore wind is to increase the distance of the turbines from the shore, and this is what many countries are doing. Sufficient distance depends on factors like the elevation of the onshore viewpoint, atmospheric conditions, and the degree of importance of the shoreline (Cape Cod Commission 2012).

Where turbines must be located near the shore, techniques similar to those suggested for land-based turbines apply. In particular, it is best to site them away from the most sensitive land features (i.e., not proximate to a headland, coves, islands, nooks, and crannies that make shorelines so visually appealing) (Cape Cod Commission 2012).

As with land-based turbines, clustering small arrays can help a project avoid overwhelming a site visually. Where offshore turbines are located at some distance from the shore, the following techniques appear to have merit (Cape Cod Commission 2012, Scottish Natural Heritage 2012):

- Arranging turbines in single lines to avoid visual overlap. Curvilinear or even circular configurations can be used to avoid monotonous grids of evenly space turbines.
- Varying the spacing between turbines, making them either tighter or looser, to better manage their visual profile.
- Choosing a color that best blends with local conditions where possible.
- As with land-based turbines, limiting night lighting where dark sky conservation is valued.
- Using intervening headlands to screen views of turbines from sensitive seascapes.
- Limiting the number of turbines to minimize the horizontal spread as viewed from shore.

7.3.4 Solar design and mitigation strategies

Successful design and mitigation strategies for commercial solar projects are more limited than for wind turbines, and depend on the type of solar technology used (see Chapter 6). As a general rule, lower profile technologies, like PV panels, have less visual impact, and more design strategies available, than solar power towers or parabolic trough facilities. To date, commercial solar facilities have been located in natural-appearing desert landscapes, such as those in the southwestern US, or the semi-natural cultural landscapes of southern Europe, but the rapid reduction in cost of solar panels and their increasing efficiency has made them more viable in cloudy northern landscapes as well.

Use topography and/or vegetation to screen facilities

Photovoltaic panels, because of their low height, can often be successfully screened from view by taking advantage of even low landforms or vegetation. In some cases, grading can be used to build screening topography. Parabolic trough facilities and associated buildings are considerably taller than photovoltaic panels, but are still much lower in height than power towers. Power towers over 400 ft in height and with highly reflective receivers at the top are, of course, much harder to screen, but it may be possible in some cases.

FIGURE 7.5 Nevada Solar One Parabolic Trough facility partly screened by a low hill.
(Photo: Robert Sullivan, Argonne National Laboratory.)

Design the perimeter of the project to blend with surroundings

Depending on scale and viewing angles, the perimeter of a solar project can be broken up to be less geometric, and can even "feather" into the surrounding, undisturbed landscape.

Blend arrays with contours to harmonize

It is possible in some cases to break up or cluster arrays of collectors, similar to what can be done with wind turbines, to better harmonize the layout with the landscape. Microtopography or other features can be incorporated into the plan to keep a site semi-natural (USDI 2013, McCarty 2013a, 2013b, 2013c).

Use lower profile solar technologies where visual issues are important

In landscapes that are highly valued for scenery, the choice of a low profile technology is more suitable than technologies that require tall towers and/or highly reflective mirrors (see Figure 7.6) (USDI 2013, McCarty 2013a, 2013b, 2013c).

Color treat the back of panels

For some facilities, the back sides of panels or mirrors are exposed to the viewer. They can be color treated to match the surrounding landscape and reduce visual contrast (see Figure 7.7) (USDI 2013, McCarty 2013a, 2013b, 2013c).

FIGURE 7.6 Panorama from Joshua Tree National Park. The dark horizontal stripe in the center of the photo is a very large PV array.

(Photo: Mark Meyer, US National Park Service.)

FIGURE 7.7 Color was used on the back of these collectors at Nevada Solar One to reduce contrast from a distance.

(Photo: John McCarty.)

Make solar power towers interesting architecturally

Architects "celebrate" tall structures rather than try to hide them. In the case of 400 to 800 ft tall towers with significant glare, it may be best to accept their visual dominance and make them interesting to look at (McCarty 2013a, 2013b, 2013c).

Avoid offsite glare through siting

Offsite glare visible from sensitive locations can sometimes be reduced or avoided by careful facility siting, for example, siting facilities far enough away from roads to avoid glare in the driver's direct line of sight (USDI 2013).

Screen solar arrays and other glare-causing elements using fencing, vegetation, or earthen berms

In some cases, screening may be used to avoid glare at sensitive viewing locations; however, for large facilities this could be prohibitively expensive, and is not effective for high-elevation viewpoints (USDI 2013).

CASE STUDY 7.1 SOUTH FORK VALLEY PV SOLAR PROJECT

A 500-acre PV solar project was proposed in the southern Sierra Nevada foothills in Kern County, California. South Fork Valley is named for the river that bisects it: the south fork of the Kern River. Upstream from the valley the river passes through federal wilderness and is federally designated as a Wild and Scenic River. The river corridor through this broad valley has an impressive riparian habitat component that is home to a California Audubon preserve considered an important bird habitat. The valley is situated between the mountainous Sequoia National Forest on the north and Bureau of Land Management (BLM) lands to the south. The valley's situation between these mountainous landscapes affords it memorable views of natural-appearing landscapes that rise dramatically above the valley floor, including the tapestry of colors created by springtime wildflower displays on the mountain slopes. This impressive adjacent scenic quality is rare in an inhabited place in the southern Sierra Nevada Mountains.

Culturally, the valley has a string of small towns along the road that follows the river's meandering course; each is unique and has its own historical context in its ranching and mining history. Families who live in villages and on the rural ranches that dot the landscape have been resident for generations. Commercial services are rare, and places of worship also serve as centers of civic congregation.

When the Kern County Planning Department issued a Notice of Preparation for the solar project's environmental review, the local residents were engaged, passionate, and vocal in their opposition to an industrial development in their bucolic valley. Visual intrusion and community character incompatibility were the stakeholders' primary concerns during the county's scoping process.

The project proponent's technology provider was surprised by the opposition to a project they designed. They had similar projects in operation in the US and Europe and considered themselves stewards of environmental responsibility. The project designer took the unusual approach of contacting the county's visual resource consultant to assist in revising the project to make it more visually acceptable to the stakeholders. County approval of this collaboration was required by California environmental law. The county acknowledged the concerns of its constituents and approved having their consultant identify mitigation measures that would reduce the project's perceived visual intrusion into a landscape that stakeholders consider to be their inheritance.

Working together, the project designer and the visual specialist developed landscape-level visual mitigation measures that would soften the industrial nature of the proposed project. This included identifying sensitive visual receptors and developing mitigation strategies specific to individual user groups on a micro-scale.

As the project was originally designed, a turn-of-the-century church and its community center would be surrounded on two sides by the project's solar arrays. The church had a unique architectural fenestration: there was a window behind the altar that allowed the congregation to gaze northward to the towering pyramidal forms of the Sierra Nevada Mountains over the proposed project's solar arrays. Visual mitigation for this historic resource included measures to triple the setbacks so the project's mass would be much less apparent.

The main thoroughfare through the valley meets the criteria for nomination as a state or federal scenic highway designation. To reduce the impact to the largest viewer group in the valley, the PV arrays were moved an additional 50 ft away from the travel way. This distance was considered acceptable because of extensive perimeter screening proposed in a concept landscape plan as an important visual mitigation measure. Endemic species would be planted around the entire perimeter, with dense mature planting at sensitive locations. The inclusion of specimen-success criteria would assure successful maturity of the vegetative screening. This measure also buffered the previously mentioned church. Additionally, the mitigated project site plan moved the electrical substation and operations and maintenance (O&M) building from the travel way's edge to a location in the center of the panel array. This move placed it as far away as possible from all sensitive visual receptors.

Kern County acknowledges the importance of residential viewsheds. The project as proposed was less than 100 ft from the largest residential area in South Fork Valley. At many residences, mature landscaping already obstructed views of a century-old historic pasture. Where solar panel arrays were adjacent to residential land uses, the setback was more than doubled to 200 ft. The mitigated project landscape plan further screens the low-profile project, making it more visually compatible.

Native Americans reside on a federally non-recognized tribal reservation in the Sierra foothills north of the river and the project. Their topographically superior landscape perspective rendered the screening provided by the riparian corridor ineffective, affording views of the full extent of the project. The perimeter landscape planting would soften the edges of the project but more effective mitigation was infeasible. The dark-colored solar panels would create distinct contrasts with the surrounding ranch lands.

Additional design considerations included:

- Typically, utility-scale solar facilities have a perimeter security chain-link fence that is 6 ft tall and topped with three strands of barbed wire. This feature was deemed an incongruent element in this landscape. The project proponent agreed to using ranch-style fencing that replicates existing features in the landscape.
- Security lighting is typical of these facilities as well; dusk-to-dawn security lighting is not uncommon. Light trespass and sky glow were stakeholder concerns; dusk-to-dawn lighting is rare in the valley and dark skies are valued. The mitigated site plan proposed only lighting at the O&M building and on-site substation. These lights would use motion detectors that were on timers so that they would only be on during maintenance activities.

- The plan also proposed to leave the existing grasslands in place. This mitigation measure was intended to reduce fugitive dust that would be an issue if the loamy soils of the valley were exposed. However, this design criterion had the secondary effect of reducing visual contrast of the light-colored soils, making the project more visually consistent with its surroundings. The biomass of the pasture was to be controlled by grazing sheep, which is a congruent landscape element.

Although it is difficult to establish a working relationship between designers and visual resource professionals, this case study demonstrates that if the parties involved are proactive to pursue a visually superior alternative, the collaborative approach would result in a project that is less visually intrusive to stakeholders.

Peter Langenfeld

7.3.5 Geothermal design and mitigation strategies

Geothermal facilities require extensive pipeline systems for geothermal fluid, water, and/or steam transport, and these pipelines may create strong line, color, and texture contrasts with the surrounding landscape. The nature, extent, and visual complexity of the pipeline system are dependent on the type of geothermal technology employed, as well as the size and configuration of the individual facility. The pipelines generally run parallel to, and very close to, the ground, so moderate-height vegetation or topographic relief may provide effective screening. Site planning should use topography and vegetation to screen pipelines from view.

During the operations phase for geothermal plants, the well field is developed; roads are built and/or improved; well field equipment, including piping, is installed; transmission lines are built; and well work-overs, repairs, and maintenance occur. All of these activities create visual contrasts. Drilling and flow-testing activities associated with geothermal energy production often are conducted 24 hours per day, and would require night lighting during operations.

Some geothermal facilities will at times produce visible plumes from cooling towers, depending on the geothermal technology used. Two other types of plumes may be visible. These consist of bore head plumes and natural vents. Bore head plumes are pressurized steam, and are quite dense in comparison to those emitted from cooling towers. The bore head plume has greater vertical scale and generally is visible during all weather conditions. Steam plumes from natural vents vary in size and density.

Use dry-cooling technology

Dry-cooling technology should be used (where appropriate) instead of wet-cooling technology so that visual impacts from water vapor plumes can be avoided.

Screen pipelines from roads and other sensitive viewpoints

Pipelines should be set back from roadways and other sensitive viewpoints, and should be sited to take advantage of available topographic or vegetative screening to conceal them from view.

Paint or coat aboveground pipelines

Aboveground pipelines should generally be painted or coated to match darker colors in the natural surroundings; however, appropriate color selection is dependent on a careful analysis of site- and project-specific factors.

Minimize drill rig and well test facility lighting

All drill rig and well test facility lighting should be limited to that required to safely conduct operations. Lighting should be shielded, and should direct light to focus on the immediate work area, except as may be required to comply with safety requirements.

7.3.6 Transmission towers and lines

We include a brief discussion of transmission line best practices, because new lines are nearly always associated with renewable energy projects. These lines create additional visual impacts to the renewable energy facilities themselves.

Transmission lines are linear features that may stretch for hundreds of miles across a landscape. The gently curving lines of the conductors are punctuated at regular intervals by the tall forms and lines of transmission poles, which may utilize either lattice or monopole construction. While transmission lines lack the "visual mass" of wind and solar facilities, their length and the need for a cleared right-of-way in forested landscape can sometimes make them conspicuously visible. The following recommendations are drawn primarily from *Best Management Practices for Reducing Visual Impacts of Renewable Energy Facilities on BLM Administered Lands* (USDI 2013), and represent common practices to reduce visual impacts.

Site right-of-way crossings to minimize impacts on linear KOPs

Locations for right-of-way (ROW) crossings of roads, waterways, trails, and other sensitive linear KOPs should be chosen to avoid the most visually sensitive points or segments along the linear KOP. Crossings should be chosen to minimize the duration of visual exposure for observers traveling on the road, trail, or waterway. As a general rule, transmission lines should cross these viewpoints at right angles whenever possible.

Collocate transmission lines in existing ROWs or corridors

Transmission lines should be collocated within or adjacent to an existing corridor to minimize impacts from clearing, access, and maintenance roads, and to utilize shared infrastructure. However, this practice can add to cumulative impacts from the existing facility.

Avoid siting transmission lines in the centers of valley bottoms and along ridgetops

The eye follows strong natural lines in the landscape, and these lines and associated landforms tend to focus views on particular landscape features. For this reason transmission lines should avoid following the centers of valley bottoms and ridgetops (i.e., routing the ROWs perpendicular to and over ridgelines).

Avoid skylining transmission lines

Transmission lines should not be placed on ridgelines, summits, or other locations where they would be silhouetted against the sky. Skylining draws visual attention to the project elements and can greatly increase visual contrast. In some cases towers can straddle a ridgeline to avoid placing the towers in the most prominent position.

Site transmission lines along natural lines within the landscape

Siting of transmission lines should take advantage of natural lines within the landscape (e.g., natural breaks in the landscape topography, the edges of clearings, or transitions in vegetation). Siting towers on steep slopes should be avoided where alternatives are available.

Use monopole, guyed, and lattice electric transmission towers appropriately

Lattice or guyed towers are less visually obtrusive in some rural landscapes than monopoles, especially if they are 0.5 mi (0.75 km) or more distant from viewpoints, and seen against a landscape backdrop. When transmission towers are placed within a half mile or less from viewpoints, monopoles occupy a smaller field of view than lattice towers. Monopoles are often more appropriate within built or partially built environments, while lattice or guyed towers tend to be more appropriate for less developed, rural landscapes, where the latticework is more transparent against natural background textures and colors.

Color treat transmission towers to reduce contrasts with existing landscape

In visually sensitive areas, electric transmission towers should be color treated to reduce contrasts with the existing landscape. Monopole towers should have a low-reflectivity treatment in the most suitable color. Corten (rusting) steel is a good choice in some landscapes like the desert southwest, where reds, tans and browns are the dominant colors. Lattice towers should receive a non-specular treatment, typically a darkened gray color. Electricity transmission facilities should utilize non-specular conductors and non-reflective coatings on insulators.

Use air transport to erect transmission towers

In areas of the highest visual sensitivity, air transport capability should be used to mobilize equipment and materials for clearing, grading, and erecting transmission towers in order to avoid additional impacts from road construction.

Design vegetative openings to mimic natural openings

Openings in vegetation for facilities, structures, and roads, should be feathered and shaped to repeat the size, shape, and characteristics of naturally occurring openings. Too often the clearing is linear, which reinforces an unnatural-appearing landscape line.

Use partial ROW clearing and feather edges of transmission ROWs

The vegetation management plan for transmission ROWs and other features in forested areas should incorporate partial ROW clearing where feasible, including topping rather than removing trees that exceed the allowable height and leaving irregular edges within the ROW. Trees that would not present a safety or engineering hazard or otherwise interfere with operations should be left on the ROW. The plan should also specify feathering of ROW edges (i.e., the progressive and selective thinning of trees) combined with mixing of tree heights to create an irregular vegetation outline.

7.4 Summary and conclusion

As we have discussed, many renewable energy facilities are simply too large and too conspicuous to be completely hidden from view, especially in the open and relatively flat landscapes in which they are often located. Good site design and mitigation must be used to minimize the impacts, recognizing that limited success is the best that can be hoped for in many cases, and thus a key strategy is to avoid siting these facilities in visually sensitive landscapes.

Design and mitigation include intelligent placement of facilities and measures taken to avoid, reduce, or compensate for the visual impacts of a facility. Intelligent strategies rely on a thorough understanding of the visual characteristics of both the project and the existing landscape, the latter of which is achieved through the landscape character assessment. The designer's knowledge of topography, vegetation, the visual character of the landscape, land uses, and visually or culturally sensitive locations within the landscape is used to site the facility to minimize visibility (where feasible) and to harmonize the proposed facility with the existing setting to the extent possible.

There are many mitigation measures, some specific to particular renewable energy and electric transmission technologies, that can be used to avoid or reduce the visual impacts, though many may have only a small effect on the facilities/overall impacts. Again, a detailed knowledge of the visual characteristics of both the project and the existing landscape is needed to identify those mitigation practices that will be most effective in a given situation. An encouraging development is the recent publication of renewable energy-specific guides to best management practices for reducing and avoiding visual impacts of these types of facilities.

Key to both successful design and successful mitigation of the visual impacts of renewable energy projects and associated electric transmission facilities is to actively consider them *before* the project is sited and designed. Too frequently, landscape-sensitive facility design and mitigation are not considered until either public opposition to a project has surfaced, or an impact assessment has determined that a project has undue or unanticipated impacts, at which point it may be too late to remedy them.

Bibliography

Cape Cod Commission. 2012. Visual Impact Assessment (VIA) Methodology for Offshore Development. Cape Cod Commission.

Carver, S. & Markieta, M. 2012. No high ground: Visualising Scotland's renewable energy landscapes using rapid viewshed assessment tools. Conference proceedings, *GISRUK 2012*, Lancaster. www.geos.ed.ac.uk/~gisteac/proceedingsonline/GISRUK2012/Papers/presentation-49.pdf

Carver, S. & Washtell, J. (2012). Real-time visibility analysis and rapid viewshed calculation using a voxel-based modelling approach. Conference proceedings, *GISRUK 2012*, Lancaster.

Carver, S., Comber, A., McMorran, R., & Nutter, S. (2012). A GIS model for mapping spatial patterns and distribution of wild land in Scotland. *Landscape and Urban Planning 104*(3), 395–409. doi: 10.1016/j.landurbplan.2011.11.016.

Carver, S., Tricker, J., & Landres, P. (2013). Keeping it wild: Mapping wilderness character in the United States. *Journal of Environmental Management 131*, 239–255. doi: 10.1016/j.jenvman.2013.08.046.

Cownover, Brad, Robert Thayer, Robert Ryan, Stephen Sheppard. 2007. Wind Energy on the Horizon. Panel Presentation for Scenic America, *American Society for Landscape Architects Conference*, Oct. 2007, San Francisco.

McCarty, John. 2013a. Visual resource inventory basics. PowerPoint for training purposes in Reno, NV.
McCarty, John. 2013b. America's energy resources on the public land. PowerPoint for training on solar energy visual impacts. Reno, NV.
McCarty, John. 2013c. Wind power on the horizon: The new energy landscape. PowerPoint presentation for BLM land managers.
Pasqualetti, M.J. & Gipe, P. 2002. *Wind Power in View: Energy Landscapes in a Crowded World*. Academic Press.
Scottish Natural Heritage. 2012. Offshore Renewables – Guidance on Assessing the Impact on Coastal Landscape and Seascape: Guidance for Scoping and Environmental Impact Statement. Scottish Natural Heritage.
Simonds, J. O. 1961. *Landscape Architecture*. McGraw-Hill.
Sullivan, R. G., and Abplanalp, J. M. 2013. Utility-Scale Solar Energy Facility Visual Impact Characterization and Mitigation: Project Final Report. Argonne National Laboratory.
Sullivan, R. G. & Abplanalp, J. M. 2015. Visibility and Visual Characteristics of the Ivanpah Solar Electric Generating System Power Tower Facility. Environmental Science Division, Argonne National Laboratory, IL.
Sullivan, R. G. and Meyer, M. 2014. The Scenery Conservation System: Incorporating Historic and Cultural Values into Scenic Inventories. Argonne National Laboratory, US Department of Energy.
Sullivan, R. G., Kirchler, L. B,. Cothren, J., Winters, S. L. 2013. Offshore wind turbine visibility and visual impact threshold distances. *Environmental Practice 15*(1), 33–49. doi 10.1017/s1466046612000464.
Sullivan, R. G., Kirchler, L. B., Lahti, T., Roche, S., Beckman, K., Cantwell, B., and Richmond, P. D. 2012. Wind turbine visibility and visual impact distances in western landscapes. Conference proceedings, *Science, Politics and Policy: Environmental Nexus*, 37th Annual National Association of Environmental Professionals Conference, May 21–24, 2012, Portland, OR, pp. 897–942.
Tetra Tech and Nixon Peabody LLP. 2008. *Wind Energy Siting Handbook*. Prepared for the American Wind Energy Association.
Thayer, Robert. 2007. Twenty Five Points about Wind Energy for Landscape Architects. PowerPoint Presentation.
University of Newcastle. 2002. Visual Assessment of Wind Farms Best Practice. Scottish Natural Heritage Commission Report F01AA303A.
US Department of Agriculture, Forest Service (USDA). 1973. National Forest Landscape Management, Volume 1 (Agriculture Handbook Number 434). US Government Printing Office.
US Department of the Interior, Bureau of Land Management (USDI). 1986. BLM Manual Handbook H-8431-1, Visual Resource Contrast Rating, Release 8-30. US Department of the Interior, Bureau of Land Management.
US Department of the Interior, Bureau of Land Management (USDI). 2013. Best Management Practices for Reducing Visual Impacts of Renewable Energy Facilities on BLM Administered Lands. US Department of the Interior, Bureau of Land Management.
Vissering, Jean. 2011. A Visual Impact Assessment Process for Wind Energy Projects. Prepared for the Clean Energy States Alliance.

8

MEASURING SCENIC IMPACTS OF RENEWABLE ENERGY PROJECTS

Louise Kling, James Palmer, and Richard Smardon

Highlights

- Presents basic Visual Impact Assessment terminology and regulatory frameworks as used in North America.
- Covers steps for determining project visual characteristics, resources to be considered, sensitivity of these resources plus determining the assessment area.
- Viewshed analysis, baseline conditions, use of photo simulations, and setting VIA criteria.
- Processes for considering cumulative VIA for renewable energy projects, plus potential onsite and offsite mitigation strategies.

8.1 Introduction

This chapter draws primarily from North America to provide an overview of select Visual Impact Assessment (VIA) procedures used to assess potential impacts to scenic and visual resources as a result of utility-scale renewable energy projects.

The history of VIA development is provided in Chapter 2, the regulatory overview is provided in Chapter 3, the European wind energy VIA practice in Chapter 4, the discussion of visual perception and social acceptability of renewable energy projects is provided in Chapter 5, visual characteristics of renewable energy projects in Chapter 6, and visual simulations are discussed in Chapter 9. Recognizing that variation in specific VIA methods exists, this overview focuses on the salient themes of a VIA, with emphasis on practical applications that yield more reliable and defensible conclusions.

8.1.1 What is a Visual Impact Assessment?

A VIA is described as the "analysis of the visual impacts of a proposed project" (Sullivan & Meyer, 2014). The analysis is typically triggered by an existing regulatory framework requiring disclosure of impacts to visual or scenic resources that may result from a proposed action. For example, in the United States, the National Environmental Policy Act of 1969 (NEPA)[1] stipulates that all federal agencies consider aesthetic impacts of projects for which there is a federal nexus (Smardon & Karp, 1993). Likewise, approximately half of US states are governed by legislation requiring consideration of aesthetic impacts of state-funded or permitted activities (see Chapter 3). In Europe, the EIA Directive[2] established in 1985 that an environmental assessment is required for development projects having significant environmental effects, and discretionary for other types of projects as required for member states. The EIA Directive has been amended several times; the most recent clearly states that:

> In order to better preserve historical and cultural heritage and the landscape, it is important to address the visual impact of projects, namely the change in the appearance or view of the built or natural landscape and urban areas, in environmental impact assessments.[3]

In addition, the European Landscape Convention (ELC)[4] proposes to "conserve and maintain the significant or characteristic features of a landscape," meaning any "area, as perceived by people, whose character is the result of the action and interaction of natural and/or human factors." The intent is, to protect the special character of all landscapes, not just the character of a few special landscapes. This is described in further detail in Chapter 4.

In some cases, the regulatory or management framework aimed at the protection of visual or scenic resources also includes VIA guidance. For example, two public land management agencies in the US Department of the Interior's Bureau of Land Management (BLM) and the United States Department of Agriculture's Forest Service (FS) administer lands according to specific management standards for scenic/visual resources. Management standards are developed through the land use planning process, in consideration of baseline visual and scenic resource data collected for the planning area (USDI BLM, 1986a; USDA FS, 1995). For projects sited on lands administered by BLM, there is a generally accepted path forward to determine impact mechanisms, change in existing visual quality and distribution of scenic values across the planning area, and conformance to existing land management standards (USDI BLM, 1986a, 1986b; USDA FS, 1995). More recently, the National Research Council in the US established recommendations for assessing visual impacts of utility-scale windfarm development (NRC, 2007, pp. 146–151). Sullivan and Meyer (2014) also provide a systematic framework by which to determine completeness of a VIA.

In the UK, *Guidelines for Landscape and Visual Impact Assessment* developed by the Landscape Institute and the Institute of Environmental Management and Assessment is in its third edition (LI & IEMA, 2013). It provides a systematic and consistent approach to Landscape and Visual Impact Assessment (LVIA) for all types of development. However, there is also substantial VIA guidance focusing on offshore windfarm development and potential impacts to coastal landscape and seascape (Scottish Natural Heritage, 2012b) and this is described further in Chapter 4. Australia and New Zealand also have guidance for onshore windfarm development within some of the states, which can be seen in Chapter 3.

It is also very common for a VIA to be required for projects sited on private or public lands that are not managed for visual or scenic resources, and where no guidance is provided despite a regulatory trigger to implement VIA. As discussed in Sullivan and Meyer (2014), no industry-wide standard for content of a VIA exists, resulting in considerable variation in approach, level of detail, and criteria used to determine level of impact or significance.

The debate continues over methods used to describe existing scenic quality and landscape character, and how to objectively identify impacts that may result from construction and operation of a project. For example, in the UK, landscape assessment evaluates how the project will affect landscape integrity, with a separate visual assessment used to evaluate how people will react to those effects. Common methods in the US, including the US Army Corps of Engineers procedure (Smardon et al., 1988) and USDI BLM (1980), develop visual impact determination through a process that combines impacts to landscape integrity and expert assessment of those impacts.

The ongoing debate over the "objectivist paradigm," considering landscape quality as an intrinsic physical attribute, and the "subjectivist paradigm," which considers landscape quality as an attribute derived from the eye of the beholder is ongoing. A discussion of a generalized framework to determine impacts is provided below. This discussion does not resolve this debate; however, it does highlight where in the assessment process each approach can be considered.

8.2 Visual Impact Assessment framework

- Determine scope, objectives
- Implement viewshed analysis
- Establish baseline conditions
- Analyze direct and cumulative impacts
- Identify mitigation

Recognizing the overarching paradigm of scenic quality assessment and the variation that exists in VIA approaches, the following sections provide a general VIA framework to yield defensible conclusions regarding the level of impact to visual/scenic resources that may result from construction and operation of a utility-scale renewable energy project. This approach draws from commonly used VIA methods, categorizing standard outcomes of these methods as indicators of magnitude/intensity, geographic extent, duration, and context. Though rooted in "expert approach," the approach draws from concepts reviewed by Tveit et al. (2006) who advocate characterizing the landscape as an object (i.e., focus on intrinsic value), while accounting for observer experience in assessing potential impacts (i.e., perceived change). Each indicator provides information needed to understand the specific impact mechanism and how it manifests on the landscape. This information is critical to developing mitigation and/or design

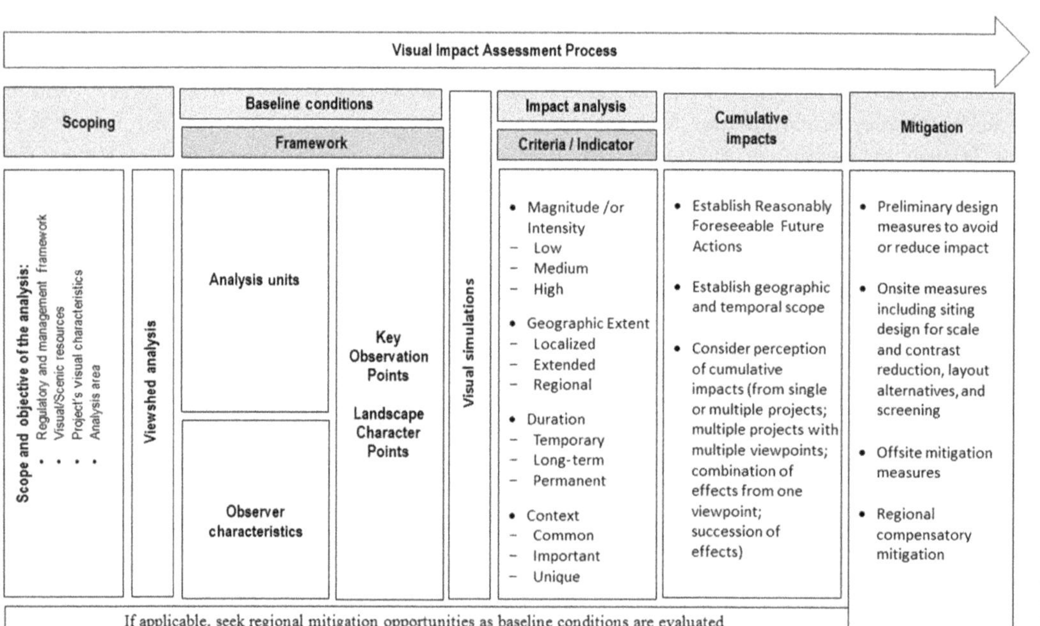

FIGURE 8.1 Visual Impact Assessment framework.

solutions to reduce impacts to visual resources. A flowchart of the impact assessment framework is provided in Figure 8.1. Examples from current VIA guidance from the US are provided where applicable. As follows, an effort is made to draw distinctions among these basic tenets of VIA, listed below, to clarify those judgments that are more subjective (value) versus those that are more quantitative or objective (magnitude of change) (LI & IEMA, 2013; Sullivan & Meyer, 2014). The procedural steps discussed below include determining the scope and objectives of the assessment, implementing a viewshed analysis, establishing baseline conditions, analyzing direct and cumulative impacts, and identifying design and mitigation opportunities. These steps can be applied to projects located in a variety of settings, land ownership, and land management regime in a manner that is commensurate with the size and scale of the project.

8.3 Scope and objectives of the Visual Impact Assessment

- Establish regulatory and management framework
- Determine the project's visual characteristics
- Define the visual/scenic resource(s) considered in the analysis
- Describe the potential sensitivity of the landscape
- Determine the analysis area
- Clearly state the scope and objectives

The scope and objectives of the VIA are typically defined at the outset, as this process determines the level of detail provided in the assessment and the specific approach used, and ensures information contained in the VIA addresses relevant issues and supports land management conformance determinations, if applicable. This process is intended to be completed as a desktop exercise, in coordination with a lead regulatory or managing agency and interdisciplinary project team. The scope and objectives of the VIA is determined by the following steps:

8.3.1 Establish the regulatory and management framework

The regulatory setting of the project, in many cases, will dictate how the assessment is structured and the level of detail contained in the analysis. It is typically commensurate with the expected level of impact, the restrictiveness of the management standard, and/or the expected level of controversy. Each of these considerations is informed by a management plan review, and internal or external scoping processes.

For example, projects located in the United States with a federal nexus and judged likely to have substantial environmental impact under NEPA would then trigger an environmental assessment and/or an Environmental Impact Statement (EIS). As part of the EIS there would be a VIA covering all lands affected by the proposed action, regardless of jurisdiction. For projects located on US federal lands, such as BLM-administered lands, the assessment would be required to address goals of the Federal Land Policy Management Act (FLPMA), which – as discussed below – would trigger a planning-level assessment of potential impacts to visual resources.

Regulatory frameworks may also exist at the state level. For example, the California Environmental Quality Act (CEQA) requires a determination of impacts to scenic vistas, resources, character and quality, including those that may result from new sources of light and glare. Though this guidance directs the assessment to consider certain types of impacts, the analyst must determine what exactly is considered in the assessment.

In the State of Oregon, certain energy projects must obtain a site certificate prior to construction.[5] As part of the site certification process, the project must meet specific standards designed to protect scenic values identified as "significant or important in local state land use plans, tribal land management plans or federal land management plans."[6] In this case, the assessment would be directed toward those scenic resources identified in planning documents and consequently not open for interpretation by the analyst. See Chapters 3 and 4 for other such examples of regional to local guidance.

8.3.2 Determine project's visual characteristics

The importance of understanding a proposed project's visual characteristics early in the process is two-fold. For one, it will inform the project team of the expected level of direct impacts to visual resources. Projects that are not expected to result in impacts to scenic quality or landscape character may be dropped from the assessment. Conversely, projects expected to result in major or significant impacts may receive elevated consideration during the development of project alternatives.

Second, understanding of a project's visual characteristics provides an opportunity to consider adjustments in project design, if such changes are expected to influence public acceptance of the project and/or successful permitting. Though details of project-related impacts may not be wholly understood, it is reasonable at this stage to identify conspicuous impact mechanisms of the proposed action, such that potential design solutions can be integrated into the proposal. Common impact mechanisms pertaining to renewable energy development are discussed in detail in Chapter 6; however, they generally pertain to deviations in form, line, color, texture, scale, lighting, motion, and glare.

8.3.3 Define the visual/scenic resource(s) considered in the assessment

The visual or scenic resources considered in a VIA may vary based on the regulatory or land use plan goals and objectives, prevailing land use patterns, or its social setting. In order to finalize the study area or specific VIA approach, the scenic resources being considered must be distinguished as inherent landscape features or visual amenities (LI & IEMA, 2013). For example, impacts to inherent landscape features are addressed in projects to be located on BLM-administered lands through the assessment of scenic quality impacts on the planning area.

Other locations may be considered important because landscape character or quality is considered integral to the human experience of that landscape or valued by the local community – i.e., a visual amenity. Examples of these areas include recreational trails and scenic vistas, and also iconic features of the built environment that may contribute to the existing landscape character. It should be noted that some landscape features have national or regional significance or have protective designations whereas some local scenic vistas may not have specific protective designations.

8.3.4 Describe potential sensitivity of the landscape

Some landscapes are assumed to have associated visual sensitivities by virtue of national, regional or local designation and so have some sensitivity because of these associated values. Sensitivity is also assessed by prevailing land use, as scenic quality may play an important role in establishing recreational setting, community identity, socioeconomic, or tourism value. Examples include a highly used beach area, public park, hiking trail, or river.

Field inventories also are necessary to document descriptive characteristics of the view. Inventories normally focus on areas of public use within a 10-mile radius of a project. These include public roads, recreation areas, trails, wilderness and natural areas, historic sites, village centers, and other important scenic or cultural features identified in planning documents or in public meetings.

Interdisciplinary coordination also plays a role with sensitivity as some culturally historic areas may not have obvious physical landscape attributes but have significant historical significance. In the UK, a key step is also checking with local planning authorities to determine sensitivity.

8.3.5 Determine the assessment area

The assessment area is generally defined by a specified radius surrounding major project elements. This distance should be determined based on the scale of these elements and an understanding of typical distances at which they can be detected or recognized on the landscape. For example, Sullivan et al. (2012) determined that wind energy facilities ranging in hub height from 61.4 meter (201.4 feet) to 88 meter (289 feet) in Wyoming and Colorado were visible at distances of >36 kilometers (23 miles). Facilities were judged to be noticed by casual observers at up to 19 kilometers (12 miles). Based on these results, Sullivan et al. (2012) recommend an assessment area defined by a radius of 48 kilometers (30 miles) surrounding wind energy facilities of this size. Sullivan and Abplanalp (2013) found a 1.6 square kilometer (400 acre) parabolic trough solar facility easily visible at a distance of greater than 23 kilometers (14 miles). A large solar tower facility was conspicuously visible at a distance exceeding 56 kilometers (35 miles). For solar facilities such as these, the assessment area should be broad enough to accommodate these anticipated visibility thresholds, particularly if observer positions surrounding the facilities are elevated with respect to these facilities.

The radius surrounding the project can also be fixed based on applicable regulations or guidance. For example, the State of Maine requires potential visual impacts of wind energy facilities to be assessed to a distance of 5 miles.

Also, in Appendix D "A Visual Impact Assessment Process for Evaluating Wind-Energy Projects" of the NRC report they suggest a 10-mile radius for viewshed assessment and this is mainly for wind turbine projects (2007, p. 350). However, the rationale for using fixed distances such as these may be questioned as visibility thresholds increase with expanding height and scale of renewable energy projects.

Scoping for cumulative effects should also be considered. See Section 8.7 for a discussion of assessment procedures for cumulative effects.

8.4 Viewshed analysis

The viewshed analysis is used to refine the geographic scope of the detailed analysis by determining locations where the proposed project may potentially be visible. This area is commonly referred to as the "seen area," viewshed, or Zone of Visual Influence (ZVI). The viewshed analysis is typically completed using Geographic Information System (GIS) software to model potential visibility based on the relationship between viewshed-limiting factors, such as topography, vegetation, structures, project components (such as wind turbines), curvature of the Earth, atmospheric refraction, and the average viewer eye level (USDI BLM, 2013b). The accuracy of a viewshed analysis is greatly affected by the quality of the data used. For example, the 10-meter resolution Digital Elevation Model (DEM) will fit the terrain more accurately than a 30-meter DEM. Likewise, first-reflective surface measured by light detection and ranging (LiDAR) and related remote sensing methods provide more accurate measurements for the canopy or building height than inference from land cover data. The parameters of the data used for the viewshed analysis must be reported to enable reviewers to evaluate the visibility analysis results.

Viewshed analysis can be used to determine potential visibility of certain project components. For example, when assessing a wind energy facility, it is helpful to understand which turbines or which portions of the turbine are potentially visible from a specific viewpoint. For instance, is the whole turbine visible, or the nacelle and rotating blades, or just the tip of an upright blade?

The results of the visibility analysis are typically displayed on a map of potentially sensitive viewer locations or management areas. This overlay can also be used to identify from which portions of linear scenic resources (e.g., Scenic Byways, Wild and Scenic Rivers, or recreational trails) a proposed project will be visible. Alternative project designs can also be compared by quantifying the proportion of the scenic resources that overlap the viewshed of each alternative.

Even with the highest quality data, an important limitation of the viewshed map is that it only represents the *potential visibility* of the project. The actual visibility of the project will depend on accuracy of the visibility model, plus a variety of viewing factors, including the visual contrast of project components against the backdrop or the horizon, existing lighting, the degree of atmospheric haze, and viewer characteristics (USDI BLM, 2013b; Felleman, 1986; USEPA, 1980).

A case study describing the Sinclair-Thomas Matrix is presented below. This approach has been used for assessing visibility impacts from wind turbines in several places around the world.

CASE STUDY 8.1 SINCLAIR-THOMAS MATRIX – USING VIEWSHED ANALYSIS AND THRESHOLD DISTANCES TO SUMMARIZE IMPACTS

By the mid-1990s planners in the UK had sufficient experience to begin generalizing an objective approach to describe and evaluate the visual impacts associated with wind energy development. In 1996 Gareth Thomas described an ordinal scale of visual impacts. He then identified the empirically observed relation between distance and visual impact for the turbines being installed at the time. He recommended that the ZVI be set at 15 kilometers. As the size of the turbines grew, Geoffrey Sinclair employed the same procedure of empirical observations to extend the usefulness of what has become known as the Sinclair-Thomas Matrix, shown in Table 8.1.

The contribution of the Sinclair-Thomas Matrix has been to clearly present an empirically based objective method to assess the potential visual impact of proposed wind energy projects. In particular, its descriptive ordinal scale provides a greater degree of differentiation than was common in the debate over visual impacts. It also recognizes that the visual impact changes as the size of the turbines increases, and it provides a guideline for the size of the VIA study area. One caution is that the visual prominence of wind energy projects is affected by atmospheric visibility, and perhaps other landscape factors. Sullivan et al. (2012) conducted an empirical observation study similar to that employed by Thomas and Sinclair, but in the arid western US. They found that the visibility distances were substantially greater than those observed in the UK, though the wind turbines may have been somewhat larger too.

While simplicity is one of the Sinclair-Thomas Matrix's strengths, it also is a weakness. For instance, it is based on the size of a single turbine and does not take into account the horizontal extent of the wind energy development. Nor does it take into account the sensitivity of surrounding landscape context and potential viewers. Others have made revisions to correct these shortcomings (e.g., Hassell, 2005), but as the method becomes more complex it can less be empirically validated. Nonetheless, the Hassell Matrix has found some acceptance outside Australia.

A final weakness of the "matrix" approach is that it does not attempt to include the perceptions of the affected public.

TABLE 8.1 The Thomas and Sinclair-Thomas Matrices to estimate the potential visual impact of different sizes of wind turbines

		Thomas Matrix				
		Original	*Revised*	*Sinclair-Thomas Matrix*		
Suggested Zone of Visual Influence (km)		*15*	*18*	*20*	*25*	*30*
Height of turbine hub (meters)		*41–45*	*41–48*	*53–57*	*72–80*	*95*
Descriptors	*Band*	*Approximate distance range (km)*				
Dominant impact due to large scale, movement, proximity and number	**A**	0–2	0–2	0–2.5	0–3	0–4
Major impact due to proximity: capable of dominating landscape	**B**	2–3	2–4	2.5–5	3–6	4–7.5
Clearly visible with moderate impact: potentially intrusive	**C**	3–4	4–6	5–8	6–10	7.5–12
Clearly visible with moderate impact: becoming less distinct	**D**	4–6	6–9	8–11	10–14	12–17
Less distinct: size much reduced but movement still discernible	**E**	6–10	9–13	11–15	14–18	17–22
Low impact, movement noticeable in good light: becoming components in overall landscape	**F**	10–12	13–16	15–19	18–23	22–27
Becoming indistinct with negligible impact on the wider landscape	**G**	12–18	16–21	19–25	23–30	27–35
Noticeable in good light but negligible impact	**H**	18–20	21–25	25–30	30–35	35–40
Negligible or no impact	**I**	20	25	30	35	40

(*Source:* Campaign for the Protection of Rural Wales (1999).)

8.5 Baseline conditions

- Regional landscape (context)
- Landscape Analysis Units (analysis)
- Key Observation Points (KOPs)
- Observer characteristics
- Perception of change
- Sensitivity to change

The assessment of baseline conditions sets the stage for the VIA by defining the context of the project, and providing the reference point from which to assess potential change in scenic quality and landscape character. As discussed above, the scope of the VIA should be commensurate with the anticipated project impacts and the goals of the assessment. Regardless of the scale of the project or the level of detail contained in the assessment, baseline data should be sufficient to meet the objectives of the assessment, which generally includes measures of direct and cumulative impacts. Indirect impacts to visual resources may also result from a proposed action. For example, clearing of a new right-of-way could increase access and subsequent proliferation of off-road vehicles trails. Those impacts that occur later in time or in a separate geographic location from the project should also be considered.

The following sections describe baseline data collection for a project. This approach evaluates baseline conditions at two spatial scales: the regional landscape setting, in which the landscape is divided into a framework of analysis units; and fine-scale, in which the project-area is assessed from standardized locations known as Key Observation Points (KOPs) (USDI BLM, 1986a), Key Viewpoints (van Haren, 1999), Key Viewing Areas (Litton, 1973), or worst-case viewpoints. For the purpose of this chapter, we will refer to these analysis locations generally as KOPs. The purpose of the multi-scale framework is to understand potential impacts in a manner that is scalable, with KOPs used to identify specific impact mechanisms and analysis units used to evaluate how these impacts manifest within a larger scale context.

8.5.1 Regional landscape

The first step in the baseline assessment involves characterizing the regional landscape setting. The regional landscape setting assessment provides context to the project within the prevailing physiographic region or planning area. The purpose of this step is to understand how the affected landscape fits within the current condition and trend of the regional landscape. For example, is the affected area considered scarce, or valued as an amenity within this larger geography? This information will also support the cumulative impact assessment by providing a baseline for this larger geographic area.

The regional landscape assessment typically entails a review of existing spatial data (e.g., shaded relief, land cover, imagery) to describe existing landscape quality, land use, and built environment. In the US, land management plans prepared for public lands may contain information that is useful to this analysis. For example, Resource Management Plans (RMPs) prepared for BLM-administered lands may provide a summary of scenic values for the planning area. Though planning area boundaries may not coincide with natural physiographic boundaries, they nonetheless provide comparative baseline information for a large geographic area. Information obtained from these spatial data sources may be augmented by field review.

To support the assessment, the regional landscape is divided into analysis units. These units provide a framework to describe existing scenic quality and landscape character at a finer resolution than that provided by the regional landscape. These units are defined based on changes in prevailing landforms, vegetation, water bodies, and development, and are established regardless of land ownership, regulatory setting or management framework. Depending on the land management context of the project, analysis units may or may not exist. For example, on BLM-administered lands in the US, analysis units have been established as part of the land use planning process. These units, referred to as "Scenic Quality Rating Units," are defined as part of the agency's Visual Resource Inventory (VRI) process. This planning-level exercise includes a scenic quality assessment and subsequent ranking of scenic quality for each unit. The results of this inventory provide a map of scenic quality (and component attributes of landform, vegetation, water, color, adjacent scenery, scarcity, and cultural modifications) for the portion of the larger regional landscape defined by the planning area. These data can be used as the baseline from which to determine potential impacts at the scale of the analysis unit, and also to understand how these impacts manifest at a larger spatial scale. As stated above, this framework also provides for a more systematic cumulative impact assessment.

Though VRI data collected as part of BLM's planning process provide an existing framework from which to structure the impact assessment, the same method can be applied to areas outside BLM's jurisdiction. For example, the USACE procedure uses this approach to define "Similarity Zones," within which scenic quality is evaluated (Smardon et al. 1988).

The approach described above defines analysis units based on intrinsic landscape character. Where it is important to understand impacts to "perceived" landscape values, analysis units may be refined or created to reflect a particular viewer-based scenic resource, such as a scenic byway or scenic vista. In this case, the analysis unit is defined by the viewshed of the scenic resource. The scale at which analysis units are defined is critical to the assessment, regardless of the analysis goal or the specific resource being assessed: large units could result in an underestimation of impacts, and small units could overestimate impacts to scenic quality and/or landscape character.

8.5.2 Key Observation Points

The fine-scale assessment of baseline conditions is completed at KOPs. KOPs represent observer locations or locations considered representative of prevailing landscape character. For worst-case viewpoints one should consider, (1) the magnitude of visibility relative to other areas (perhaps defined by distance or relative scale), (2) the sensitivity of viewers at this location, and (3) the number of viewpoints.

KOPs are classified as point-, area-, or corridor-based, depending on the type of observer experience or exposure to the project.

- *Point*: Point-based KOPs represent specific locations, such as designated vistas or interpretive signs, residences, or community bus-stop where the viewer experience is typically stationary and experienced from a single vantage point. Views from these locations may be directional (e.g., toward a focal point) or not (e.g., a 360 degree panoramic).
- *Areas*: Area-based KOPs represent geographic areas where visual resources could be experienced from a variety of locations. Views from these locations are typically transient, and experienced by viewers moving through the area (e.g., dispersed recreation, residential area). The likelihood of viewers standing in the same spot during repeated visits is low, or the area may include a number of different vantage points relative to the project. The degree of variability of views experienced will depend on the complexity of the landscape or potential views. It is expected that more than one KOP would be required to adequately address an area-based viewer experience.
- *Corridors*: Corridor-based KOPs (or linear KOPs) represent linear viewing experiences in which scenic attributes are experienced as a continuum. They may be focal or directional (e.g., leading toward a noteworthy natural feature or entrance way), and/or transient (e.g., passing through a landscape). It is expected that more than one KOP would be required to adequately address a corridor-based viewer experience.

Note that KOPs may also be established to represent common landscape features within an analysis unit, and not be tied to any specific observer position.

CASE STUDY 8.2 CAPE COD COMMISSION VISUAL IMPACT ASSESSMENT GUIDANCE FOR OFFSHORE DEVELOPMENT

The following is a brief overview of the VIA methodology for offshore development developed by the Cape Cod Commission (CCC) (Rooney et al., 2012). This VIA methodology addresses impacts associated with offshore structures within three nautical miles of the Massachusetts coast, the limit

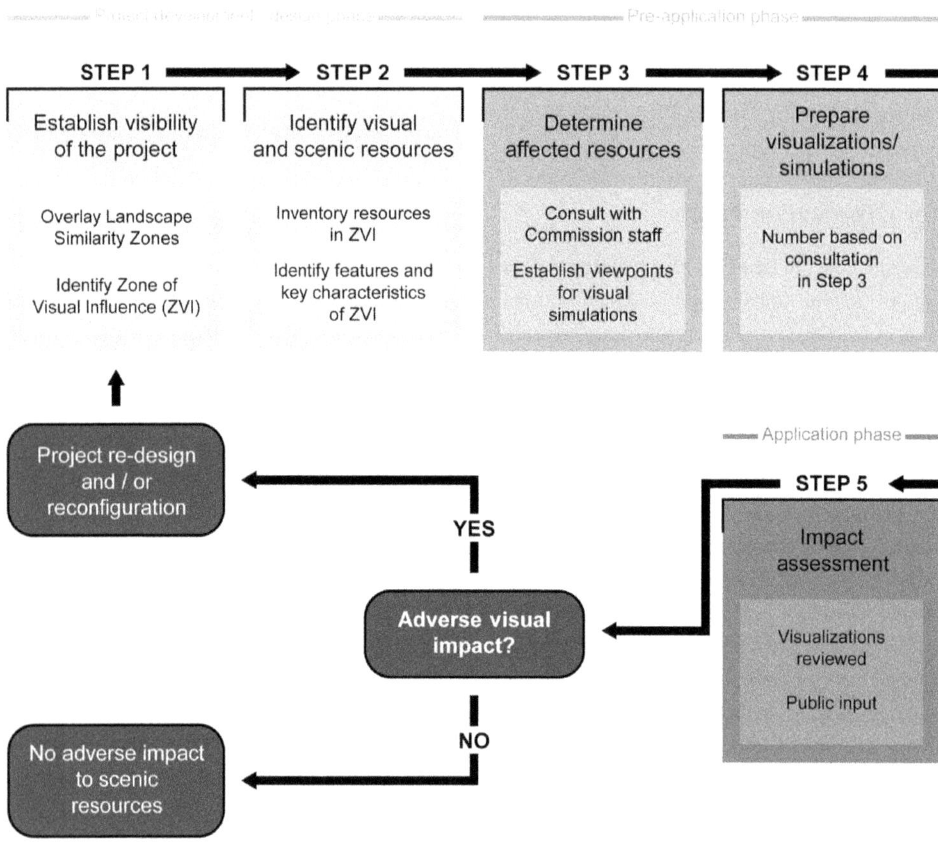

FIGURE 8.2 Five-step Cape Cod Commission VIA process.
(Rooney et al., 2012)

of state jurisdiction. This methodology was developed, in part, in response to the controversy (Williams & Whitcomb, 2007) arising from the proposed siting of the Cape Wind facility in federal jurisdictional waters off Cape Cod.

This methodology has five basic steps as shown in Figure 8.2, which include:

- Establishing *visibility of the project* with ZVIs or views where the project will likely be visible;
- Identifying *visual and scenic resources* utilizing established databases plus onsite inventory;
- Determining *affected resources* from the information developed in the previous step;
- Preparing *visualizations/simulations* from key viewpoints; and
- Conducting *an impact assessment* to determine if there is *adverse visual impact.*

What is unusual about this methodology is that:

- Under Step 1 the CCC has already developed landscape similarity zones for all of Cape Cod land area and shoreline that need to be overlaid with the ZVIs as shown in Plates 18 and 26;

- Under Step 2 the CCC already has GIS databases of scenic and historic landscape features. The CCC also has standard landscape inventory forms to be used by all project applicants. These forms contain specific sections for metadata, physical landscape factors, sketches for specific viewpoints plus matching photos, activities and functions within the viewed area, as well as ranking of landscape quality, value, and absorption capability of each viewed landscape;
- The Step 5 impact assessment includes three basic factors affecting magnitude of impact: (1) compatibility with landscape/seascape surroundings, (2) scale compatibility, and (3) dominance. There are very detailed indicators for magnitude of change for each of these factors.

If there is adverse visual impact from an offshore structure, the technical bulletin has mitigation guidance including: (1) project siting layout and design, (2) project screening options, and (3) offset mitigation. This is a fairly rigorous regionally based assessment procedure that has the advantage of using already established databases for visibility, landscape/seascape characteristics, plus prescribed detailed inventory procedures.

On the other side of the Atlantic there have been detailed procedures for assessing existing landscape character. In the UK, the landscape character assessment describes the defining or essential landscape characteristics and is not concerned with perceived scenic quality as seen above. In this context, landscape character relates to the "visual expression of spatial elements, structure, and pattern in the landscape" (Ode et al., 2008, p. 90), or a "distinct, recognisable and consistent pattern of elements in the landscape that makes one landscape different from another" (Scottish Natural Heritage, 2012b, p. 4). Please see chapter 4 for more detail regarding this approach.

8.5.3 Observer characteristics

Whereas analysis units are used to describe intrinsic landscape elements, observer characteristics describe how those elements are perceived. Understanding observer characteristics is critical to the assessment of perceived impacts, as it will inform the extent to which potential impacts are detected (perception of change), and the degree to which potential impacts are perceived as adverse (sensitivity to change). However, few studies have addressed the relationship of primary recreation activity and reaction to visual impacts (Palmer, 1999). There have been surveys in the UK and US that ask visitors if they would visit coastal recreation areas again given the existence of new windfarms (see Chapter 5, Section 5.4 for a summary of these studies).

Several existing methods provide guidance to determine perception of change. USDI BLM (1986a), for example, considers "Environmental Factors" when assessing impacts, such that the relationship of the observer to the potential impact is central to the assessment. Examples of key factors include observer angle, duration, and geometry (superior, inferior, or at grade). USDI BLM (1986a) also provides criteria to assess visual sensitivity by considering type of user, amount of use, public interest, adjacent land use, and special landscape management areas, e.g., Columbia Gorge National Scenic Area. In general, perception of change tends to be tied more to the position of the viewer in relation to the project, whereas the viewer's sensitivity pertains more to the viewer's activity, engagement and expectations of the landscape's visual experience. Table 8.2 provides a summary of the attributes of observer perception and sensitivity to change.

TABLE 8.2 The attributes of observer perception and sensitivity to change

Observer perception of change	*Observer sensitivity to change*
Type of view Is the view experienced while stationary or in motion? Is the landscape panoramic or focal?	*Primary activity* Viewers may be more sensitive to change in residential, recreational, or historic property viewsheds, or in areas associated with their primary activity.
Observer position Is the viewer position relative to the project elevated (superior), at grade, or below grade (inferior)?	*Engagement* Viewers may be more sensitive to change in areas where they are directly engaged with the landscape.
Viewer exposure Is viewer exposure limited to a short duration or limited number of viewpoints, or is exposure prolonged and/or experienced from multiple viewpoints?	*Experience* Viewers may be more sensitive to change in areas where the visual experience is paramount to their activity.
	Expectation Viewers may be more sensitive to change in areas where high quality or undisturbed landscapes are expected.

CASE STUDY 8.3 VIEWER INTERCEPT SURVEYS

We all know that there is no sound when a tree falls in the woods if there is no one to hear it. Does it then follow that there is no scenic impact if there is no one to see it? If scenic impact relies on the experience of viewers in the affected landscape, then how can it be measured without involving the affected viewers? Fortunately there are valid and reliable objective procedures to measure landscape perceptions (Churchward et al., 2013). While their use is not common in impact assessment, the State of Maine[7] in the US requires all wind energy projects to evaluate:

> E. The extent, nature and duration of potentially affected public uses of the scenic resource of state or national significance and the potential effect of the generating facilities' presence on the public's continued use and enjoyment of the scenic resource of state or national significance.

The law is quite specific – only impacts to designated state or national scenic resources (Scenic Resources of State or National Significance – SRSNS) can be considered, and the criteria for determining if a scenic impact is unreasonable must consider the expectations, continued use, and enjoyment of the users at the affected SRSNS. In practice this has consisted of the use of photo simulations and surveys of people using the affected SRSNS.

The intercept survey is typically conducted at or near the viewpoint for a photo simulation so that contextual validity is maintained. This is enhanced by introductory questions about why visitors came to the SRSNS and the nature of their experience. Respondents are asked to evaluate the scenic quality of a photograph of the existing condition, which is similar to the surrounding visual condition. This is followed by rating the scenic quality of the simulated proposed condition. Scenic impact is the difference between the scenic value of the proposed and existing conditions. Of greater relevance to Maine's law, they are also asked how this visual change will affect their enjoyment and continued use of the SRSNS.

Intercept surveys have been conducted for 10 projects between 2003 and 2013; 20 viewpoints have been evaluated resulting in 1,397 comparisons of existing and proposed scenic value (Palmer, 2015). These surveys provide results that are typically statistically significant and reliable, even though the number of users (and therefore respondents) at many SRSNS is normally rather low. However, statistical significance is not the same as importance, and is not an indicator of whether the impact is so unreasonable that the permit should be denied. It is more appropriate to use the effect size of the change to make this interpretation, following the guidelines proposed by Stamps (1990). Where the effect size (i.e., Hedges' *g* or Cohen's *d*) is between 0.0 and -0.2 the impact would possibly go unnoticed; between -0.2 and -0.5 it is noticeable but not adverse; between -0.5 and -1.1 it is adverse; and when it is beyond -1.1 the impact is unreasonably adverse.

A summary of the importance of the visual change from these wind projects is presented in Table 8.3. The three criteria are shown, and the results indicate that users of these SRSNS found the scenic impact to be adverse, and even unreasonable, while the effect on enjoyment is expected to be noticeable but not particularly adverse, and the respondents expect there to be little overall effect on their continued use – in some cases the wind development had a slight positive effect.

TABLE 8.3 How criterion affects the strength of impact as measured by Hedges' *g*

	Strength of impact (Hedges' g)		
Criterion	*Worst*	*Best*	*Weighted mean*
Scenic impact	–1.840	–0.352	–0.942
Effect on enjoyment	–0.725	–0.014	–0.304
Continued use	–0.528	0.423	–0.108

In the context of Maine's law, it is effect on enjoyment and continued use that are determinant, along with consideration of "extent, nature and duration of potentially affected public uses." The result has been that most wind projects have been permitted, even though the direct evaluation of their scenic impact is very high, and perhaps unreasonable.

8.6 Visual Impact Assessment

The following section includes a step-by-step description of VIA, including methods to develop conclusions on magnitude, geographic extent, duration, and context of the impact.

8.6.1 Photosimulations

The impact assessment is typically supported by visualizations that depict the appearance of the project from KOPs. When done correctly, visualizations – whether static-frame or video – provide a tremendous amount of information about how the project could appear within the landscape. Because few project budgets can support an unlimited number of simulations, the selection of views to simulate should be deliberate, and based on clear criteria. Such criteria might include "worst-case" potential visual impact, focusing on locations where perceived change may be greatest. This approach would assess each KOP based on criteria including, but not limited to, the number of potential viewers, distance from viewer to the project area, time of day or season of the year. Criteria that consider the type of renewable energy development may also be considered in this context. For example, a solar PV facility may result in strongest perceived visual contrast when viewed from a superior or higher elevation viewer position relative to the project. In this case, KOPs that demonstrate the appearance of the project from both a superior viewer position and one situated at a grade relative to the project would provide valuable information to stakeholders and decision makers regarding how perceived impacts may occur across the landscape.

Detailed guidance on the production of graphic simulations is covered in Chapter 9.

8.6.2 Visual Impact Assessment

There are several common impact assessment methods used in North America to determine potential impacts of renewable energy projects. Although there are differences between the approaches, the overarching goal is to determine potential impacts to visual resources that may result from construction, operation, and decommissioning of a proposed project in a way that is logical, repeatable and defensible. To be defensible, the VIA should provide detailed information on the impact mechanism, the manner in which impacts manifest across the scale of the landscape, and the extent to which impacts influence trends in regional landscape character. The assessment framework presented in this chapter builds on baseline data collected as part of the existing conditions assessment to develop a summary impact or significance determination based on the assessed magnitude/intensity, geographic extent, duration, and context of the impact. Example criteria for this determination are provided in Figure 8.3; however, criteria should be established in a manner that is relevant to the project type, affected landscape, and its geographic context.

Magnitude/intensity of the impact

The magnitude/intensity of the impact will determine the change in resource condition, and provide some measure by which to assess the continued function of that resource. *Change in resource condition* is defined as the reduction in scenic quality or change in landscape character. *Change in resource function* is defined as the extent to which the resource continues to provide the visual/scenic values expected by the observer.

The change in resource condition indicator measures impacts to scenic/visual resources by assessing change in scenic quality and landscape character from KOPs. As shown in Figure 8.3, impacts are measured by assessing the level of visual contrast and scale dominance, and the geographic extent and juxtaposition of the impact. Visual contrast is determined by implementing the visual contrast rating at each KOP (USDI BLM, 1986b). This methodology is a widely used tool to identify the mechanism of change in visual resources that may result from a proposed action by evaluating the extent to which basic elements of form, line, color and texture of the proposed project contrast with the existing landscape. This method assumes that visual contrast between the project and the existing landscape character contributes to an adverse visual impact and it is not a measure of the project's overall attractiveness (USDI BLM, 1986b). The results of the visual contrast rating assessment inform project design refinements and/or mitigation by isolating which portion(s) of the proposed action result in the strongest visual contrast, and therefore may have the greatest influence on overall impacts.

Change in resource condition can be measured by estimating the geographic extent and juxtaposition of moderate to strong visual contrast created by the project. This step is completed by establishing impact thresholds based on environmental factors that influence visual contrast. For example, if *distance* from the project is a primary factor in the attenuation of visual contrast, an impact threshold that corresponds to a specific distance could be applied to the viewshed map to demonstrate the spatial extent of impacts. This step informs our understanding of how impacts manifest across the larger scale of the resource, measured at the scale of the analysis unit(s). Several existing methods provide criteria to determine the magnitude or intensity. For example, on BLM-administered lands, this approach can be used to determine the overall change in scenic quality within each affected *scenic quality rating unit* (SQRU). Likewise, the US Army Corps of Engineers Visual Resources Assessment Procedure (Smardon et al., 1988) incorporates the contrast rating into a quantitative index of visual *impact severity*. The Cape Cod Commission VIA methodology (Rooney et al., 2012) for offshore development documents the existing landscape quality of each similarity zone then assesses *magnitude of change* from dominant to very small with lists of indicators and examples. It is similar to the latest offshore visual impact assessment guidance for offshore wind energy development from the UK (Scottish Natural Heritage, 2012b, p. 25), which focuses on "judgments on *magnitude, sensitivity* and *significance*." Impacts measured at the analysis unit scale can also be incorporated into the cumulative effects assessment and other landscape-level analyses.

As discussed above, the *change in resource function* indicator measures the extent to which the resource continues to provide the visual/scenic values expected by the observer. This indicator evaluates the *change in resource condition* indicators within the context of the observer's perception and sensitivity to change, and requires establishing a threshold beyond which visual/scenic values are no longer provided. To date, there is no established procedure by which to establish visual/scenic value thresholds. Some studies have determined this based on visitor surveys (Palmer, 1999, 2015; Ode et al., 2008), during which visitors are asked how the proposed visual change would affect (1) enjoyment, or (2) sense of place. For example, see the viewer intercept survey example on pages 210–211.

Geographic extent of impact

As shown in Figure 8.3, determining the geographic extent of the impact supports the impact determination by providing a summary metric of the spatial extent of impacts. An example of potential thresholds is provided; largely based on criteria developed by the BLM and FS as part of an EIS-level VIA for a transmission project (USDI BLM, 2014c). This indicator focuses on moderate-high intensity impacts, and how those impacts are manifested on the landscape. Like the assessment of magnitude/intensity, the determination of geographic extent can be assessed in terms of both the resource and the perceived impacts. To understand how impacts manifest within the resource, the spatial extent of impacts is assessed in terms of the proportion of the resource impacted, the extent to which the resource is fragmented as a result of those impacts, and the extent to which those impacts alter the distribution of scenic values across a larger geographic area. The geographic extent of impacts is classified as localized, extended, or regional. This metric can also be assessed in terms of observer perception, whereby the geographic extent of exposure is quantified in terms of overall percentage of travel time along a linear resource, and proportion of the field of view from stationary viewer platforms. Examples of visibility distance indicators are relevant to large-scale western US landscapes; these criteria may be adjusted to better suit the geographic context of the proposed action.

Duration of impact

The duration of impacts addresses the length of time project-related impacts would occur. This indicator includes temporary, long-term, or permanent impacts. This metric allows the length of time an impact would occur to be considered as part of the overall impact determination, and is of particular importance in separating impacts pertaining to construction or operational phases of the project. This metric also prompts the assessment to disclose how successful post-project reclamation efforts are expected to be.

Context of impact

Information on the context of the impact informs the overall significance determination by considering the potential scarcity or distinction of the resource in the overall impact determination. A basic assumption of this metric is that impacts are assumed to be greater when they affect scarce resources. Another approach is to consider the "value" of existing landscape quality and use. Landscapes with intrinsic value may be more valuable when they are scarce. Landscape qualities that are common now may not be in the future.

Magnitude / Intensity	Geographic extent	Duration	Context
LOW: Visual contrast of project components is WEAK; project components are not apparent or are visually SUBORDINATE; impacts would be limited geographically to the FG/MG distance zone (3–5 miles). **MEDIUM:** Visual contrast of project components is MODERATE; project components would be visually PROMINENT; impacts would extend to the background distance zone (15 miles). **HIGH:** Visual contrast of project components is MODERATE to STRONG; project components are DOMINANT; impacts would extend beyond the background distance zone (>15 miles).	**LOCALIZED:** Moderate-Strong visual contrast affects 0–15% of the visual/scenic resource. Though juxtaposition of the impact relative to the resource may reduce the size of the resource, it does not fragment the resource. Key factor used to rank scenic quality in affected analysis units could be changed; however, no change in overall scenic quality classification of affected analysis unit would result. **EXTENDED:** Moderate visual contrast affects 15–45% of the visual/scenic resource. Juxtaposition of the impact relative to the resource may reduce the size of the resource and result in fragmentation of the resource. Change in overall scenic quality classification for at least one analysis unit would occur. **REGIONAL:** Moderate visual contrast affects <45% of the visual/scenic resource. Juxtaposition of the impact relative to the resource may reduce the size of the resource and result in fragmentation of the resource. Change in overall scenic quality classification for more than one analysis unit would occur. **LOW:** Observers are exposed intermittently to Medium to High magnitude impacts for 5–20% of the total travel time in a linear resource. Relative degree of exposure to Medium to High magnitude impacts from stationary views is <25 degrees for non-focal views and 90 degrees for focal views. **MODERATE:** Observers are exposed to Medium to High magnitude impacts for 20–40% of the total travel time in a linear resource. Relative degree of exposure to Medium to High magnitude impacts from stationary views is 25–90 degrees for non-focal views and <70 degrees for focal views. **HIGH:** Observers are exposed to Medium to High magnitude impacts for 40–100% of the total travel time in a linear resource. Relative degree of exposure to Medium to High magnitude impacts from stationary views is 90–124 degrees for non-focal views and 70–124 degrees for focal views.	**TEMPORARY:** Impacts would be intermittent, infrequent, or for the duration of a discrete activity, such as construction. **LONG-TERM:** Impacts would extend from several years to the life of the project. **PERMANENT:** Impacts would cause a permanent change in visual/scenic resources that would last beyond the life of the project, even if the actions that caused the impacts were to cease.	**COMMON:** The affected visual/scenic resource is considered *usual or ordinary* in the local or region context; it is not rare and is not protected by legislation. The resource does not fill a distinctive role within the locality or physiographic region. **IMPORTANT:** The affected visual/scenic resource is *scarce and may be protected by legislation* or similar designation. The resource *fills a distinctive role* within the locality or physiographic region. **UNIQUE:** The affected visual resource is *protected by legislation*, or is *scarce or unique* within the locality or the region. The resource fills a *distinctive* role within the locality or the physiographic region.

FIGURE 8.3 Criteria for determining impact significance.

8.7 Cumulative visual impact

In the US, the Council of Environmental Quality (CEQ) regulations for implementing NEPA define cumulative effects as:

> the impact on the environment which results from the incremental impact of the action when added to other past, present, and reasonably foreseeable future actions regardless of what agency or person undertakes such other actions.[8]

Cumulative impacts to visual resources as a result of renewable energy can occur in situations where multiple projects are sited in proximity to other development, thereby resulting in impacts to the regional landscape. The purpose of cumulative effects analysis is to document the consideration of the context and intensity of the effects of a proposed action, particularly whether the action is related to other actions with individually insignificant but potentially cumulatively significant impacts.[9] This information is critical to forecasting future land use patterns, providing a platform for agencies and local governments to engage in the discussion of development trends, and identifying mitigation or design solutions. In order to arrive at a conclusion that is useful to planners, the analysis needs to move beyond simply listing the Reasonably Foreseeable Future Actions (RFFAs), and instead discuss how these actions – both alone and together – will influence the overall health and trajectory of the visual resource and whether potential impacts will exceed the capacity of the resource to remain productive and sustain itself (i.e., maintain scenic value in perpetuity) (CEQ, 1997).

Cumulative effects are categorized as follows: (1) additive – the effects of the actions add together to make up the cumulative effect, (2) countervailing – the effects of some actions balance or mitigate the effects of other actions, and (3) synergistic – the effects of the actions together is greater than the sum of their individual effects (CEQ, 1997). These concepts, and how they apply to visual resources, are described well in a Draft Environmental Impact Statement (DEIS) prepared for a large-scale transmission project in the southwestern United States:

> In general for visual resources in context with transmission line projects, if the Project is collocated with other existing or future linear projects, cumulative effects are synergistic because the form, line, color and texture of the features of the facilities (i.e., access roads, earthwork, and facilities) are similar and form a cohesive, united feature in the landscape. If new projects are located outside the context of existing corridors, landscapes (i.e., scenic quality rating units) would be further divided and the cumulative effect becomes strongly additive. In terms of effects on views, the larger corridor could be more dominant based on site-specific viewing conditions, separation of facilities, and other elements. However, if the projects are sited in context with existing transmission lines and they are similar in scale, form, line, color and texture, the cumulative effect would be reduced due to the synergistic relationship of the Project's features.
>
> (USDI BLM, 2014b, p. 4.275)

Despite its importance in land use planning, no single approach to cumulative effects analysis exists. Alternative approaches were identified by a study group convened by Maine's Office of Energy Independence and Security (2012). In the UK there is some guidance for assessing cumulative visual impacts (LI & IEMA, 2013), especially for multiple offshore wind energy projects (Scottish Natural Heritage, 2012a). In the US, the CEQ provides an eleven-step process structured on scoping, determining affected environment, and analyzing environmental consequences, citing the need to develop an analysis framework that provides a "general causal model of cumulative effects that incorporates information on the causes, process, and effects involved" (CEQ 1997, p. 50). A discussion of common methods used to assess cumulative impacts is provided below, with reference to existing published examples where applicable.

The first step of the cumulative effects analysis entails identifying the temporal and geographic scope of the analysis. The temporal scope of the analysis typically extends for the life of the proposed action. The spatial scope of the analysis typically extends beyond that used to analyze direct and indirect impacts of the proposed action. The analysis area may also vary depending on the type of project analyzed. For example, the analysis area used to assess cumulative effects for a proposed 500 kilovolt transmission project was defined as the viewshed of the proposed transmission line within 10 miles. This distance was selected based on the assumption that this distance would be near the limit of visibility of skylined transmission towers, and beyond that distance the proposed action and alternatives would have negligible if any contribution to cumulative visual resources impacts (USDI BLM, 2014c; Sullivan et al., 2014). Another similar project in the US established the geographic scope of the analysis as all analysis units (in this example, SQRUs) that were located within 3 miles of the routes, and areas within the viewshed of residences and other viewing areas (USDI BLM, 2014b). A cumulative impact for a wind project located in Mohave County – a non-linear project – was assessed within a 20-mile radius surrounding the project boundary (USDI BLM, 2013b). The analysis focused on the viewshed of the proposed project within the Detrital Valley; however, areas outside the viewshed were considered if a clear nexus with direct or indirect impacts of the proposed project would exist. In this example, the focus was on a geographically discrete area defined in large part by the surrounding topography. Of importance in these examples was the ability to complete cumulative impact analysis within the context of analysis units, thereby providing conclusions regarding the larger scale trajectory of the resource quality and quantity.

Regardless of the specific configuration of the analysis area, the geographic scope should be sufficient in size to address how the project and RFFAs would contribute to current land use trends.

The list of RFFAs is typically identified by review of agency documentation, NEPA analyses, reports and resource studies, peer-reviewed literature, and consultation with regional and local planning staff. The goal is to ensure the list of RFFAs is complete, and includes a broad suite of expected land use actions. As indicated in Plate 19, RFFAs should be mapped to demonstrate the spatial relationship with sensitive areas (e.g., wilderness areas, proposed wilderness, National Recreation Areas, BLM-administered Areas of Critical Environmental Concern, Indian reservations, and national parks). Existing residential areas and travel ways are also shown.

Once the RFFAs are identified and mapped, it is time to think critically about each in terms of its contribution to cumulative impacts, and how those impacts would manifest in the context of current resource trends. One approach is to consider the expected viewshed of each RFFA. This can be accomplished by running a viewshed model for each RFFA using estimated project specifications, as project specifications are likely not known. An alternative and simpler approach would be buffering each RFFA by its expected visual impact threshold distance. Regardless of the approach used, the outcome would be a spatial representation of potential impacts from the proposed project and RFFAs. This approach can also help to determine the extent to which the proposed action contributes to the overall cumulative impact. The cumulative impact map can also be overlaid on analysis units used in the assessment of project-related impacts. The juxtaposition of cumulative impacts within the larger land management framework will inform the understanding of larger-scale resource trends.

As with direct impacts, cumulative impacts can also be assessed based on how they are perceived by overlaying KOPs on the cumulative impact map. This KOP-based assessment can provide a framework to consider perceived impacts that may result from a combination of impacts, a succession of impacts, or a sequential experience, as described below.

- *Combination* from one (static) viewpoint where more than one development can be seen within the observer's arc of vision at the same time;
- *Succession* from one (static) viewpoint, i.e., where the viewer has to turn to see a number of developments around him/her;
- *Sequential effects* on a journey, i.e., where more than one wind development can be seen one after another over a period of time by an observer moving through the landscape.

One of the key issues in cumulative impact assessment is projecting future land use development patterns, not controlled by regulation or review, in combination with renewable energy development.

8.8 Mitigation of visual impacts

Ideally the results of the visual impact analysis and assessment should lead directly to proposed visual mitigation measures for wind, solar, or geothermal renewable energy projects. Mitigation through project design is covered in Chapter 7. Detailed menus of mitigation measures for wind, solar, and geothermal development are covered in *Best Management Practices for Reducing Visual Impacts of Renewable Energy Facilities on BLM-Administered Lands*, produced by BLM (USDI BLM, 2013a). The UK Department of Trade and Industry (2005) also suggest a menu of mitigation measures for offshore wind development. Detailed offshore mitigation measures from the Cape Cod Commission technical document (Rooney et al., 2012) include siting, design and layout measures, design measures, screening, and offsets. Other detailed onsite mitigation measures are covered in Chapter 7. Another set of mitigation tools includes offsets and offsite mitigation measures, such as the regional compensatory offsite mitigation process described below.

8.8.1 Offsite regional approach to mitigation[10]

Some US land management agencies are considering a landscape-scale approach to natural resource management and broadening the concept of how impacts associated with localized land use decisions would be mitigated in a *regional context*. Mitigating impacts to an acceptable level onsite will remain a priority; however, there are circumstances when onsite mitigation alone may not be sufficient. Managing at a regional scale enables the resource manager to look at the greater context of the ecosystem and account for the role certain impacted resources may play.

This effort to shift the focus away from a permit-by-permit mitigation planning perspective to a more proactive regional-scale mitigation planning perspective places an emphasis on attaining the highest mitigation benefit regardless of ownership. Setting mitigation criteria that holistically consider co-benefits between all resources impacted at a regional scale will lead to sustaining more complete and functioning ecosystems, as opposed to only recovering parts and pieces.

Such a policy could strengthen mitigation practices to effectively offset impacts of large development projects and explore all possible measures including *landscape-level planning*, *resource banking*, and *in-lieu fee arrangements*. It would also support the development or adoption of *regional mitigation plans* as a part of landscape-level land use planning to address mitigation for multiple resources, such as biological, ecological, cultural, and scenic resources, and consideration of socioeconomic factors, as appropriate.

Some US federal land management agencies have existing visual resource management systems that support baseline inventory of such resources. The BLM Visual Resource System (USDI BLM, 1986a, 1986b) and USDA Forest Service Scenery Management System (USDA FS, 1995) both describe processes for inventorying the inherent scenic quality of the natural landscape. The inventory procedures delineate the surface area boundaries of inventory units and assign a *discernable value* (scenic quality, sensitivity, visibility, integrity, etc.) to the units. As new development occupies the surface, these values are subject to change. Measuring the cumulative degree of change over time compared to the remaining intact natural landscape can serve as a baseline for assessing the need to offset continued unavoidable reduction in scenic values associated with new development.

Unavoidable impacts to visual resources are those that cannot be adequately mitigated onsite by avoidance and/or the implementation of design features meant to minimize impacts, but that still leads to a loss or reduction in inventoried visual values (USDI BLM, 2014a). Sustaining visual resource quality at a regional scale, can be achieved by visually restoring areas currently damaged by inactive land use practices (e.g. abandoned mines, unauthorized resource surface damage) or retrofitting visual resource best management practices into active existing facilities (e.g. transmission towers, fluid mineral facilities, etc.) that have unmitigated visual impacts, in parallel with new development activity (such as solar, wind, and geothermal energy). Implementing these practices leads to some questions:

1. What is the *unit of measurement* for the exchange in value?
2. Is the *exchange* based on acres of disturbance, or acres of viewshed from which the disturbance is viewed?
3. What is an acceptable *level of complexity*, keeping in mind that the more complex the process the less likely it is to be implemented?

As an example, USDI BLM's process (USDI BLM, 2014a) for assessing the occurrence of unavoidable impacts to visual resources and identifying those impacts that may warrant regional mitigation consists of four steps:

1. Baseline information is used to evaluate the visual resource condition and trends within the ecoregion, and how proposed actions will contribute to the condition and trends.

2. The scarcity of visual values vulnerable to impact is assessed. The visual resource inventory provides a quantified appraisal of visual values (scenic quality, public sensitivity, and distance zones) that can be used to determine the relative scarcity of a particular value within the region.

3. The resiliency of the visual resource is assessed. The first metric concerns the ability of the landscape to *visually absorb* the visual change imposed by development. The second resiliency metric addresses the landscape's *ability to be returned* to a naturalistic, visually intact appearance after decommissioning the developed operation.

4. The visual resource importance is determined. A review is conducted of decisions regarding management of the visual resource and how project-level amendments to those decisions would affect the balanced management of visual values within the region.

In conclusion, offsite compensatory mitigation of impacts to visual resources at the regional scale presents unique challenges relating primarily to difficulties in monetizing visual resource values. Methods can be devised that do not attempt to monetize the value of scenic resources, but instead use existing visual resource inventory and management metrics to establish non-arbitrary visual resource *exchange of values* that help guide mitigation decision-making.

CASE STUDY 8.4 DRY LAKE SOLAR ENERGY ZONE OFFSITE MITIGATION

The following is an example of such mitigation as part of the BLM's solar program for solar development in Dry Lake, Nevada. In addition to the visual resource inventory and management direction for visual resources, more consideration is being given to visual impacts from development that is positioned within viewsheds of *Special Designated Areas* (SDAs). SDAs include a variety of types of areas that have received recognition or designation because they possess unique or important resource values (e.g., wilderness areas, historic sites, and national forests). SDAs may also be designated and managed by states and localities. While these areas would not be available for large-scale renewable energy development, they could be located near and be affected by such development. Stakeholders are increasingly raising concerns regarding dominating visual changes to landscapes that are outside, but are within viewshed of SDAs.

The Solar Programmatic Environmental Impact Statement (PEIS) (USDI BLM & DOE, 2012) serves as a comprehensive example of addressing the proximity of SDAs to proposed Solar Energy Zones (SEZs), and analyzing potential impacts that may occur within SDA viewsheds. While an SEZ may be in full conformance with the visual resource management objectives, it was acknowledged there may be overriding concerns regarding visual encroachments into valued views as seen from the SDAs.

Several SDAs are within varying distance ranges of the Dry Lake SEZ, varying from less than a mile to 20 or more miles. Solar facilities would likely be less visually dominant the further away they are from observation locations. Landscapes that seem most suitable to host solar energy development are also low in visual variety and have low visual resilience. Implementing the visual resource design features within the Solar PEIS would substantially reduce certain aspects of the visual dominance of SEZ development.

The Solar PEIS analysis of visual impact from SEZ development on SDAs is a preliminary assessment of potential scale of impact ranked as weak, moderate, or strong contrast, and based on proximity, acreage and percentage of the SDA that is within view of the SEZ. When conducting the Dry Lake SEZ assessment, unavoidable visual impacts were limited to those SDAs that were listed as being exposed to moderate or strong levels of contrast in the Final Solar PEIS. The Dry Lake SEZ evaluation found some solar technologies presented unavoidable impacts and restricting these technologies eliminated those impacts, while still being able to achieve the renewable energy purpose of the SEZ (USDI BLM, 2014a). For example, if technologies applied in the Dry Lake SEZ were limited to photovoltaic only, the larger visual impacts associated with very tall power-tower facilities would not need to be considered.

Determining visual impacts that may warrant regional mitigation should also include the human cultural context of the landscape in question, the degree of visual change, and the level of visual dominance. It also is important that the impact assessment evaluates not only the specific project, but also how the project fits into the context of the full viewshed, and in association with other existing built features already in place within the viewshed. Both the degree of visual change and how the visual change will dominate the view can be determined following the steps in the BLM's *Visual Resource Contrast Rating Manual*, including the 10 environmental factors outlined in the handbook (USDI BLM, 1986b).

A final list of SDA regional mitigation recommendations can be assembled based on the results of the analyses for all potentially affected SDAs. Recommendations may include restoration or enhancement of other impacting facilities within view of the impacted SDAs, or within view of other SDAs located within the ecoregion. Considering co-benefit opportunities should remain a part of the integral process for determining regional mitigation recommendations. Comparing visual resource alternatives to the regional mitigation strategies for other resources and identifying those that overlap for multiplying the value (not cost) of the mitigation could lead to a more positive outcome.

8.9 Summary

This chapter has reviewed visual impact procedures for assessing renewable energy development and defined key impact terms, plus it offers an outline for general visual impact assessment process development. The methodology is still emerging for assessing new forms of renewable energy developments, especially offshore energy development and cumulative impacts. The positive aspect is that such development has caused professionals and academics to develop logical sound and defensible VIA procedures in consultation with affected parties. The next chapter addresses the use and review of visualization tools for simulating visual impacts.

Notes

1 42 U.S.C. §§4321-4370h.
2 85/337/EEC.
3 Directive 2014/52/EU.
4 Council of Europe Treaty Series no. 176.
5 ORS 469.300.
6 OAR 345-022-0080.
7 State of Maine 2008 (35-A MRSA §3452) *Wind Power Development*.
8 40 CFR §1508.7 (see https://ceq.doe.gov/nepa/ccenepa/exec.pdf).
9 40 CFR §1508.27(b)(7).
10 Portions of this section are adapted from McCarty et al., 2014.

References cited

Campaign for the Protection of Rural Wales. 1999. The "Sinclair-Thomas Matrix" in Evidence to the House of Lords European Communities Committee, Sub-Committee B-Energy and Transport. Appendix 1: The Potential Visual Impact of Wind Turbines in Relation to Distance. www.parliament.the-stationery-office.co.uk/pa/ld199900/ldselect/ldeucom/18/18a04.htm

Churchward, C., Palmer, J. F., Iverson Nassauer, J., & Swanwick, C. A. 2013. Evaluation of Methodologies for Visual Impact Assessments. Washington, DC: Transportation Research Board.

Department of Trade and Industry (DTI). 2005. Guidance on the Assessment of the Impact of Offshore Wind Farms: Seascape and Visual Impact Report. Department of Trade and Industry, Publication no. 80666, UK.

Felleman, J. P. 1986. Landscape visibility. In R. C. Smardon, J. F. Palmer, & J. P. Felleman, *Foundations for Visual Project Analysis*. New York: Wiley. Retrieved at www.esf.edu/via

Finavera Wind Energy. 2011. Tumbler Ridge Wind Energy Project Landscape Visual Assessment. Appendix E. Vancouver, BC: Finavera Wind Energy.

Hassell. 2005. Taralga Wind Farm Development, Landscape Visual Assessment. Final Draft. Sydney, NSW: Department of Infrastructure, Planning and Natural Resources.

Landscape Institute & Institute of Environmental Management and Assessment (LI & IEMA). 2013. *Guidelines for Landscape and Visual Impact Assessment*, 3rd ed., London: Routledge.

Litton, R. B. Jr. 1973. Landscape control points: a procedure for predicting and monitoring visual impacts. USDA Research Paper PSW-91, Berkeley CA: Pacific Forest and Range Experiment Station.

McCarty, J. H., Sullivan, R. G., & Hartmann, H. M. 2014. When and how visual resources fit into regional mitigation of onsite natural resource impacts. Presented at the 39th Annual National Association of Environmental Professionals Conference, *Changing Tides and Shifting Sands Conference Proceedings*, April 7–10, 2014, St. Petersburg, FL.

National Research Council (NRC). 2007. *Environmental Impacts of Wind Energy Projects*. Washington, DC: National Academy Press.

Ode, A., Tveit, M. S., & Fry, G. 2008. Capturing landscape visual character using indicators: Touching base with landscape aesthetic theory. *Landscape Research 33*(1), 89–117. doi: 10.1080/01426390701773854.

Palmer, J. F. 1999. Recreation participation and scenic value assessments of clearcuts. In H. G. Vogelsong (Ed.), *Proceedings of the 1998 Northeastern Recreation Research Symposium*, (pp. 199–203). Gen. Tech. Rep. NE-255. Radnor, PA: USDA, Forest Service, Northeastern Forest Research Station.

Palmer, J. F. 2015. Effect size as a basis for evaluating the acceptability of scenic impact. *Landscape and Urban Planning 140*, 56–66. doi: 10.1016/j.landurbplan.2015.04.004.

Rooney, S., Dascomb, P., Korjett, S., & Smardon, R. C. 2012. Technical Bulletin 12-001, Visual Impact Assessment Methodology for Offshore Development. Barnstable, MA: Cape Cod Commission.

Scottish Natural Heritage. 2012a. Guidance: Assessing the Cumulative Impact of Offshore Wind Energy Developments. Glasgow: Scottish Natural Heritage. www.snh.gov.uk/docs/A675503.pdf

Scottish Natural Heritage. 2012b. Offshore Renewables: Guidance on Assessing the Impact on Coastal Landscape and Seascape. Glasgow: Scottish Natural Heritage.

Sinclair, G. 2003. The Potential Visual Impact of Wind Turbines in Relation to Distance: An Approach to the Environmental Assessment of Planning Proposals. Narberth, Pembrokeshire, UK: Environment Information Services.

Smardon, R. C. & Karp, J. P. 1993. *The Legal Landscape: Guidelines for Regulating Environmental and Aesthetic Quality*. New York: Van Nostrand Reinhold. Available at www.esf.edu/via

Smardon, R. C., Palmer, J. F., Knopf, J., Grinde, K., Henderson, J. E., & Peyman-Dove, L. D. 1988. Visual Resources Assessment Procedures for US Army Corps of Engineers. Instruction Report EL-88-1. Vicksburg, MS: USACOE, Waterways Experiment Station. Available at www.esf.edu/via

Stamps, A. E. 1990. A paradigm for distinguishing significant from nonsignificant visual impacts: Theory, implementation, case histories. *Environmental Impact Assessment Review* 17(4), 249–293. doi: 10.1016/S0195-9255(97)00008-5.

Sullivan, R. G., & Abplanalp, J. M. 2013. Utility-scale Solar Energy Facility Visual Impact Characterization and Mitigation Study: Project Report. Chicago: Environmental Science Division, Argonne National Laboratory.

Sullivan, R. G. & M. Meyer. 2014. Guide to Evaluating Visual Impact Assessments for Renewable Energy Projects. Natural Resource Report NPS/ARD/NRR-2014/836. Fort Collins, CO: USDI, National Parks Service.

Sullivan, R. G., Kirchler, L. B., Lahti, T., Roché, S., Beckman, K., Cantwell, B., & Richmond, P. 2012. Wind turbine visibility and visual impact threshold distances in Western landscapes. In *Science, Politics and Policy: Environmental Nexus*. Proceedings of the 37th Annual National Association of Environmental Professionals Conference, May 21–24, 2012, Portland, OR, pp. 897–942.

Tveit, M., Ode, A., Fry, G. 2006. Key concepts in a framework for analysing visual landscape character. *Landscape Research 31*(3), 229–256. doi: 10.1080/01426390600783269.

US Council on Environmental Quality (CEQ). 1997. Environmental Justice Guidance under the National Environmental Policy Act. Washington, DC: Council of Environmental Quality.

USDA, Forest Service (USDA FS). 1995. *Landscape Aesthetics: A Handbook for Scenery Management*. Agricultural Handbook No. 701. Washington DC: USDA, Forest Service. Available at www.esf.edu/via

USDI, Bureau of Land Management (USDI BLM). 1980. Visual Simulation Techniques. Washington, DC: US Government Print Office.

USDI, Bureau of Land Management (USDI BLM). 1986a. Visual Resource Inventory. Manual H-8410-1. www.blm.gov/style/medialib/blm/wo/Information_Resources_Management/policy/blm_handbook.Par.31679.File.dat/H-8410.pdf

USDI, Bureau of Land Management (USDI BLM). 1986b. Visual Resource Contrast Rating. Manual 8431. www.blm.gov/style/medialib/blm/wo/Information_Resources_Management/policy/blm_handbook.Par.79462.File.dat/8431.pdf

USDI, Bureau of Land Management (USDI BLM). 2013a. Best Management Practices for Reducing Visual Impacts of Renewable Energy Facilities on BLM-Administered Lands. Cheyenne, WY: USDI, Bureau of Land Management.

USDI, Bureau of Land Management (USDI BLM). 2013b. Mohave County Wind Farm Project Final Environmental Impact Statement. Phoenix, AZ: USDI, Bureau of Land Management. www.blm.gov/az/st/en/prog/energy/wind/mohave/reports/FEIS.htm

USDI, Bureau of Land Management (USDI BLM). 2014a. Regional Mitigation Strategy for the Dry Lake Solar Energy Zone. Technical Note 444. Washington, DC: USDI, Bureau of Land Management.

USDI, Bureau of Land Management (USDI BLM). 2014b. Energy Gateway South Transmission Line Project. Volume 1b. Cheyenne, WY: USDI, Bureau of Land Management. www.blm.gov/wy/st/en/info/NEPA/documents/hdd/gateway_south/DEIS.html

USDI, Bureau of Land Management (USDI BLM). 2014c. Boardman to Hemingway Transmission Line Project Draft Environmental Impact Statement. Vale, OR: USDI, Bureau of Land Management. http://boardmantohemingway.com/documents.aspx

USDI, Bureau of Land Management and US Department of Energy (USDI BLM & DOE). 2012. Final Programmatic Environmental Impact Statement (PEIS) for Solar Energy Development in Six Southwestern States. FES 12-24, DOE/EIS-0403. Washington, DC: USDI, Bureau of Land Management and US Department of Energy.

US Environmental Protection Agency (USEPA). 1980. Protecting Visibility: An EPA Report to Congress. Report number EPA-450/5-79-008, US Environmental Protection Agency. Research Triangle Park, NC: Office of Air, Noise and Radiation; Office of Air Quality Planning and Standards.

van Haren, F. 1999. Visibility Monitoring Data Analysis for the Columbia River Gorge National Scenic Area. Columbia River Gorge Commission Publication no. 99-202 and Washington State Department of Ecology. https://fortress.wa.gov/ecy/publications/SummaryPages/99202.html

White Consultants. 2013. Pembrokeshire and Carmarthenshire: Cumulative Impact of Wind Turbines on Landscape and Visual Amenity Guidance. Final Report for the Carmarthenshire County Council, Pembrokeshire Coast National Park Authority and Pembrokeshire County Council. www.pembrokeshire.gov.uk/objview.asp?object_id=8887

Williams, W. & Whitcomb, R. 2007. *Cape Wind: Money, Celebrity, Class, Politics, and the Battle for Our Energy Future.* New York, NY: Public Affairs.

9

VISUALIZING PROPOSED RENEWABLE ENERGY PROJECTS[1]

Robert Sullivan

Highlights

- Spatially accurate and realistic visual simulations of renewable energy projects are an important tool for visual impact assessment.
- Simulations are not substitutes for actual views of a project; they are not the same as "being there."
- Photomontages are currently the most widely used simulation medium for environmental impact assessments.
- The preparation of visual simulations is a complex technical undertaking, prone to several kinds of errors that can be difficult to detect.
- Video montages and 3D computer models of proposed projects are increasingly being used for visual simulations.

The old adage that "A picture is worth a thousand words" is particularly apt when assessing visual impacts. Visual impact assessment (VIA) preparers use visual simulations to depict projects before they are built, both for impact assessment and stakeholder communication. This chapter introduces the major types of simulations currently used in VIAs for utility-scale energy projects, describes how the simulations are developed and presented, provides guidance on evaluating the quality of simulations, and presents important limitations and common problems associated with the use of simulations. This discussion is focused on techniques generally employed in the U.S. While the basic practices are generally the same elsewhere, some countries, such as the U.K., have stringent requirements for how simulations are prepared and presented.

9.1 Introduction

In the broadest sense, visual simulations include a wide range of representations of proposed projects, including maps, aerial and ground-level photographs, paintings, engineering and architectural drawings, solid models, computer generated still images and animations, etc. (Zube et al. 1987). Simulations (in the form of sketches and drawings) have been used to depict projects prior to construction since the time of the ancient Greeks, at least (Zube et al. 1987).

In the context of utility-scale energy project VIAs, visual simulations are visualizations (typically computer-generated) of the proposed project and surrounding landscape that are used to depict the overall appearance of a proposed project after it is operational. In properly prepared VIAs, spatially accurate and realistic simulations are used to depict views of the proposed project from important viewpoints, which are often referred to as key observation points (KOPs). These simulations may be supplemented by other types of visualizations, such as wireframe images or sketches.

Simulations are an important component of VIAs for renewable energy projects. In VIAs, their primary use is to help the impact analyst assess the visual contrasts created by the project and to help stakeholders visualize and respond to development proposals. Simulations can also be useful in the project design process, as a tool to assess the potential impacts associated with particular project siting strategies. They are also used to evaluate the effectiveness of potential mitigation measures and best management practices to avoid or reduce impacts.

In the U.S., simulations are not technically required to be used in VIAs, but it is common practice to use them in impact assessments for utility-scale renewable energy projects, which typically involve large visual impacts. In some countries, applications for utility-scale wind facilities must include simulations prepared to exacting specifications (Scottish Natural Heritage 2014). While visual simulations are not specifically called for in all European countries, there is guidance on best practice for visualization process (Landscape Institute 2011; Scottish Natural Heritage 2014). Further, there have been specific publications addressing the technical aspects of producing photographs and visual simulations (Benson 2005; Macdonald 2012). Visual simulations are not the same as "real life views" but are useful tools to assist in the VIA process (NZILA Education Foundation 2010; Landscape Institute 2011; Scottish Natural Heritage 2014). Although simulations have important limitations, it is difficult to understand and assess the potential visual impacts of a proposed project without them.

The development of simulations for a VIA is a complex technical undertaking. Guidelines for the development and presentation of visual simulations are presented in Section 9.2. A summary of the development process for photomontages, currently the main method for producing simulations for VIAs, is presented in Section 9.3, and the use of animations for simulations is discussed in Section 9.4. Even the best simulations have important limitations in terms of accuracy and realism. These limitations are discussed in Section 9.5. While the technology to create visual simulations constantly improves, the process is subject to errors and inaccuracies that may have important effects on the quality of the finished product. Sources of error and inaccuracy in simulations are presented in Section 9.6. Other types of simulations used to supplement photomontages are discussed in Section 9.7.

9.1.1 Responsibility for simulation preparation

Preparation of simulations is normally the responsibility of the project proponent; however, they are usually prepared by the VIA team or another contractor that specializes in preparing simulations. The preparation of high-quality simulations is a specialized skill that should be undertaken only by trained, experienced professionals with adequate resources and time to develop a high-quality product. For controversial projects, additional visual simulations sometimes are prepared by project opponents. They may be of poor quality, or misleading in showing impacts, and should not be used for impact determination unless properly prepared by trained personnel.

9.1.2 Types of simulations

Although there are many types of visual simulations, the "workhorse" of VIAs is the photomontage (also referred to as a photosimulation), a still image of a highly realistic three-dimensional (3D) model of the proposed facility superimposed onto a photograph of the existing landscape. Other types of simulations include computer-generated schematic wireframe drawings, completely synthetic landscape renderings that do not involve the use of photographs, sketch graphics, and animations. These types of simulations may be very useful for depicting particular aspects of a proposed facility or for illustrative purposes, but they are generally used to supplement photomontages in VIAs. These techniques can be quite sophisticated, especially synthetic landscape renderings, which in some cases can be generated by high-end visualization software integrated with computer-aided design (CAD) and/or geographic information systems (GISs), such that structures, vegetation types and height, growth rates, and other environmental factors can be modeled to develop synthetic landscapes that "evolve" over time (3D Nature LLC 2015). This chapter focuses primarily on photomontages, because they are the most commonly used simulation format for VIAs. Because movement is an important aspect of the visual experience of wind energy facilities, and to a lesser extent solar facilities, the use of animation in simulations is also discussed.

9.2 Guidelines for producing and evaluating simulations

The following guidelines present best practices for simulation preparation and use in VIAs.

9.2.1 General principles

The following general principles for producing and presenting visual simulations are adapted from Sheppard (2005) and serve as high-level guidelines for visual impact simulation:

- *Simulations should be spatially accurate and realistic.* Simulations should simulate the actual or expected appearance of the landscape and project as closely as possible, according to the data available at the time.
- *Simulations should be representative.* Simulations should represent the important and typical range of views which would be experienced with the actual project and provide viewers with a range of viewing conditions, including typical worst-case scenarios.
- *Simulations should be visually clear.* Simulations should be properly prepared and displayed, such that project components and the surrounding landscape are depicted clearly and with sufficient detail to serve as a sound basis for impact assessment.
- *Simulations should be engaging without entertaining.* Simulations should engage and hold the interest of viewers, while avoiding the use of presentation techniques that are entertaining or impressive without adding significant information value.
- *Simulation methodology should be defensible and documented.* The methodology used to produce simulations should follow procedures that will withstand scientific scrutiny, are consistently followed, and are thoroughly documented. The documentation should include clear descriptions of known and potential sources of error and uncertainty.

A VIA that includes simulations that do not adhere to the above criteria is seriously deficient.

9.2.2 Single-frame vs. panoramic simulations

Simulations may be single-frame or panoramic, or both. The choice should be determined on a case-by-case basis. Simulations should show enough of the surrounding landscape to show the project in the appropriate spatial context. In most cases, the presentation of both a single-frame and panoramic image for each simulated view is greatly preferable to showing either one alone. Single-frame images may be needed to show adequate details of the objects in the simulation to depict it realistically, but they may not show as much of the facility as would be seen in "real world" views. Also, a single-frame image may not show enough of the surrounding landscape to portray the visual context properly, or to show the scale of the project in relation to its surroundings (Landscape Institute 2011). Panoramic images show more of the project and its surroundings; this is particularly important for the depiction of renewable energy facilities in open landscapes, as the facilities tend to be very large. The human horizontal field of view is approximately 124° (NZILA Education Foundation 2010), and this entire field of view can be shown in a panoramic image. However, in some cases, panoramic images may lack details that would be visible in reality, especially when they are reproduced at relatively small sizes, which is common in practice. Showing both types of images for each simulated view allows proper depiction of facility details, while showing the facility in the proper visual context and better depicting its scale in relationship to its surroundings.

9.2.3 Selecting KOPs, season, time of day, and lighting conditions for simulations

The selection of appropriate KOPs and the season, time of day, and lighting conditions for simulations requires careful consideration of a number of factors. Typically, simulations should capture seasons and time periods of highest visitor use at KOPs, and typically under sunny conditions with good visibility for the specified location. At a minimum, simulations should depict conditions with the greatest reasonable visual impact scenario for KOPs. For example, if a solar project is expected to produce glare (annoyingly bright reflections) at KOPs, at least some of the simulations for the project should depict the glare as it would be seen from KOPs. Similarly, simulations of wind facilities should not be limited to depictions of the facility with the turbine blades parallel to the line of sight, such that they show the least amount of visible surface area in the simulations. Where projects are expected to have substantial night-sky impacts, these impacts should be portrayed in night-sky impact simulations.

In more sensitive situations, it may be desirable to depict additional seasons, times of day, and lighting conditions in simulations; however, high-quality simulations are expensive and time-consuming to produce. A reasonable balance must be achieved between producing enough simulations to show the range of impacts from all important KOPs and other typical viewing situations, and expending excessive money and time either producing or redoing simulations that do not add significantly to an understanding of the impacts. For example, where potential KOPs are closely spaced and have very similar views of the project, it may be preferable to develop one representative simulation rather than developing multiple simulations that would show essentially the same view and visual contrasts.

9.2.4 Preparing accurate and realistic simulations

Simulations should be based on accurate project design specifications and on accurate spatial information (particularly elevation data), and must account for Earth curvature and screening vegetation and structures. Simulations should be spatially accurate; they should show all of the project elements and other project development effects that would be visible, at the correct size and location, and in the correct visual perspective. They should be realistic (i.e., they should look like the facility would actually look when built, as seen from the relevant KOP). They should be illuminated, colored, and shaded realistically, with cast shadows that are depicted accurately and realistically. Any known errors or uncertainties in accuracy or realism should be clearly stated so that viewer expectations are appropriate. Visibility factors, including screening and lighting are discussed in detail in Chapter 6.

9.2.5 Simulation image size and viewing distance

Simulations should be reproduced at a size large enough to be comfortably viewed from the appropriate specified viewing distance that accurately depicts the apparent size of the facility as it would be seen in the real landscape. For printed simulations, foldouts should be used, if needed, to enlarge the simulation so that adequate detail can be seen while preserving a large horizontal field of view. The proper viewing distance for the simulation (the distance that the simulation should be held from the viewer's eye) at its intended reproduction size should be specified. If panoramic images are shown at large sizes in public meetings or similar settings, they should be mounted on curved surfaces so that all parts of the image can be viewed at the correct viewing distance. If this is not possible, they should be displayed in a space that allows viewers to move to a position directly in front of the portion of the image they are viewing. Note, however, that recent guidance from Scottish Natural Heritage (2014) has relaxed these requirements while noting that doing so will result in the wind facilities not being seen at the size they would appear in reality.

9.2.6 Simulation output quality

High-quality prints are usually desirable for impact assessment purposes and for public information presentations. Professional printing services should be used whenever possible. Night-sky impact simulations should preferably be projected using a high-resolution projector and a projection screen (not a wall) in a darkened room to simulate the real night sky viewing situation to the extent possible. Correct color balance is also important, and color correction hardware should be used if possible. Higher end color correction hardware will take ambient light readings to assist in determining the correct brightness for screen display.

9.2.7 Simulation presentation lighting

Simulations should be presented under adequate lighting to show fine details in all parts of the simulation. Lighting should not be so bright, however, as to cast obscuring glare onto the image. Night-sky impact simulations present special problems for display, because much of the image is very dark, and the lights visible within the simulation may not be bright. Viewing night-sky impact simulations in a well-lit room makes it very difficult to judge accurately the effect of the lights on dark night skies, and the lights depicted in the simulation may be hard to see at all. To the extent possible, night-sky impact simulations should be projected in a room that is totally dark.

9.2.8 Labeling and supplementary information

Supplementary information should be provided to indicate the project, KOP location, and alternative depicted in the simulation. This information should be provided on a separate sheet that accompanies the simulation and is always displayed with the simulation. Simulations should include a variety of supplementary information, including:

- Geographic coordinates for the camera/KOP location;
- Date and time of the photograph;
- View direction;
- Weather conditions;
- Lighting condition (frontlit, backlit, sidelit);
- Camera and lens make and model;

- Focal length used for photograph (for film SLR cameras) or 35-mm equivalent focal length for digital SLR cameras;
- Horizontal and vertical width of field depicted in the simulation;
- Distance to nearest and farthest visible portions of facility (e.g., the nearest and farthest visible wind turbines);
- Proper viewing distance for the simulation in the presented format; and
- An inset or supplementary map (with a legend) that shows the location of the KOP, the facility boundary and major components, nearby features such as roads and populated places, and that depicts graphically the horizontal field of view shown in the simulation.

9.2.9 Simulation methodology documentation

In addition to the simulation-specific supplementary information discussed above, the methods used to develop simulations should be thoroughly and clearly described in the VIA. Documentation should describe the rationale for KOP selection and KOPs eliminated from consideration, parties identifying KOPs, photographic and simulation production methods and equipment, as well as measures taken to ensure simulation accuracy. Base photographs, field logs, and other intermediate products such as wire-frames should be available on demand.

9.3 Photomontage production summary

Photomontages developed for modern VIAs use specialized computer software to superimpose a computer-generated image of the proposed project facility onto a digital photograph of an existing view of the proposed project site. The photograph of the existing landscape (referred to as a base photograph)

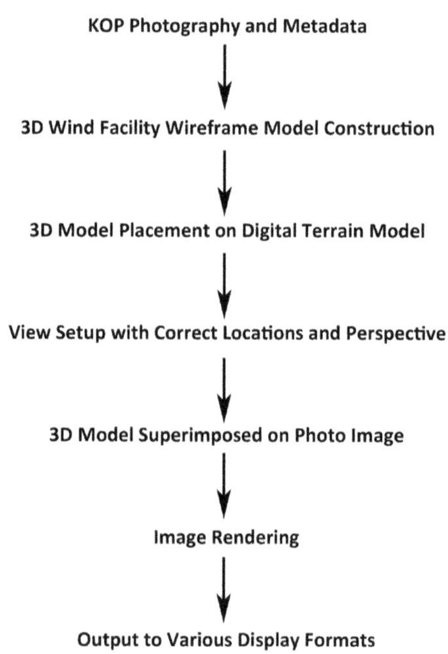

FIGURE 9.1 Photomontage production process.
(Credit: Robert Sullivan, Argonne National Laboratory.)

is taken from a desired viewpoint (generally a KOP) and captures the project site and the surrounding landscape. The image of the facility is derived from a spatially accurate and realistically rendered 3D model of the facility that is placed in the base photograph so that it appears at exactly the same position as it would be seen from the KOP, is the same size as it would be seen from the KOP, and in the same visual perspective.

The typical process used for preparation of high-quality photomontages for use in VIAs is shown in Figure 9. 1, and can be summarized as follows:

- After KOPs for the proposed project have been identified, one or more KOPs are selected for use in preparing simulations.
- Using high-quality digital cameras and lenses, and under carefully controlled conditions, the project location is photographed from each KOP for which a simulation will be prepared. The precise locations of the camera, the point at which the camera is aimed, and other control points in the surrounding landscape visible within the photo are recorded, as is the lens focal length and other photo data. The date and time are also recorded, in order to calculate the position of the sun in the sky. A photograph of the project area for an offshore wind facility used as the base photograph for a photomontage is shown in Plate 20.
- Using appropriate visualization software, a spatially accurate 3D wireframe model of the project is built and placed on a spatially accurate 3D model of the project landscape (the *terrain model*), derived from elevation data for the Earth's surface.
- A "view" of the 3D model is set up in the software using the identical lens focal length, camera, and project coordinates recorded for the photograph, so that the model is shown in the same location, at the same size, and in the same visual perspective as the built project would be seen from the KOP.
- The wireframe terrain model and the 3D model are superimposed precisely onto the photograph using the control points to assist with accurate alignment of the terrain model and the photograph. The 3D model is thus shown in the correct location, at the correct size and aspect, and in the correct visual perspective in the photograph. Minor adjustments are often necessary because of imperfections in the elevation and/or project data. Plate 21 shows appropriately scaled 3D wireframe models of offshore wind turbines accurately superimposed onto the base photograph shown in Plate 20.
- The terrain model is stripped away, and the model is *rendered*, that is, its surfaces are colored and shaded in the computer program to match the properties of the facility elements of the real project. The sun position data as well as information on cloud cover and weather patterns recorded at the time the photograph was taken are used by the visualization software to calculate lightness and darkness of shading and the correct positioning of shadows cast by model elements. Minor adjustments to the rendering may be made using photo editing software, typically to achieve greater realism (e.g., blending of sharp edges to mimic the softening of outlines with distance). Image editing tasks may also include removal of trees and other structures in the existing landscape, re-insertion of foreground elements that would be seen in front of the project, adjustments to landforms, and other edits related to changes to the landscape associated with project construction. Plate 22 shows a spatially accurate and realistic simulation of an offshore wind facility.
- The simulations are prepared for presentation in one or more output formats. Preparation includes supplementary information for the simulation to aid simulation viewers in interpretation of the simulation. This supplementary information should be provided on a separate sheet that accompanies the simulation (see Plate 23). Providing the supplementary information separately avoids showing distracting graphical elements when the simulation is presented, and affords as

much room as possible to show the project and surrounding landscape in the simulation. Output formats typically include large-format, high-resolution prints; they may also include digital files for computer projection and smaller format prints for use in reports and other communication pieces.

- If the simulation is a panoramic simulation (see Section 9.2.2), additional steps are needed that involve taking additional photos to capture a larger portion of the landscape, the digital "stitching" of multiple photos to create the panoramic view, and correction of distortions that may be introduced into the stitching process.

CASE STUDY 9.1 VISUALIZATION STUDY FOR OFFSHORE NORTH CAROLINA

Project location

The Outer Banks, North Carolina

Project purpose

Two highly scenic National Seashores span 40% of North Carolina's coast, including some of the darkest night-sky areas in the eastern United States. When the Bureau of Ocean Energy Management (BOEM) organized a North Carolina Task Force to explore offshore wind energy development, the National Park Service (NPS) became concerned about scenic impacts. Ordinarily, the development process begins prior to consideration of scenic impacts. The NPS was concerned that decisions would be made without the benefit of information about the visual scope and scale of potential wind energy development.

In response to this concern, BOEM contracted T. J. Boyle Associates of Burlington, Vermont to create highly "accurate and realistic visual representations of offshore energy facilities" under various day and night conditions using static and dynamic media. In addition, the contractor facilitated public outreach events presenting and explaining the visual effects to the public.

Project description

Each simulation portrays an array of 200 turbines and a service platform with 1,000-m spacing and irregular project boundary.

The simulated scenarios represent one condition from each of the following four parameters:

- 18 viewpoints (from Corolla Lighthouse to Sunset Beach, NC),
- Up to four lighting conditions (morning, afternoon, starlit, and misty night),
- three distances from shore (10, 15, and 20 nautical miles [nm]),
- two wind turbine models (Siemens 3.6 MW, and Vestas 7 MW),

The project produced:

- 234 single-frame simulations,
- 21 immersive panoramic photomontages,
- 48 animated photographs,
- six videos.

Role of landscape architect

All substantive work for this project was conducted by landscape architects, including:

1. Project scoping and technical justification

It is typically assumed that a wind turbine will not be visible at 20 nm – however a member of the project's Technical Review Panel had documented clear visibility of existing offshore turbines at over 20 nm. The Panel required technical justification for our approach to the following issues:

- The effect of Earth curvature and atmospheric refraction on visibility;
- Reduction in visibility from fog and other meteorological conditions;
- Visibility of Federal Aviation Administration (FAA) and Coast Guard warning lights;
- Human visual acuity relating to the perception of wind turbines;
- Photographic resolution requirements;
- Media resolution requirements for complete representation.

Meeting these requirements as well as achieving a high level of spatial accuracy and realism pushed the limits of currently available simulation technology and current best professional practices.

2. Desk studies

Preliminary desk studies included:

- Preparing aerial photo maps of all 18 viewpoints at 30, 100 and 500 scale to guide the field work;
- Developing and testing the protocol for photography and record keeping;
- Developing viewshed maps and 3D computer visualizations using the project geographic information system;
- Incorporation of simulation viewpoint locations provided by BOEM;
- Incorporation of offshore turbine locations provided by BOEM.

3. Fieldwork and photography

The primary task was to photograph the views toward the wind project scenarios from 18 sites in the early morning, late afternoon, and starlit night. Challenges addressed in this process included:

- Establishing registration markers for repeat photography and simulation scaling;
- Panoramic photography required a minimum of three rows of seven or more photographs, each at a specified bearing and azimuth in order to capture the proposed turbine layout;
- Magnetic disturbance from iron lighthouses to the compass used for orienting photographs occasionally required the use of aerial maps for orientation.

Nighttime observation and photography of existing FAA warning lights was conducted to verify the visibility and apparent size of the lights at different distances.

4. Static and dynamic simulations

Four different types of simulations were prepared:

- *Single-frame view:* The "normal" lens has a 37° horizontal angle of view. When printed it is 10.75" by 16" (27 cm by 41 cm) and is properly viewed at 23.5" (60 cm).
- *Immersive panorama:* A cylindrical projection is applied to multiple single-frame photos that are stitched together to form a 124°-by-55° panorama. When plotted it is 26" by 58" (66 cm by 147 cm) and is properly viewed at 27" (69 cm).
- *Animated photograph:* The turbines of a single-frame simulation are animated and saved as a 1080p high definition video. Animation adequately represented FAA warning lights if viewed in a dark room.
- *Video simulation:* Movement and sound in the landscape is captured as well as the turning turbine blades. Though the resolution is lower, the public accepted the video simulations as most realistic.

The basic approach is the same for each type of simulation. A computer graphics image of the project is matched to the field photography as shown in Plate 24. All simulations accounted for the Earth's curvature and atmospheric refraction.

5. Public outreach

An initial public meeting was held with the North Carolina Task Force to discuss the results and implications of the project's findings. Subsequent open house meetings were held in Kitty Hawk and Wilmington, NC. The information was presented at individual stations hosted by one of the project staff to assist the public in how to properly view the simulations and answer questions. The stations included:

- Overall purpose of the project, the scenarios, and how the simulations were created;
- 234 single-frame day and nighttime simulations;
- 21 immersive panoramas displayed in a curved stand or mounted flat on a wall;
- 11 daytime animations;
- six videos;
- 16 nighttime animations presented in a dark room.

A court recorder was available to transcribe public comments on the simulations and the potential visual impacts of projects that might be proposed. Based on these comments and staff observations, it is clear that the public responded to all of the media, but that they each had their own strengths and weaknesses. The single-frame simulations provided a convenient way to review a great number of simulations, but they did not provide as realistic an experience as the other media. The nighttime impacts of FAA lighting were poorly represented by the print media, but were convincingly portrayed by an animation displayed in a darkened room after the audience was given several minutes to adjust to the light level. The public appreciated the greater realism of the immersive panoramic display; however it could be viewed effectively by only one person at a time. It seemed that the public thought portraying the rotation of the turbine blades was important, but that the static nature of the landscape in the animations seemed artificial. Overall, the public seemed to prefer the videos as the most effective way to portray a wind energy project, however there are resolution limitations that effect the simulation of distant objects.

Special factors and project significance

This was the first time that BOEM has considered scenic impacts *before* an action was proposed – in this case before publishing the Notice of Intent for commercial wind leasing and site assessment activities.

This project was the result of a partnership between two federal agencies with contrasting mandates – BOEM is tasked with managing the development of offshore energy resources and the NPS with protecting scenery of National Park Units for future generations. As with many environmental conflicts, the key to successful communication lay in well-documented information that both parties could accept as accurate and realistic.

James Palmer

9.4 Animations

If desired, with the use of appropriate software and techniques, a sequence of still images (e.g., photomontages) can be combined to create an animation that shows movement of either project components or the viewer, or change in the appearance of the project and surrounding landscape over time. For example, animations are particularly effective for showing wind turbine blade motion, the synchronized flashing of wind turbine aviation obstruction lighting at night, changes in position and reflectivity of solar collectors/reflectors over the course of a day, and changes in lighting conditions as the sun rises and sets or clouds pass in front of the sun. Animations can also depict the change in the viewed landscape and visual contrasts as the viewer moves (e.g., while hiking a trail, floating on a river, or driving a scenic byway). Animations used for visual simulations typically run from a few dozen seconds to a few minutes in length, and they are shown in a variety of digital movie formats on computer monitors or as larger projected images.

The visual characteristics of renewable energy facilities are such that the viewing experience is dynamic. The appearance of wind and solar facilities, in particular, may change dramatically in the course of a day. Because change over time and movement of both the viewer and objects in the landscape are both inherent and essential aspects of the human visual experience, the use of animation in visual impact simulation can increase the realism of the simulations and is particularly useful for applications where movement is an important characteristic of either the project or the viewing experience.

Despite the advantages of animations for showing movement and change over time, still-image photomontages currently are much more widely employed for visual impact simulation than animations. Animated photomontage sequences are more expensive, time consuming, and difficult to create than still-image photomontages, and the output resolution of the animations is not as high as the resolution of high-quality digital images and prints. Also, the technology needed to display animations limits their use for some applications, such as use in the field to compare the simulation with the real view of the landscape. Despite these limitations, the use of animations for visual impact simulation is increasing and may be appropriate for particularly sensitive or important simulations.

9.5 Limitations of simulations

Although high-quality visual simulations are a useful tool for assessing visual contrast, for a variety of technical reasons, even when properly prepared, simulations are limited in their ability to realistically portray the full range of the visual experience of utility-scale renewable energy facilities, particularly wind and solar facilities. It must be kept in mind that they are based on photographs or videos, and, ultimately, what they simulate is a photograph or a video of the proposed project, not the actual visual experience a viewer would have in a real landscape looking at the real project (NZILA Education Foundation 2010; Horner & Maclennan and Envision 2006). Because of limitations inherent in the photographic medium, simulations are approximations of what the project would look like and are not the same as "being there." Indeed, observations made by Benson (2005) suggest that simulations of proposed windfarms in VIAs often underestimated the impacts compared with field observations of the built projects, in part, because "the windfarm often looked nearer, more visible, and more conspicuous than the photomontage predicted." The following sections discuss some of the major limitations of visual simulations based on photography, including photomontages.

9.5.1 Loss of dynamic visual experience

Although everyone is familiar with photographs and accepts them as a close approximation of visual reality, still photographs are static, limited representations of the view from one location at an instant in time, and the view is chosen by the simulation creator, not the viewer. In reality, the human visual experience is dynamic, changing constantly as the viewer turns his/her head or moves through the landscape as he/she desires. The visual environment changes constantly too, as the sun's position changes in the course of the day and as clouds pass overhead. And, of course, elements of the project itself may move or change their appearance dramatically over time. The synchronized motion of wind turbine blades is a very important part of the visual experience of wind energy; solar collectors/reflectors may change their orientation and their reflective characteristics substantially in just a few hours or even minutes; and vapor plumes from cooling towers and boilers are in constant motion. Movement of objects in the landscape usually increases their visibility and may increase the visual contrast that is perceived by viewers (Benson 2005; Bishop 2002; Sullivan et al. 2012, 2013), so it is possible that failure to depict movement through the use of animation in simulations may result in lowered perceptions of visual contrasts associated with a project. Animations can depict viewer and object motion as well as change over time, but what is actually seen in the animation is still controlled entirely by the simulation creator, not the viewer, and animations are subject to output quality limitations as noted above.

9.5.2 Limitations to contrast range

Because photomontages are based on photographs, they are subject to the limitations of photography. A camera cannot capture the same range of visual contrast as the human eye (Horner & Maclennan and Envision 2006), and although the images can be manipulated to enhance color saturation and contrast slightly, simulations typically appear to be somewhat more flat and dull than real-life views of the landscape. The lower contrast shown in simulations than in the field may contribute to a lower perception of visual impacts than actually is observed when projects are built (Benson 2005). While this limitation affects all simulations based on photography, it is an especially important limitation for simulations depicting glare such as the reflections from solar collector arrays and other facility components. Because of the extreme variation of brightness between glare and the surrounding landscape and the relatively limited dynamic range of contrast of the various output media such as prints and computer screens, the glare is typically depicted as being less bright than actually observed in the field.

9.5.3 Limits to field of view

Photographs also have a limited and predetermined field of view. They only show what is shown "within the frame," and the visual context provided by that part of the landscape that would be visible in real life but is outside of the field of view of the photograph is lost (Sheppard 1989, 2005). The visual context may in some cases be very important to determining the full effect that the addition of the project to the visible landscape would have. Panoramic simulations are often used to expand the field of view of simulations in order to show more of the surrounding landscape; however, unless the image size at which the simulation is displayed is dramatically increased, the use of panoramic images may result in a loss of detail in the image. Furthermore, correct viewing of panoramic images is more complicated than for "normal" views, in part, because of apparent distortions that are observed when a panoramic image is projected onto a flat surface such as a screen or a printed page.

9.5.4 Limited viewpoints

Simulations developed for views from KOPs can only depict the views from the specified KOPs and omit potential views of the project from many locations within the viewshed that are not KOPs or close to KOPs (Horner & Maclennan and Envision 2006). A typical VIA for a utility-scale renewable energy project will include simulations representing views of the project from a relatively small part of the total area from which the project would be visible. While well-selected KOPs may cover all viewpoints of major concern to stakeholders, it is important to remember that the project will be visible from many locations not represented by KOPs, and that a substantial portion of the total visual impact of the project may be experienced from these locations.

9.5.5 Viewing distance requirements

Finally, in order to see the project components in a simulation at the same size as they would be seen from the camera viewpoint in a real view of the landscape, the simulation must be viewed at a specific viewing distance (NZILA Education Foundation 2010; Horner & Maclennan and Envision 2006). The correct viewing distance for a photograph (or a simulation based on a photograph) is a function of the camera photographic sensor width, the focal length of the lens used to take the photograph, and the size at which the photograph is reproduced, for example, an 8½"-by-11" page. The formula for determining the correct size of the image in relation to the distance viewed is as follows (National Academy of Sciences 2007):

> Distance from viewer = Width of image /(2 • tan (HFOV/2)), where HFOV = the horizontal field of view

Viewing the photograph at an incorrect viewing distance will result in objects within the photograph appearing larger or smaller than they would actually be seen in the real landscape. For visual simulations based on photographs, viewing the simulation from the incorrect viewing distance may result in the project appearing to be larger or smaller than would be observed in the field, which could result in an over- or under-estimation of the project's visual contrast (Sheppard 1989).

9.6 Sources of error and inaccuracy in simulations

As noted, the development of spatially accurate and realistic simulations to support a VIA is a complex technical process that requires a high degree of skill, appropriate technology, accurate data, and rigorous methods. If improperly selected or prepared, simulations may be misleading with the errors not apparent to casual observers. The following discussion examines major sources of errors and inaccuracy in simulations.

9.6.1 Improper selection of KOPs and simulation parameters

Improper selection of KOPs may limit viewers' knowledge of the full range of visual impacts from a project, or worse still, may bias the impact assessment by not showing or fully disclosing important impacts. Sources of error include:

- Not selecting enough KOPs for simulations;
- Not selecting KOPs that represent the "worst-case scenario";
- Selecting so many KOPs for simulations that viewers are bored or confused by viewing large numbers of simulations without significant differences;
- Omitting KOPs for simulation that are important to stakeholders; and
- Selecting KOPs for simulations that as a group over- or under-represent expected impacts.

Similarly, systematic selection of lighting conditions (as determined by date, time of day, weather, and atmospheric conditions) and other simulation parameters that over- or under-represent expected impacts may result in bias in the impact assessment. A common example is the exclusion of night-sky impact simulations for solar and wind facilities, which are known to have potentially substantial night-sky impacts.

9.6.2 Spatial inaccuracy in simulations

Spatial inaccuracy in simulations results from omitting elements that would be visible in the real landscape; showing elements that would not be visible; and showing objects in the wrong locations, at the wrong sizes, or in the wrong visual perspective. There are many potential sources of spatial inaccuracy in simulations. Some may introduce potentially large errors in contrast assessment; others typically result in minor errors in the assessment.

Inaccuracies potentially resulting in significant over- or under-estimation of visual contrast include:

- Changes to the project design after the simulations are prepared.
- Incorrect locations for the KOP or project elements, resulting in potentially significant location errors; however, large errors would likely be corrected prior to simulation development.
- Incorrect setup of viewing parameters in the visualization software, such as incorrect focal length specification.
- Errors in elevation data used to develop the simulation that occur near the KOP, potentially resulting in large-scale incorrect concealment or exposure of landforms and project elements.
- Failure to account for screening elements, such as vegetation and structures, that would be present or would be removed if the project were built.
- Use of incorrect or incomplete models of facility components.
- Project element models not oriented properly with respect to the viewer.

Inaccuracies typically resulting in minor over- or under-estimation of visual contrast include:

- Errors in elevation data used to develop the simulation that occur near the project location, potentially resulting in small-scale incorrect concealment or exposure of landforms and project elements.
- Failure to incorporate Earth curvature and atmospheric refraction in simulation development, resulting in incorrect concealment or exposure of landforms and project elements. The effect of Earth curvature is continuously variable, but at 10 mi, the elevation difference is approximately 55 ft; that is, at 10 mi, about 1/8 the height of a moderately sized wind turbine (measured to blade tip) would be hidden by Earth curvature. At 5 mi, about 14 ft of the turbine would be hidden from view.
- Minor distortions inherent in the base photograph which will not be reflected in the 3D terrain model.
- Improper registration of terrain and project models with the base photograph.

Google Earth, SketchUp, and similar free or inexpensive terrain and model visualization software can be used to preview project element location and orientation. The developer or visualization contractor should be able to provide facility models, or simple models can be easily constructed based on available project data. Models for wind turbines and transmission towers are frequently available at no cost through the Internet. Visualizations produced using Google Earth, SketchUp, and primitive models of facility elements will lack precision and realism, and will not account for screening by vegetation or structures but may be useful for getting a general idea of the size and aspect of a project before accurate simulations are provided.

9.6.3 Lack of realism in simulations

A simulation may be spatially accurate but not realistic; that is, the project elements do not look the way they would in a real view of the project. For example, the project elements may be the wrong color, look cartoonish, have overly sharp visual edges, or not blend seamlessly with the background. Incorrect depiction of atmospheric haze effects can result in misperception of the distance from the viewer to the project. Errors in realism may be very subtle and difficult to detect but can change the perceived contrast of project elements, causing bias in impact assessment. Potential sources of a lack of realism in simulations include:

- Insufficient contrast, as discussed in Section 9.5.2. This is particularly important for depictions of glare from solar facilities and for lighting at all types of facilities.
- Over- or under-incorporation of distance and atmospheric effects, manifested as the project elements appearing too dull or hazy, or too clear and bright to match the surroundings.
- Improper coloring and shading of project elements, resulting in incorrect colors for project elements, a flat appearance, or an overemphasis on three-dimensionality.
- Improper blending of model edges with the background photograph, resulting in overly sharp or sawtooth edges on objects, or "mushy" blurred edges that are inconsistent with the edges of other objects in the base photograph.
- Incorrect lighting and shadow casting, often resulting from incorrect time specification for sun positioning that causes illumination and shadowing of the project elements that do not match the lighting and shadows in the base photograph.

Detecting subtle problems with realism in simulations can be aided by having a photograph of a similar facility in a similar setting available for simultaneous viewing. Also, free or inexpensive mobile phone applications and websites provide easy-to-use tools to calculate sun positions, lighting directions, and shadow directions and length for any location at any time of year.

9.6.4 Improper display of simulations

Simulations must be displayed properly in order for accurate impact assessments to be made, and also for stakeholder information purposes. Potential problems with simulation display include:

- Improper viewing distance, as discussed in Section 9.5.5. It is common for simulations to be reproduced at a size small enough that they cannot be comfortably viewed from the required viewing distance. In this situation, viewers will usually view the simulation from a comfortable distance that is farther than the required distance, resulting in the project appearing smaller than it would in reality.
- Lack of necessary detail in the displayed simulation. If simulations are presented at a small size, there may be insufficient detail to see project elements that may be important to accurate assessment of impacts. Detail may also be lost if low-quality reproductions are made, for example if a printed image is photocopied or a digital image is down-sampled to reduce file size for email transmittal.
- Improper lighting. If lighting is overly bright and improperly positioned, there may be glare on printed simulations that makes details difficult to see. Insufficient lighting may also make details difficult to see, especially in darker images; however, as noted above, night-sky simulations should be viewed in dark conditions or else the lights depicted in the simulations will not be as bright as they would be seen in a real landscape.
- Lack of supplementary information. A variety of information (listed in Section 9.2.8) is needed to fully understand what is being shown and to understand important limitations and other important information about the simulation; however, this information is sometimes lacking.

The simulations included in typical paper copies of VIAs downloaded from websites and printed on standard office printers are generally of insufficient quality for accurate impact assessment. They will usually be much duller in overall appearance relative to the real landscape view, will often have incorrect colors, and will usually lack details apparent in the original simulation.

9.6.5 Improper documentation of simulations and methodology

Sound methodology and supporting documentation are required for producing defensible simulations in VIAs. However, adequate documentation of the methodology used and of the simulation production process may not be provided in VIAs and may otherwise be unobtainable.

9.7 Other types of simulations

While still-image and animated photomontages are the primary type of visual impact simulation used in VIAs, there are several other types of simulations that may be used for various purposes in the course of project development, such as visualization for project siting and mitigation alternative evaluation and planning. They may be encountered as part of stakeholder involvement activities outside the formal impact assessment process. These are discussed below.

9.7.1 Wireframes

Wireframes are computer-generated line drawings of the landscape terrain and the project elements without realistic rendering and shading of the ground surface, vegetation, and project components. In essence, wire frames are "schematic" drawings of the project and the surrounding landscape. They can be developed and manipulated rapidly and are used extensively in the project siting and design process to show project elements in the proper location and at the correct size and aspect as they would be seen from a selected viewpoint. They can be useful in stakeholder involvement activities to explore alternative project designs and to develop basic visualizations of the project from different viewpoints. Wireframe images are also useful to check the spatial accuracy of simulations; that is, to determine that objects in the simulation are shown in the right location, at the right size, and in the proper visual perspective. In this role, as noted above, they are critical to photomontage development. However, the lack of surface rendering and shading in wireframe simulations results in visualizations that lack realism; the project and surrounding landscape do not look like they would in a real view of the landscape. Another limitation of wireframes is that for more complex landscapes and projects, the large number of lines involved can make the images very difficult to "read" and understand, especially for persons not familiar with this visualization method.

Because of these limitations, wireframe simulations are not suitable as finished simulations of projects and are sometimes omitted from VIAs in the U.S. despite the fact that they are used extensively in the course of project development and in the preparation of the photomontages typically used to simulate views of the project. Wireframe simulations may sometimes be included in VIAs to supplement other types of simulations. However, in other countries, e.g., Scotland, wireframes are presented along with finished photomontages when wind facility simulations are presented (Horner & Maclennan and Envision 2006, 2014).

Wireframes can also be superimposed onto photographs; typically this involves superimposition of the wireframe project elements onto a photograph of the existing landscape. If carefully done, so that the project elements are shown in the right location and correct visual perspective, this approach can combine the realism of the photograph with the spatial accuracy of the wireframe image to create an image that may be easier to understand than a regular wireframe image without the necessity for the complex surface rendering and shading needed to create a high-quality photomontage. While useful, the inclusion of the wireframe project elements still results in a lack of realism in the final simulation that makes these wireframe/photo hybrids inferior to well-executed photomontages for use in VIAs.

9.7.2 Synthetic landscape renderings

Synthetic landscape renderings are computer-generated drawings of the landscape terrain and the project elements with realistic rendering and shading of the ground surface, vegetation, and project components. Unlike photomontages, synthetic landscape renderings are not modified photographs; the computer generates the entire viewed scene and thus can depict the landscape and project from any location. Sophisticated high-performance software, and often hardware, is typically used to develop synthetic landscape renderings and is capable of depicting surface textures, lighting, shadows, and atmospheric effects on visibility with a high degree of realism. High-quality synthetic landscape renderings may be nearly indistinguishable from photographs of the real landscape, at least in some settings. Multiple synthetic landscape rendering images can also be combined to create animated sequences.

A powerful capability of synthetic renderings that cannot be duplicated using photomontage is the integration of dynamic landscape modeling to build complex 3D landscape simulations that incorporate dynamic landscape processes. The incorporation of a digital terrain model frees the simulation from the constraints of relying on a photograph taken from a predetermined viewpoint, so simulations can be created of any desired view from any desired viewpoint. The incorporation of geospatial data for terrain and other landscape features integrated with predictive modeling allows ecological data such as vegetation types and height, growth rates, and other environmental factors to be factored in to the automated simulation process to develop synthetic landscapes that "evolve" over time (3D Nature LLC 2015). For example, a GIS can be used to develop polygons representing different vegetation types distributed throughout the landscape. The system stores data on physical and visual characteristics of the vegetation, such as trunk diameter, height, morphology, and color, as well as the growth rate of the vegetation. Using the geospatial and other attributes, the visualization system can develop simulations of different future states of the landscape, for example, at different times in the future, or assuming different stand compositions. These systems incorporate high-end rendering systems that can accurately account for visibility factors, such as sun angle, lighting and shadows, the reflectivity of surfaces, and atmospheric effects. CAD data for project elements such as wind turbines, solar arrays, transmission towers, roads, and buildings can be imported and accurately placed within the terrain model.

These high-end visualization systems offer truly powerful and exciting capabilities to open new possibilities for project visualization, both for impact assessment and other purposes, for example the simulation of multiple sense modalities, as was done by linking a spatially accurate 3D model of a windfarm and surrounding landscape in Switzerland to wind turbine noise that modulates in volume according to viewer position (Manyoky et al. 2014), and a similar effort undertaken in Greece (Papakonstantinou et al. 2013). However, the visualization of the millions of fine details and complex surfaces found within a real landscape setting is computationally complex and enormously demanding on computing resources. The development of synthetic landscape renderings requires advanced computing capabilities and skilled operators, and may take hours or even days to render highly realistic simulations. The use of synthetic landscape renderings has historically been limited in VIAs, in part because of the lack of suitable software/hardware to achieve high levels of realism in the simulations in a reasonable amount of time, and a lack of well-trained users of the software within the visual impact simulation profession. Even modern visualization systems may have difficulty rendering highly realistic complex foreground elements, especially vegetation, to the extent that the images fail to meet the "photorealistic quality threshold." This has resulted in limited use of synthetic landscape renderings for VIAs, though they are sometimes used to supplement photomontages. As high-speed computing systems advance, barriers to the use of synthetic landscape renderings will be reduced, and, eventually, synthetic renderings are likely to replace photomontages in some applications.

9.7.3 Hand-drawn sketches, "cut and paste" images, and generic facility photographs

Hand-drawn sketches are rarely used in VIAs. While they may be useful for presenting conceptual information, they lack both the accuracy and realism necessary for judging visual impacts. Similarly, "cut and paste" images where a photograph, model, or sketch of the facility is overlaid onto a background photograph of the existing project setting without accurate registration and realistic rendering are unsuitable for use as simulations in formal impact assessments, though they may be useful in planning exercises.

Generic photographs of existing facilities that are similar to the proposed project may be valuable for readers to understand the general appearance of the proposed facility, and they can be quite useful for general communication purposes. However, they are not a substitute for preparing spatially accurate and realistic simulations of the proposed project as it would be seen from KOPs, which are necessary for gaining a more realistic idea of the project's likely visual impacts.

9.8 Summary

Visual simulation is a key tool for visual impact assessment but can never exactly simulate the real appearance of a project. A variety of simulation types exist, though currently photomontages remain the primary means for realistically and accurately portraying projects prior to their construction. Photomontage development is complex and exacting work; and photomontages and other simulations must not only be correctly prepared but also properly displayed in order to be effective. Because of the many subtle nuances of both the actual appearance of objects in the landscape and the development of realistic and accurate simulations, there are many possible sources of error and inaccuracy in visual simulations, some of which may not be apparent to the untrained eye. Care is needed to evaluate the quality of simulations and to interpret them correctly. Renewable energy projects have dynamic visual qualities that are well suited to depiction through animations. As both computer hardware and software increase in speed and sophistication, animations and synthetic landscape renderings may replace photomontages as the primary means of visual simulation for renewable energy facilities.

Note

1 This chapter is based in part on Chapter 5 of Sullivan and Meyer (2014).

References

3D Nature LLC. (2015). *Brief History of Computer 3D Landscapes and 3D Nature.* Retrieved from the 3D Nature Studio website: www.3dnature.com/index.php/about/company-history/

Benson, J. (2005). The visualization of windfarms. In I.D. Bishop and E. Lange (Eds.), *Visualization for Landscape and Environmental Planning: Technology and Applications* (pp. 184–192). Oxford: Taylor & Francis.

Bishop, I.D. (2002). Determination of thresholds of visual impact: The case of wind turbines. *Environment and Planning B: Planning and Design 29*(5): 707–718. doi: 10.1068/b12841.

Horner & Maclennan and Envision. (2006). *Visual Representation of Windfarms: Good Practice Guidance.* Perth, Scotland: Scottish Natural Heritage.

Landscape Institute. (2011). Photography and Photomontage in Landscape and Visual Impact Assessment. Landscape Institute Advice Note 01/11.

Macdonald, A. (2012). *Windfarm Visualization: Perspective or Perception?* Dunbeth, Scotland: Whittles Publishing.

Manyoky, M., Wissen Hayek, U., Heutschi, K., Pieren R., and Grêt-Regamey, A. (2014). Developing a GIS-based visual-acoustic 3D simulation for windfarm assessment. *ISPRS Int. J. Geo-Inf.* 2014(3), 29–48. doi:10.3390/ijgi3010029.

National Academy of Sciences. (2007). *Environmental Impacts of Wind-energy Projects.* Washington D.C.: National Academies Press.

New Zealand Institute of Landscape Architects (NZILA) Education Foundation. (2010). *Best Practice Guide: Visual Simulations BPG 10.2.* Wellington, New Zealand: NZILA.

Papakonstantinou, A., Kontos, T., Christodoulou, E., and Soulakellis, N. (2013). 3D Geovisualisation of Noise and Visual Impact of a Proposed Wind Farm Development Using a GIS-based Visual-acoustic 3D Simulation. Paper presented at the 26th International Cartographic Conference, Aug. 25–30, Dresden, Germany.

Scottish Natural Heritage. (2014). *Visual Representation of Windfarms: Version 2.1.* Perth, Scotland: Scottish Natural Heritage.

Sheppard, S. (1989). *Visual Simulation: A User's Guide for Architects, Engineers, and Planners.* New York: Van Nostrand Rheinhold.

Sheppard, S. (2005). Validity, reliability, and ethics in visualization. In I.D. Bishop and E. Lange (Eds.), *Visualization for Landscape and Environmental Planning: Technology and Applications* (pp. 79–97). Oxford: Taylor & Francis.
Sullivan, R.G. and M. Meyer. 2014. *Guide to Evaluating Visual Impact Assessments for Renewable Energy Projects.* Natural Resource Report NPS/ARD/NRR-2014/836. Fort Collins, CO: USDI, National Parks Service.
Sullivan, R.G., Kirchler, L.B., Lahti, T., Roché, S., Beckman, K., Cantwell, B., and Richmond, P.D. (2012, May). Wind Turbine Visibility and Visual Impact Threshold Distances in Western Landscapes. Paper presented at the 37th Annual National Association of Environmental Professionals Conference, May 21–24, Portland, OR.
Sullivan, R.G., Kirchler, L.B., Cothren, J., and Winters, S.L. (2013). Offshore wind turbine visibility and visual impact threshold distances. *Environmental Practice 15*(1), 33–49. doi 10.1017/s1466046612000464.
Zube, E.H., Simcox, D. and Law, C. (1987). Perceptual landscape simulations: History and prospect. *Landscape Journal 6*(1), 62–80. doi: 10.3368/lj.6.1.62.

10

ENGAGING COMMUNITIES IN CREATING NEW ENERGY LANDSCAPES

Richard Smardon and James Palmer

Highlights

- Meaningful participatory processes are key to siting commercial scale renewable energy projects.
- More inclusive and new techniques are being utilized for some projects.
- There is lack of evaluation regarding which participatory processes work and which do not.

10.1 Introduction

As we have learned from our earlier chapters, particularly Chapter 5 on public acceptability, participatory processes for renewable energy facilities are central to siting success. The absence of such participation, for whatever reason, helps explain why different stakeholders do not perceive the benefits of renewable energy as a low- to no-carbon form of energy production in the landscape (Pasqualetti et al., 2002; Selman, 2010). Key factors to consider in this regard include participating in the planning process, trust or confidence in renewable energy developers and decision makers, perceived equity in the landscape impacts, and economic benefits and costs from such facilities on ratepayers, property owners, and communities. Traditional participatory methods in landscape planning and environmental impact assessment have included facilitated group meetings or workshops, charrettes, and public hearings. The question is whether these approaches actually increase public acceptance of the new energy landscape.

This chapter presents participatory process models that address the issues that are commonly part of renewable energy planning, siting, design, and mitigation of impacts. We draw examples from Cape Cod Massachusetts, upstate Maine, upstate New York, the UK, Demark, and the Canadian Province of British Columbia. As Sorensen reminds us, the public can be involved in such projects three different ways (Sorensen et al., 2003):

- Through information about ongoing development;
- Through involvement in the decision making process; and
- Through financial involvement in the project.

Our purpose here is to review and evaluate current consultation and participation methods used as part of the renewable energy process, including public and informational meetings, public surveys, use of social media, mediation, and co-operatives. We then present how such participation techniques can be utilized within the visual impact assessment (VIA) process including utilization of visualization, the identification of landscape values, and projecting landscape futures. Last, we present a brief discussion of landscape impact equity, cumulative impacts, and mitigation of those impacts.

Thirty years ago, Pease and Smardon (1984) proposed participatory methods for stakeholder scoping of environmental impacts as part of the National Environmental Policy Act (NEPA). These participatory methods included small working group facilitation, and in some cases, mediation, plus multiple alternative generation and impact projection. In the UK there is detailed guidance on statutory and stakeholder consultation process (BWEA, 2002) for offshore wind.

Table 10.1 outlines this three-step process:

- Stage 1 identifies stakeholders, issues and processes, which correspond to scoping within the NEPA process.
- Stage 2 includes listening and learning to clarify issues, develop appropriate research, and consultation processes. This corresponds with the commissioning of the EIA and design stage.
- Stage 3 is for monitoring, evaluating, and maintaining contact with stakeholders. This stage corresponds with mitigation, monitoring, and adjustment of the project.

Specific consultation processes, all of which have been used for wind energy projects, include public hearings, information sessions, exhibitions, mediation, public surveys, and use of social media.

TABLE 10.1 Summary of stakeholder consultation and statutory processes

Stakeholder consultation process	*Environmental impact assessment and planning process*
Stage 1: Identifying stakeholders, issues, and processes	**Stage 1: Site selection and scoping**
• Create core team to advise on consultation • Identify stakeholders and issues • Establish key contacts • Draw up detailed consultation process plan • Prepare information for dissemination	• Undertake pre-feasibility studies • Site selection • Screening under the habitats directive, if appropriate • Outline environmental profile • Consideration of alternatives • Scoping exercise (identification of main environmental effects) • Production of scoping report
Stage 2: Listening and learning	**Stage 2: Commission EIA and scheme design**
• Clarify issues, expose assumptions, reduce uncertainties, build on common ground and explore ideas to resolve differences • Commission independent research and fact-finding to avoid the 'adversarial science' problem • Improve communication and relationships • Manage ongoing uncertainties • Turn new ideas into solutions • Agree changes to existing plans where necessary / possible • Develop continuing commitments • Establish monitoring and reporting procedures	• Description of the development • Description of existing environment • Description of environmental impacts • Identify residual effects • Interpretation of scale and significance of impacts • Identification of mitigation measures • Development of management systems and controls to avoid, reduce, and enable mitigation • Propose possible monitoring and reporting measures • Advertise application and lodge in public domain for review and comment

TABLE 10.1 Continued

Stakeholder consultation process	*Environmental impact assessment and planning process*
Stage 3: Monitoring, evaluating, and maintaining contacts • Reporting back to stakeholders on results of consultation • Reporting back to stakeholders on how results were used as part of decision-making processes on the development • Evaluation of consultation process • Ongoing contacts • Return to earlier stages if and when necessary	**Stage 3: Post-granting of consents** • Implementation of mitigation or compensation and control measures • Monitoring and reporting • Continual adjustment where monitoring reveals undesirable results

(*Source:* BWEA, 2002, p. 11.)

10.2 Consultation and participation methods

Public hearings are formal meetings that have an official transcript taken by a court recorder or stenographer. People sign in and are called to present their comments, often with a time limit. They may also present written statements to accompany their oral presentations. Our experience suggests that citizens are most often ill trained to contribute to the discussion using this process. They tend to talk 'from the heart,' which can include impassioned declarations, but usually has no relation to the law or regulations. Non-governmental organizations (NGOs) tend to address the issues in a more professional or 'savvy' representation. Regardless of which method is used, however, our experience has been that neither has much effect on the process or the decision. Often there is little or no two-way communication in a hearing.

Public meetings are informal, and there often is no official record, just personal notes and perhaps an untranscribed recording. Again, people sign up for time and talk from the heart. The purpose is only to gather public comment; typically no information is given to the public other than a short introduction to the project in the US. In the UK, on the other hand, there have been instances where comment cards and other forms of interchange have been utilized (Devine-Wright et al., 2001).

Informational meetings are informal and typically organized by the developer, though they may also be offered by a public agency or NGO. The State of Maine requires that wind developers have an information meeting within 90 days of filing for a permit. The public is encouraged to ask questions and the developer is expected to provide a response. Agency representatives may or may not be required to be present. In the UK informational meetings have been organized as *exhibitions* in combination with comment cards and other feedback mechanisms (Devine-Wright et al., 2001). In the US the information-meeting format has been utilized as a set of presentations about impacts of the Cape Wind project, offshore from Massachusetts. Richard Smardon presented background on visual impact assessment of windfarms in 2002 (Raab Associates, 2002).

Public surveys, particularly if they use a probability sample of the affected public, are a form of public participation. Surveys of users of scenic resources of state or national significance have been used effectively in Maine. The Cape Cod Commission used surveys (see Plate 24) to help establish distance thresholds for their approach to the assessment of offshore wind projects (Rooney et al., 2012). Palmer (2015) shows how to use effect size statistics to evaluate the strength of scenic resource users' response to wind projects. On occasion, survey forms have been developed as part of the consultation process for UK windfarm siting.

Social media and the Internet have been used to make information more accessible to the public. There are many cases where this has happened with multiple websites providing information from different perspectives for large renewable energy projects. One example is the Cape Wind project off Cape Cod Massachusetts where both project proponents and detractors utilized the web to present information and critique the other side's information. Permitting agencies have adopted the Internet as the primary vehicle to transparently make the public record for a proposed renewable energy project available. For instance, England's Planning Inspectorate, the agency responsible for reviewing nationally significant infrastructure projects, has presented thousands of documents associated with the Navitus Bay Wind Park near the Isle of Wight.[1] "Challenge Navitus," an opposition group, has effectively used traditional media and the web to press its claim that the developers misrepresented the project's scenic impacts as well as the impacts on tourism and recreation.[2] The Ivanpah Solar Electric Generating System is the largest solar thermal power station. The Bureau of Land Management (BLM) has used the Internet to make available the public record,[3] while the Basin and Range Watch citizens' opposition group has used the Internet and social media to document the effect of this project on a visually sensitive and fragile ecosystem.[4] It is common for such groups to express their fears and frustrations through dark humor, as shown in Figure 10.1.

FIGURE 10.1 It is common for citizen opposition groups to use social media to share dark humor about the consequences of large renewable energy projects.

(Source: Basin and Range Watch.)

Mediation is increasingly used to address disputes regarding renewable energy siting issues. Mediation is useful in uncovering key factors affecting acceptance of siting a renewable energy project and working through what would be mutually acceptable conditions for siting such projects with a neutral facilitator. Examples of this approach in the US include the following:

- Professor Larry Susskind and his Consensus Building Institute teach a course on Facilitating Wind Energy Siting for involved parties.[5]
- Clifford C. Rohde, Esq. advertises as a lawyer/mediator for renewable energy.[6] In an article for the *Huffington Post*, David Yarnold, President of the National Audubon Society, indicates that wind developers accept mediation for potential disputes.[7]
- Union County Wind Energy in Eastern Oregon used a collaborative project for mediation supported by the Oregon Consensus.[8]
- The National Wind Watch website includes other examples of mediation.[9]

The Middlegrunden Wind Project in Copenhagen utilized an extensive mediation-like consultation process to change the project's gridded layout to a sweeping curved line of fewer turbines (Sorensen et al., 2003). Figure 10.2 shows the dramatic sweep of turbines that have become the iconic symbol of Copenhagen, much the same way that the Golden Gate Bridge has for San Francisco and the Eiffel Tower has for Paris.

FIGURE 10.2 Middlegrunden wind turbine arc array near Copenhagen, Denmark.
(Source: Photo is public domain, through Wikipedia: https://commons.wikimedia.org/wiki/File:DanishWindTurbines.jpg)

Co-operatives that provide local residents the opportunity to share in both the expenses and income from wind power have played a major role in acceptance where they have been tried. The Middlegrunden wind project cooperative has 8,500 local shareholders and "is the world's largest turbine cooperative" (Sorensen et al., 2003, p. 2). In 1996 the Copenhagen Environment and Energy Office took the initiative to organize the project. As mentioned above, an extensive public consultation process resulted in significant design changes as well as the project's financial model. Because of this process, the Danish Society of Conservation of Nature changed its opposition to the project because the "8500 members generated widespread understanding and social acceptance of the chosen location and layout of the farm" (Sorensen et al., 2003, p. 3).

10.3 Participatory process evaluation

Two papers have critiqued the siting of selected renewable energy projects where meaningful citizen participation was not used. Devine-Wright et al., (2001) assessed the use of community involvement strategies in three UK wind energy projects. Sorensen et al., (2003) reviewed public involvement strategies for offshore wind projects in Denmark and Sweden. For the UK windfarms, two failed to get approval and one succeeded. In the successful case, "the developer adopted a more flexible approach, meeting with the community not to inform them of a scheme that was already devised by experts, but a series of options that they could choose from" (Devine-Wright et al., 2001, p. 137). The Swedish case of Karlskrona Vindkraf offshore wind was characterized by Sorensen et al., (2003) as focused on "management of dissent," but use of the Internet has helped to promote a more dynamic dialogue. Both critiques emphasize a more open consultation process and dialogue but stop short of recommending any specific process guidance.

10.4 Visual impact assessment and the consultation process

Additional methods have been proposed for citizen involvement in environmental decision-making. Some of them could possibly be utilized for renewable energy siting, including: developing future landscape scenarios (Dockerly et al., 2005; Laison & Gunnarsson-Ostling, 2009; Loibl & Walz, 2010; Shaw et al., 2009; Tress & Tress, 2003), landscape planning charrettes (Girling et al., 2006; Lennertz & Lutzenhiser, 2006), multi-criteria analysis (Sheppard & Meitner, 2005), and mediated modeling (Montalto et al., 2013).

10.4.1 Visualization to assist in understanding

In addition, computer assisted 2D and 3D visualization tools (see Plate 26) have been steadily developing to assist with landscape value mapping, visibility analysis and simulation of landscape futures (Bishop & Lange, 2005; Kwartler, 2005; Lewis & Sheppard, 2006; Miller et al., 2011; Perkins & Barnhart, 2005; Salter et al., 2009; Sheppard, 2001, 2005, 2006). The applications for future scenario development with visualizations include: projecting future climate change landscape impacts (Dockerly et al., 2005; Laison & Gunnarsson-Ostling, 2009; Loibl & Walz, 2010; Shaw et al., 2009; Sheppard, 2005, 2012), regional landscape planning (Steinitz et al., 2003; Valencia et al., 2009), community and neighborhood planning (Girling et al., 2006; Loibl & Walz, 2010; Salter et al., 2009; Sheppard, 2006), as well as coastal zone planning (Jude et al., 2006). Probably Stephen R.J. Sheppard from the University of British Columbia has done more work than any other scholar on the use of scenario development and visualization for use in engaging the public and helping them understand future landscape changes (Sheppard 2001, 2005, 2006, 2012).

10.4.2 Identifying landscape values

One of the key issues affecting siting renewable energy facilities is that place-based landscape values often go unrecognized, are ignored, or are improperly assessed. Examples of these deficiencies include the proposed Cape Wind project between Cape Cod and Martha's Vineyard, Massachusetts (Firestone & Kempton, 2007; Firestone et al., 2009), geothermal development in Hawaii (Edelstein & Kleese, 1995), and offshore development near scenic shorelines in the UK (Devine-Wright & Howes, 2010) and the Greek Islands (Dimitropoulos & Kontoleon, 2009). In the US, federal lands in the western United States are inventoried for scenic values by the US Forest Service and BLM. For non-federal lands in such states as California, Maine, and New York, environmental laws protect significant scenic areas from unreasonable scenic impacts. In some cases, there may be locally significant landscapes or seascapes that may not be considered as part of the environmental assessment and siting process. Ignoring such locally significant scenic landscape areas may lead to vigorous opposition from local communities.

In Massachusetts, the Cape Cod Commission (CCC) VIA process (Rooney et al., 2012) for offshore structures in state jurisdictional waters solved this issue by mapping and classifying the entire Cape Cod land and shore area. This method documented some 31 key viewpoints looking out toward the Atlantic Ocean from the Cape Cod shoreline. Cape Cod is one of the earliest areas settled by Europeans coming to the New World. It is also a major recreation destination, and many recognize that the Cape Cod landscape has unique historic and natural scenic qualities. The CCC's landscape classification is not based solely on expert judgment, but is also grounded in residents' perceptions of scenic quality and landscape change on Cape Cod over a 20-year period (Palmer, 1983, 1997, 2004). Public opposition to the Cape Wind development provided strong motivation for CCC to develop its VIA procedure. The CCC's procedure included creating photo simulations (Plate 24) of windfarms from three different settings (a beach, a marsh, and a harbor) at four different distances (0.3, 1.5, 3.0, and 5.0 nautical miles) from the shore that were then evaluated by both recreational beach tourists and residents with very consistent results between the two groups.

10.5 Projecting landscape futures and alternatives

One challenge in renewable energy development is projecting future landscape conditions with and without a project, in order to evaluate the impacts of such facilities and their alternatives. The only established VIA methods to specifically project futures with and without the project is the US Army Corps of Engineers' *Visual Resources Assessment Procedure* (Smardon et al., 1988) for water resource development projects and the US Forest Service Scenery Management System's proposed Appendix J (USFS, 2007), which focuses on identifying ecological forest landscape futures given different anthropocentric and natural stresses like forest fires and climate change. Both of these methods have little to say about how to involve various stakeholders in envisioning these alternative landscape futures.

Comparing a future condition with the project to the existing condition ignores how the landscape will change without the project. This presents a conundrum for those responsible for portraying the visual effects of renewable energy projects – how should the visual effects of climate change be portrayed if actions are not taken?

One of the more innovative and inclusive processes has been developed by the Collaborative for Advanced Landscape Planning (CALP) at the University of British Columbia (Pond et al., 2010; Sheppard, 2012). CALP developed a *Local Climate Change Visioning and Landscape Visualization Guidance Manual* (Pond et al., 2010) and applied it in 2008–2009 to the climate change adaption project for the City of Kimberly, British Columbia.

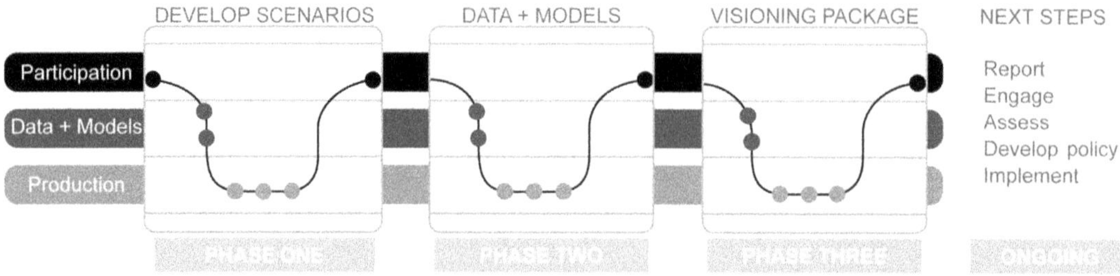

FIGURE 10.3 Three-phase scenario development process.
(Source: Pond et al., 2010, p. 5.)

This is an iterative process that contains three phases with ten specific steps. The process is illustrated in Figure 10.3. The dots at the top of the curves are points for public input in the process.

- Phase one is for participatory scenario building, which includes initiating the project, reviewing the locally relevant science, identifying local priorities and developing future scenarios.
- Phase two is the actual data production and modeling, which includes environmental, land use, impact, and spatial data. 2D and 3D visualizations of these data are prepared and reviewed in a workshop with interested parties.
- Phase three is the full visioning package, which includes refinement of the data and modeling, distribution of the full visualization and technical documents, and stakeholder presentations via an open house or similar venue.
- A planned fourth phase will include further dissemination of the results, public involvement in developing and implementing a policy for adapting to climate change.

Continuous involvement is highly desirable in order to insure the credibility of such visualization tools (Manno et al., 2008). Visualizations are very powerful ways to communicate the impacts of and differences among various alternatives. Landscape scenario development, with modeling, is also very similar to a new developing field of *mediated modeling*.[10] An example of this is a recent study where investigators from Drexel University used visualization of hydrologic models to illustrate the impact of green infrastructure to reduce urban runoff in a Philadelphia neighborhood (Montalto et al., 2013).

10.6 Landscape impact equity

Another issue raised with renewable energy, as well as other forms of development, concerns which individual or groups of individuals are impacted within the landscape fabric. Does one community benefit or suffer impacts more than another? Examples include wind power development in the scenic coastal areas of the UK, or renewable energy power transmission through a new regional interconnect power line in New York State. In Europe, several studies have investigated how renewable energy developments degrade the landscape of one region to the benefit of residents in another (Mason & Milbourne, 2014; Walker et al., 2010; Wolsink, 2007a, 2007b). This also relates to the issue of cumulative impacts from multiple renewable energy developments impacting tourism or 'industrializing' the landscape in one particular area. How should cumulative impacts from multiple renewable energy projects be considered as part of the scenic impact assessment? Certainly the visioning process as previously described above would be useful in this regard.

An example of landscape impact equity is found in the New York Regional Interconnect (NYRI) transmission line proposed in upstate New York, which would run from the middle of the state to just north of New York City. As proposed in 2010, NYRI would transport direct current (DC) electricity generated by wind projects in northern New York State some 150 miles to Orange County, New York, where it would be converted to alternating current (AC) for use in the New York City metropolitan area.

This proposal required a scenic and visual impact assessment for the entire NYRI corridor. The communities along the proposed corridor organized to oppose the project, with two major issues driving the opposition. First, were the perceived scenic and recreational impacts of the additional power line structures, and second was the absence of direct local access to the transmitted DC electricity. A different participatory approach might have altered such perceptions by relocating the transmission line away from village centers, and identifying how local communities might benefit from the project.

There are many cases in Europe where local communities co-own renewable energy production, so they more directly benefit from such power generation. A New York Times feature article (Cardwell, 2015) concerned an island community in Denmark that generates its own electricity using offshore wind power and local biomass fuel, where previously it was dependent on expensive carbon-based energy sources transported by ship.

Conducting choice-based surveys is one of the ways to engage the subject of equity with stakeholders and help them weigh the social, environmental, and economic benefits and costs of renewable energy alternatives. Visualizations can play an important role in the public's understanding of these alternatives. The choice-based surveys offer different energy production and use alternatives as part of the scenario development presented in this chapter. Therefore, this would be a form of multi-criteria assessment combined with scenario generation to gauge reactions to landscape change, economic outcomes, reduced carbon footprint, and energy independence. The critique of using choice-based surveys is that public respondents may or may not have sufficient knowledge of alternatives and what impacts could accrue. The fact is that most energy projects leave the host community (and sometimes adjacent communities) with tax revenues and perhaps other benefits. However, these are community benefits, whereas opposition is almost always from people who feel that there are unbearable costs to them individually. This contrast between how benefits and costs are distributed among individuals and the public at large is at the crux of social equity and why the opposition is sometimes so intransigent.

10.7 Mitigation of impacts

If renewable energy projects are to be built, what role does a host community have in determining how to mitigate its effects? Which forms of visual mitigation hold most promise in terms of reducing significant visual impacts? Do we rely entirely on expert judgment regarding effectiveness of visual mitigation measures or is there a way of incorporating stakeholder and public judgment in such matters?

The US Department of the Interior Bureau of Land Management (BLM, 2013) has been leading the way with visual mitigation measures for land-based wind, solar and geothermal energy development mitigation measures. In the UK there has been guidance for offshore wind development visual mitigation (Enviros Consulting, 2005) as well as the Cape Cod Commission VIA method (Rooney et al., 2012). The BLM is exploring compensatory mitigation by preserving high quality landscapes as an offset to developing lower quality landscapes for renewable energy development (Sullivan et al., 2012). The BLM has employed extensive public involvement in this process for mitigation of solar energy development. Again scenario-based visualization of mitigation alternatives may be useful. Also, use of Internet surveys such as those done by Ian Bishop (2002) to gauge reactions to offshore windfarms' distance, contrast, movement and atmospheric affects may be effective.

10.8 Cumulative impacts

One of the most difficult questions to gauge with renewable energy development is the cumulative landscape impact. This could be from multiple energy projects (possibly including both renewable and non-renewable), or other types of development. Examples include places where multiple wind and/or solar developments with their associated power lines may visually dominate the natural features in the landscape. An example is the western coastline of the UK and Scotland where multiple offshore windfarms cover the water edge horizon. Yet another example is the Wadden sea area on the Denmark shoreline, where there are proposals to replace smaller older wind turbines with those of twice the size. So how do we assess cumulative landscape and seascape impacts from renewable energy projects?

A good example of a cumulative visual landscape impact application is *Pembrokeshire and Carmarthenshire: Cumulative Impact of Wind Turbines on Landscape and Visual Amenity Guidance* (White Consultants, 2013). This approach was agreed upon by three local authorities in the UK and the overall process guidance comes from the Scottish Natural Heritage (2012). Cumulative impact for this project application is defined as, "the additional changes caused by a proposed development in conjunction with other similar developments or as the combined effect of a set of developments, taken together" (Scottish Natural

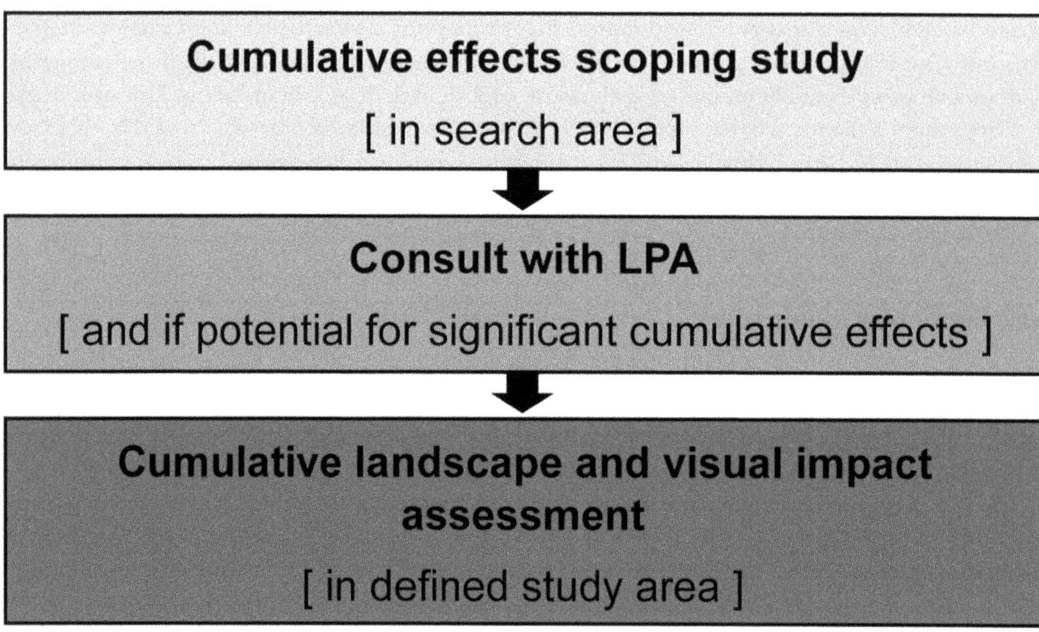

FIGURE 10.4 Three-step cumulative assessment process. The "LPA" is a local planning authority. (Source: White Consultants, 2013, p. 6.)

Heritage, 2012, p. 4). This approach is a three-step process (see Figures 10.4 and 10.5 below). Where a scoping process is used to define the search area, there is consultation with the local planning authority to see if there is potential for significant cumulative effects, and if so then a cumulative landscape and visual impact assessment is done for the defined study area. The scoping process could involve public consultation.

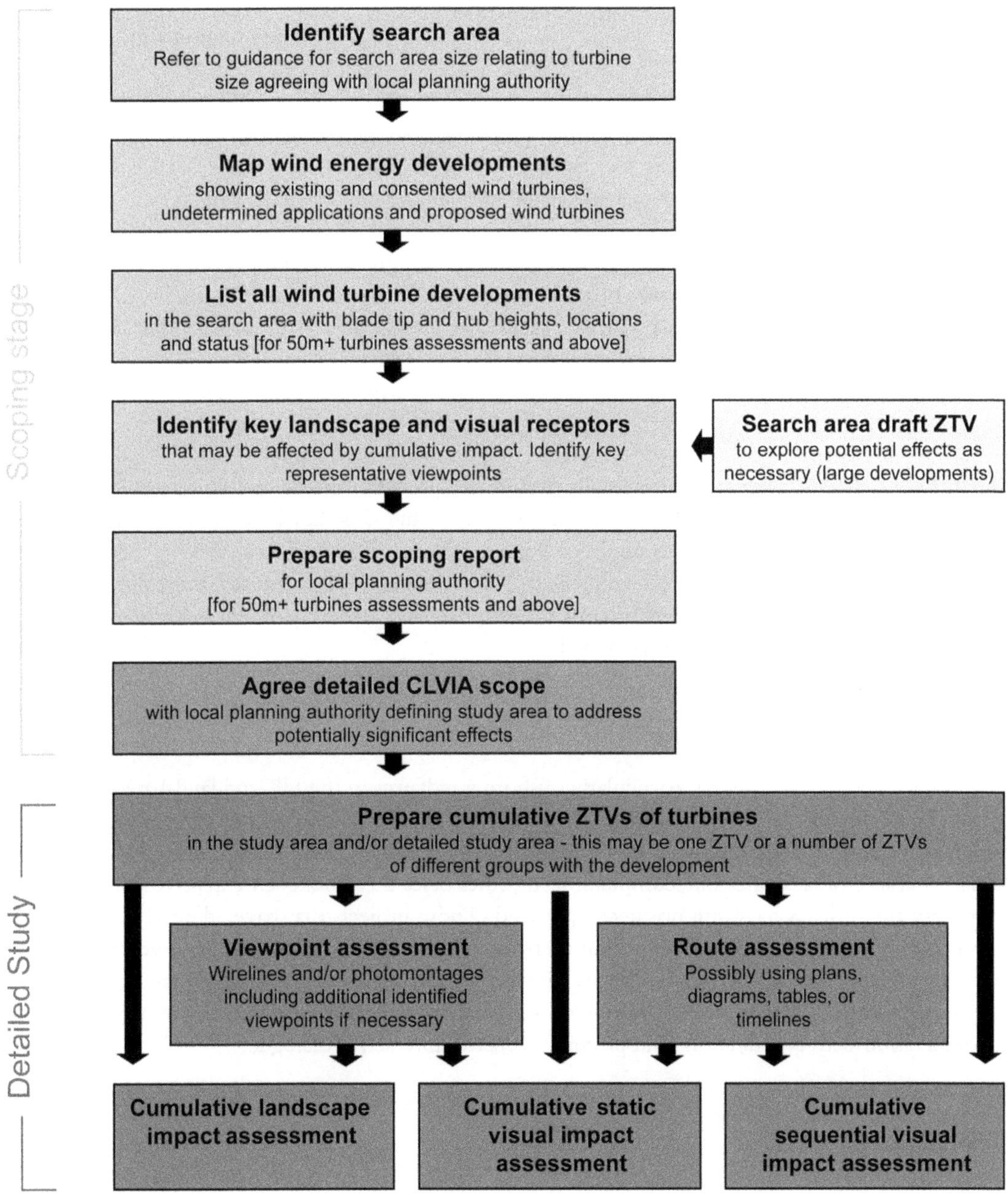

FIGURE 10.5 Overall flow chart of cumulative assessment.
(Source: White Consultants, 2013, p. 25.)

Factors that contribute to cumulative impact of wind (or renewable energy development) include:

- The distance between individual wind turbine developments;
- The distance and area over which they are visible;
- The overall character of landscape and its sensitivity to total development;
- The siting and design of the wind turbine and turbine windfarms; and
- The way in which the landscape is experienced.

(White Consultants, 2013, p. 5).

The study application goes on to describe the existing UK landscape, individual wind turbine design characteristics, existing renewable energy development, and landscape management objectives for this particular area. The detailed guidance for assessing cumulative impacts on amenity or scenic landscape includes three types of cumulative visual effects (White Consultants, 2013, p. 18):

- *Combination* from one (static) viewpoint where more than one development can be seen within the observer's arc of vision at the same time;
- *Succession* from one (static) viewpoint, i.e., where the viewer has to turn to see a number of developments around him/her;
- *Sequential effects* on a journey, i.e., where more than one wind development can be seen one after another over a period of time by an observer moving through the landscape.

The study also describes impacts to potential sensitive receptors from static or a single viewpoint versus sequential effects as one travels along a footpath or scenic route. The key here at this step is what type of outreach or participatory methods might be done to identify these key receptors.

Again, the future visioning process would be useful in this regard. It seems that there may be thresholds of cumulative visual impact as we increase the number, size, and array of renewable energy projects within a given landscape or seascape viewshed.

10.9 Summary and conclusion

Renewable energy development, particularly offshore windfarms in the UK and land-based windfarms in the US, has seen some growth in public consultation and participation processes for siting and environmental review. Some proposed renewable energy development projects have not gotten approvals partially because of lack of public consultation. In other cases there has been extensive public consultation, but the projects have still not been approved. Those projects that have adopted open, flexible consultation processes have done better, but there are no guarantees if there are sensitive landscapes or seascapes involved and sensitive populations affected. Use of such tools as multiple facilitation modes that are interactive, accurate visualization, scenario development, plus cumulative assessment methods in combination with consultation hold promise for better interactive dialogue.

Notes

1 http://infrastructure.planninginspectorate.gov.uk/projects/south-east/navitus-bay-wind-park/?ipcsection=docs
2 www.challengenavitus.org.uk/ and https://en.wikipedia.org/wiki/Navitus_Bay_wind_farm
3 www.blm.gov/ca/st/en/fo/needles/nefo_nepa.html
4 www.basinandrangewatch.org/IvanpahUpdate.html

5 www.cbuilding.org/courses/facilitating-wind-energy-siting
6 http://windpowerlaw.info/about-the-author-and-more-about-the-blog/
7 www.huffingtonpost.com/david-yarnold/new-guidelines-a-step-for_b_1375204.html
8 http://oregonconsensus.org/projects/union-county-wind-energy/ and http://oregonconsensus.org/
9 www.wind-watch.org/news/2013/08/21/wind-turbine-lawsuit-ordered-to-mediation/
10 Mediated modeling is a continuous process of stakeholder engagement from data acquisition to model construction to alternative evaluation.

References

Bishop, I.D. 2002. Determination of thresholds of visual impact: The case of wind turbines. *Environment and Planning B: Planning and Design, 29*(5), 707–718.

Bishop, I.D. & Lange, E. 2005. *Visualization in Landscape and Environmental Planning: Technology and Application*. London: Taylor and Francis.

British Wind Energy Association (BWEA). 2002. *Best Practice Guidelines: Consultation for Offshore Wind Energy Developments*. London: British Wind Energy Association. Retrieved from www.energy.ca.gov/windguidelines/documents/other_guidelines/BWEA-BPG-OFFSHORE.PDF

Cape Cod Commission. 2011. *Cape Cod Ocean Management Plan*. Barnstable, MA: Cape Cod Commission.

Cardwell, D. 2015. Green-energy inspiration off the coast of Denmark. *New York Times*, January 17. Retrieved from http://www.nytimes.com/2015/01/18/business/energy-environment/green-energy-inspiration-from-samso-denmark.html

Devine-Wright, P., McAlpine, G., & Batley-White, S. 2001. Wind turbines in the landscape: An evaluation of local community involvement and other considerations in UK windfarm development. In Martin Edge (Ed.), *Proceedings of the Environmental Research Design Association, Vol. 32. Old World – New Ideas: Environmental and Cultural Change and Tradition in a Shrinking World*, pp. 133–137. Edinburgh: Environmental Design Research Association.

Devine-Wright, P. & Howes, Y. 2010. Disruption to place attachment and the protection of restorative environments: A wind energy case study. *Journal of Environmental Psychology, 30*(3), 271–280.

Dimitropoulos, A. & Kontoleon, A. 2009. Assessing the determinants of local acceptability of wind-farm investment: A choice experiment in the Greek Aegean Islands. *Energy Policy, 37*(5), 1842–1854.

Dockerly, T., Lovett, A., Appleton, K., Bane, A., & Sunnenberg, G. 2005. Developing scenarios and visualizations to illustrate policy and climatic influences on future agricultural landscapes. *Agriculture, Ecosystems and Environment, 114*(1), 103–120.

Edelstein, M.R. & Kleese, D.A. 1995. Cultural relativity of impact assessment: Native Hawaiian opposition to geothermal energy development. *Society and Natural Resources, 8*(1), 19–31.

Enviros Consulting. 2005. Guidance on the Assessment of the Impact of Offshore Wind Farms: Seascape and Visual Impact Report. London: Department of Trade and Industry.

Firestone, J. & Kempton, W. 2007. Public opinion about large offshore wind power underlying factors. *Energy Policy, 35*(3), 1584–1598.

Firestone, J., Kempton, W., & Krueger, A. 2009. Public acceptance of offshore wind power projects in the USA. *Wind Energy, 12*(2), 183–202.

Girling, C., Kollett, R., & Johnston, S. 2006. Informing design charrettes: Tools for participation in neighborhood scale planning. *The Integrated Assessment Journal, 6*(4), 109–130.

Jude, S.R., Jones, A.P., Andrews, J.E., & Bateman, I.J. 2006. Visualization for participatory coastal zone management: A case study of the Norfolk Coast England. *Journal of Coastal Research, 22*(6), 1527–1538.

Kwartler, M. 2005. Visualization in support of public participation. In I.D. Bishop & E. Lange (Eds.), *Visualization in Landscape and Environmental Planning: Technology and Application*, pp. 251–260. London: Taylor and Francis.

Laison, K. & Gunnarsson-Ostling, U. 2009. Climate change scenarios and citizen participation: Mitigation and adaptation perspective in constructing sustainable futures. *Habitat International, 33*(3), 260–266.

Lennertz, B. & Lutzenhiser, A. 2006. *The Charrette Handbook: The Essential Guide for Accelerated, Collaborative Community Planning*. Chicago: The National Charrette Institute and American Planning Association.

Lewis, J.L. & Sheppard, S.R.J. 2006. Culture and communication: Can landscape visualization improve forest management consultation with indigenous communities. *Landscape and Urban Planning*, 77(3), 291–313.

Loibl, W. & Walz, A. 2010. Generic regional strategies from local stakeholders' scenarios: An Alpine village experience. *Ecology and Society, 15*(3), 3. Retrieved from www.ecologyandsociety.org/vol15/iss3/art3

Manno, J., Smardon, R., DePinto, J., Cloyd, E., & del Granado, S. 2008. The Use of Models in Great Lakes Decision Making. Pack Occasional Paper No. 16. Syracuse, NY: SUNY College of Environmental Science and Forestry. Retrieved from www.esf.edu/es/pack/

Mason, K., & Milbourne, P. 2014. Constructing a 'landscape justice' for windfarm development: The case of Nant Y Moch, Wales. *Geoforum, 53*, 104–115.

Miller, D., Fry, G., Quine, C.P., & Morrice, J. 2011. *Managing and Planning Landscape Change: The Role of Visualization Tools for Public Participation*, New York: Springer.

Montalto, F.A., Bartrand, T.A., Waldman, A.M., Travaline, K.A., Loomis, C.H., McAfee, C., Geldi, J.M., Riggal, G.J., & Boles, L.M. 2013. Decentralized green infrastructure: The importance of stakeholder behavior in determining spatial and temporal outcomes. *Journal of Structure and Infrastructure Engineering, 9*(12), 1187–1205.

Palmer, J.F. 1983. Assessment of coastal wetlands in Dennis, Massachusetts. In R.C. Smardon (Ed.), *The Future of Wetlands: Assessing Visual Cultural Values,* pp. 65–82. Totowa, NJ: Allenheld Osmun. Retrieved from www.esf.edu/via

Palmer, J.F. 1997. Stability of landscape perceptions in the face of landscape change. *Landscape and Urban Planning, 37*(1/2), 109–113.

Palmer, J.F. 2004. Using spatial metrics to predict scenic perception in a changing landscape: Dennis, Massachusetts. *Landscape and Urban Planning, 69*(2–3), 201–218.

Palmer, J.F. 2015. Effect size as a basis for evaluating the acceptability of scenic impacts: Ten wind energy projects from Maine, USA. *Landscape and Urban Planning, 140*, 56–66.

Pasqualetti, M.J., Gipe, P., & Righter, R.W. (Eds.). 2002. *Wind Power in View.* San Diego, CA: Academic Press.

Pease, J.P. & Smardon, R.C. 1984. The scoping concept and citizen involvement: An opportunity for revising NEPA. In S. Hart, G.A. Erick, & W. Hornick (Eds.), *Improving Impact Assessment: Increasing Reliance and Utilization of Scientific and Technical Information*, pp. 253–267. Boulder, CO: Westview Press.

Perkins, N.H. & Barnhart, S. 2005. Visualization and participatory decision-making. In I.D. Bishop & E. Lange (Eds.), *Visualization in Landscape and Environmental Planning-Technology and Application*, pp. 241–250. London: Taylor and Francis.

Pond, E., Schroth, O., Sheppard, S.R.J., Muir-Owen, S., Liepa, I., Campbell, C., Salter, J., Tatebe, K., & Flanders D. 2010. Local Climate Change Visioning and Landscape Visualizations: Guidance Manual. Vancouver, BC: Collaborative for Advanced Landscape Planning (CALP), University of British Columbia. Retrieved from http://web.forestry.ubc.ca/calp/CALP-Visioning-Guidance-Manual-V1-1.pdf

Raab Associates. 2002. Cape and Islands Offshore Wind Public Outreach Initiative. Meeting #4: December 12; Visual Impacts and Alternative Sites. Cambridge: Massachusetts Technology Collaborative.

Rooney, S., Dascombe, P., Karjoff, S., & Smardon, R.C. 2012. *Visual Impact Assessment (VIA) Methodology for Offshore Development.* Tech Bulletin #12–001. Barnstable, MA: Cape Cod Commission. Retrieved from www.esf.edu/via

Salter, J., Campbell, C., Journeay, M., & Sheppard, S.R.J. 2009. The digital workshop: Exploring the use of interactive and immersive visualization tools in participatory planning. *Journal of Environmental Management, 90*(6): 2090–2101.

Scottish Natural Heritage. 2012. Assessing the Cumulative Impact of Offshore Wind Energy Developments, Scottish Natural Heritage. Retrieved from www.snh.gov.uk/docs/A675503.pdf

Selman, P. 2010. Learning to love the landscapes of carbon neutrality. *Landscape Research, 35*(2), 157–171.

Shaw, A., Sheppard, S.R.J., Burch, S., Flanders, D., Weik, A., Carmichael, J., Robinson, J., & Cohen, S. 2009. Making local futures tangible-synthesizing, downscaling and visualizing climate change scenarios for participatory capacity building. *Global Environmental Change, 19*(4), 447–463.

Sheppard, S.R.J. 2001. Guidance for crystal ball gazers: Developing a code of ethics for landscape visualization. *Landscape and Urban Planning, 54*, 183–199.

Sheppard, S.R.J. 2005. Landscape visualization and climate change: The potential for influencing perceptions and behavior. *Environmental Science and Policy, 8*(6), 637–654.

Sheppard, S.R.J. 2006. Bridging the sustainability gap with landscape visualization in community visioning labs. *Integrated Assessment, 6*(4), 79–108.

Sheppard, S.R.J. 2012. *Visualizing Climate Change: A Guide to Visual Communication of Climate Change and Developing Local Solutions.* New York: Routledge.

Sheppard, S.R.J. & Meitner, M.J. 2005. Using multi-criteria analysis and visualization for sustainable forest management planning with stakeholder groups. *Forest Ecology and Management, 201*(1–2), 171–187.

Smardon, R.C., Palmer, J.F., Knopf, A., Grinde, K., Henderson, J.E., & Peyman-Dove, L.D. 1988. *Visual Resources Assessment Procedure for US Army Corps of Engineers*. Instruction Report EL-88–1. Vicksburg, MS: USACOE Waterways Experiment Station. Retrieved from www.esf.edu/via

Sorensen, H.C., Hansen, L.K., Hammarlund, K., & Larson, J.H. 2003. Experience with strategies for public involvement in offshore wind projects. Paper presented at the National Procedures for Offshore Wind Energy in the EU. Brussels, Belgium: Institute for Infrastructure, Environment and Innovation. Retrieved from http://cvi.se/uploads/pdf/Kunskapsdatabas%20samhalle/tillstandsprovning/opinion%20och%20acceptans/public_involvement_in_offshore_wind_projects.pdf

Steinitz, C., Arias, H., Bassett, S., Flaxman, M., Goode, T., Maddock, T., Mount, D., Peiser, R., & Shearer, A. 2003. *Alternative Futures for Changing Landscapes: The Upper San Pedro River Basin in Arizona*. Washington, DC: Island Press.

Sullivan, R.G., Cothren, J., Williamson, M., Smith, P., McCarty, J., & Kirchler, L.B. (2102). Visual impact risk assessment and mitigation mapping system for utility-scale wind energy facilities. In *Science, Politics and Policy: Environmental Nexus: Conference Proceedings*, National Association of Environmental Professionals, 37th Annual NAEP Conference (May 21–24, Portland, OR), pp. 667–692.

Tress, B., & Tress, G. 2003. Scenario visualization for participatory landscape planning: A Study from Denmark. *Landscape and Urban Planning, 64*, 161–178.

US Department of Agriculture, Forest Service (USFS). 1995. *Landscape Aesthetics: A Handbook for Scenery Management*. Agricultural handbook No. 701. Washington, DC: USDA Forest Service.

US Department of Agriculture, Forest Service (USFS). 2007. Appendix J. Recommended scenery management system refinements. [Draft appendix to USFS 1995.]

US Department of Interior, Bureau of Land Management (BLM). 2013. *Best Management Practices for Reducing Visual impacts of Renewable Energy Facilities on BLM Administered Lands*. Cheyenne, WY: USDI BLM.

Valencia, C., Flanders, D., & Kozak, R. 2009. Participatory landscape planning and sustained community development: Methodological observations from a case study in rural Mexico. *Landscape and Urban Planning, 94*(1), 63–70.

Walker, G., Devine-Wright, P., Hunter, S., High, H., & Evans, B. 2010. Trust and community: Exploring the meanings, contexts and dynamics of community renewable energy. *Energy Policy, 38*(6), 2655–2663.

White Consultants. 2013. Pembrokeshire and Carmarthenshire: Cumulative Impact of Wind Turbines on Landscape and Visual Amenity Guidance. Haverfordwest, Pembrokeshire, UK: Pembrokeshire County Council. Retrieved from www.pembrokeshire.gov.uk/objview.asp?object_id=8887&language=

Wolsink, M. 2007a. Planning of renewables schemes: Deliberative and fair decision-making on landscape issues instead of reproachful accusations of non-cooperation. *Energy Policy, 35*(5), 2692–2704.

Wolsink, M. 2007b. Wind power implementation: The nature of public attitudes: Equity and fairness instead of 'backyard motives.' *Renewable and Sustainable Energy Reviews, 11*(6), 1188–1207.

11

CONCLUSION

Policy recommendations for the new energy landscape

Dean Apostol, James Palmer, Martin Pasqualetti, Richard Smardon, and Robert Sullivan

Highlights

- Discusses importance of renewable energy.
- Summarizes issues of renewable energy and scenic impacts.
- Recommends policies and best practices.

As is now generally accepted, continued use of our present energy portfolio threatens the health of the planet. Just in time, however, lower-impact renewable resources are becoming available as economically viable alternatives to satisfy our energy needs.

However, while renewable energy resources offer many opportunities to rebalance the environment (or at least slow the accumulating damage), we must accept that they still create environmental and social impacts we must address. Because many of these impacts are new to us, we are just beginning to adjust our perceptions and our policies.

These impacts have several root characteristics. One characteristic – at least for solar, wind, and geothermal – is their site-specific nature. Every operational step prior to transmission is concentrated where the resource is located. This characteristic concentrates the associated environmental impacts and often leads to land use and visual conflicts.

A second characteristic, typically for wind power, is movement. Unavoidably, human attention is drawn to them. They are difficult to ignore. In contrast, conventional power plants often appear to be completely stationary.

A third characteristic of renewable energy that influences landscape impact is low energy densities. Neither geothermal, wind, nor solar energy is blessed with the high energy densities of the fossil fuels, let alone uranium. As a result, these renewable energy sources often disturb greater swaths of land than conventional energy resource development.

The primary goal of *The Renewable Energy Landscape* is to propose policies, procedures, and changes in attitude that can help the world accommodate a future energy portfolio that has a softer touch on the Earth's natural systems as viewed and experienced by people.

11.1 Regulatory legal and policy issues

As *Chapter 3* emphasizes, it is not surprising that development of renewable energy facilities near visually sensitive areas is often controversial. Landscape changes and aesthetic disturbances are the biggest source of that controversy, though complaints about health and wildlife often serve as surrogates since they have stronger legal protections. A critical issue influencing landscape and scenery impacts is the wide variability in guidance, regulations, and incentives that exist for developing renewable energy sources across national, provincial, state, and local jurisdictions. While controversy is commonplace, it can be reduced by careful and sensitive planning and adherence to *best practices*, applicable across a wide range of regulatory jurisdictions.

It is important to reiterate that scenery is not just some 'nice thing' to enjoy, something of no financial or intrinsic personal value. Views of attractive surroundings can unleash creativity and even reduce stress, and even contribute to health (Ulrich 1984; Tzoulas et al. 2007). Additionally, landscapes are a source of pride and identity for nations, regions, and local communities. They contribute, sometimes significantly, to economic development, particularly through tourism. As we have attempted to illustrate throughout this book, losing valued landscapes extends far beyond the mere loss of a view. We must incorporate that sensitivity within the policies and procedures we utilize as we move toward a more sustainable energy future.

11.2 Developing multiple landscape zoning

Because landscapes vary in scenic quality, and the extent to which residents and visitors value that scenery also varies, we suggest nations, sub-regions (states, provinces), and localities consider developing a basic three-part 'zoning' of lands, at least with respect to commercial scale renewable energy. This type of zoning has long been used with good success on federal lands in the US, and has been emulated elsewhere such as the Wadden Sea coast in Germany, Netherlands, and Denmark, as well as Scotland's Atlantic coast.

For example, *Zone A* could encompass lands that governments have designated as most highly valued for their scenic (and/or other) visually sensitive qualities, and within which large scale commercial development of renewable energy would not be allowed. Examples of such landscapes include National Parks, designated wilderness areas, Wild and Scenic Rivers, National Seashores, UNESCO World Heritage Sites, State and Provincial Parks, historic sites, wildlife refuges, and other areas with some official sensitivity designation. Also to be included in this category should be high value *cultural* landscapes that are not compatible with large-scale renewable energy. Some of these are already protected, but in many cases, they are not. Those with existing protection should be easier to set aside or remove as potential renewable energy sites. There are myriad examples, including the Statue of Liberty, the Golden Gate Bridge, and the Hollywood sign.

Zone B could include natural landscapes that have clear visual value (see Chapters 5 and 8), but do not hold official protection. Many thousands of potential sites could fall into this category, and many environmental websites list some. Limited renewable energy development may be compatible with these lands, but not when the presence of renewable energy facilities changes the essential appeal and value of the landscape. The definition of 'essential landscape character' requires some negotiation in many cases. But care must be taken to limit the extent of energy development, because opening the door with one project can lead to 'piggy backing': where new projects are justified by the presence of the previous one. At some point, cumulative impacts accrue and can destroy the valued qualities of the landscape. In Zone B areas, special attention should be paid to design and mitigation techniques

(Chapter 7) to reduce potential impacts, and where possible to use energy development to enhance landscape views.

Finally, *Zone C* includes lands with low scenic and/or other values, although such designation naturally can be contentious. At the most basic level, these lands could be considered to be areas that remain after all other categories of protected land have been established. Naturally, designation that opens the door to development can be aggressively debated. For example, the European Landscape Convention (Council of Europe 2000) emphasizes the importance of protecting the integrity of all landscapes, even 'ordinary' ones (see Chapter 4). Already disturbed areas are also candidates for Zone C designation. But an alternative to the 'sacrifice' of such areas is to bring them up to higher standards by repairing past damage. This is in line with the approach of US Forest Service Scenery Management System (USDA FS 1995, 2006a).

11.3 Determining visibility across landscape zones

While such zones are designated by lines on maps, we know that the eye sees across these lines, and it is difficult to place such lines with much precision. For this reason, the concept of 'viewsheds' is important in regional renewable energy planning. In particular, a good principle is that the view from a protected 'A' area should be buffered from 'C' areas by 'B' areas, where development is allowed but limited and designed in ways sensitive to the views from within the protected area. The transition between B and C lands is more problematic, since there is no obvious intermediate zone. However, we do recommend that, at the least, even C areas must receive design attention to avoid diminishing locally important views and viewsheds. For wind and solar power towers, this is often proving difficult in practice, illustrating the difficulties of preserving landscape quality while introducing industrial development, even if it is of the renewable energy variety.

Some renewable energy projects are visible for substantial distances (see Plate 25). This raises the question of where to draw the line on assessment of scenic impacts to viewsheds. Maine uses an 8-mi (13.3-km) area around wind projects. Beyond that point no analysis is done and no mitigation required. This is likely too short a distance for many locations, particularly where skies tend to be clear, and visibility nearly limitless, as in some areas of the western US. Sullivan et al. (2013b) found from the results of a visibility study conducted for offshore wind facilities that 25 mi (40 km) is a reasonable limit for impact analysis. Scottish Natural Heritage has a 20 km limit (Hornor & Maclennan and Envision 2006). The effective distance buffer will, of course, depend on the size of the project, the publically ascribed value, topography, land cover, and atmospheric conditions. Whatever the distance, various other factors will come into play as significant, including the elevation of the observer and the intensity of any possible night lighting.

11.4 Building scenic inventory baselines

Deciding which lands to protect and where development should be allowed depends on a careful and sensitive inventory of scenic resources. An important tool in scenic conservation is a regional landscape inventory that includes an evaluation of scenic value, such as those used by US federal land management agencies (USDI BLM 1980; USDA FS 1995; Sullivan and Meyer 2014), Canadian and Australian forest agencies, and in the UK and other European regions such as the Wadden Sea coast (Smardon 2009) and Scotland's Atlantic Coast (Hill et al. 2001). It is difficult to conserve what is not documented, and these inventories are the basis for establishing conservation rules. Too often, inventories are not begun until projects are already proposed, with the inventories themselves conducted by the energy developer, who has an obvious incentive to downplay scenic value.

There are good examples of landscape and seascape classifications, such as the one developed by the Cape Cod Commission by Rooney et al. (2012) (Plates 18 and 26) and in the UK for offshore energy development (Hill et al. 2001). The southwestern US states of Arizona, California, Colorado, Nevada, New Mexico, and Utah have a set of directives and a programmatic environmental impact analyses for solar development on lands managed by the US Department of the Interior Bureau of Land Management (BLM) (see http://solareis.anl.gov/ and USDI BLM 2012). These and several other western states have directives and a programmatic environmental impact analyses for wind development (see Chapter 3).

The BLM's Visual Resource Inventory process provides a good model that can be adapted to non-federal lands. It catalogues the presence of features known to contribute to scenic value, including naturalness, vegetation, landform, and presence of water. It has been adapted to many landscape types, including high mountains, hot deserts, and Great Plains grasslands (USDI BLM 1980).

Whatever the method, it is impossible to overstate the value of an inventory as a tool in helping decide on conservation actions (see Chapter 8). Most appropriate is a comprehensive inventory, as at Cape Cod, that used one set of openly reviewed criteria, rather than private-sector developers assembling an inventory piecemeal on a project-by-project basis using various criteria. In an approach that differed somewhat from the Cape Cod example, Maine inventoried and rated specific features, such as rivers, lakes, roads, etc. for their scenic and other environmental values.

Whatever system of evaluation is used, renewable energy developers must be sensitive to the impacts to natural resources that have been determined to be of significance. These could be scenery, but also other values such as solitude. In addition, such significance can be judged on an individual personal, as well as an official, level.

11.5 Integrated environmental planning for renewable energy

A central theme in this book is that renewable energy development can transform landscapes. Though we have limited our focus to scenic resources, there are also impacts to wildlife, noise, and cultural resources, and potentially to streams and water bodies, that must be accounted for. As we noted in Chapter 6, renewable energy can be land extensive, being deployed over large areas to capture low-density energy sources. However, it is often possible to incorporate renewable energy into a multiple-use working landscape. Examples of this approach include integrating wind energy development with commercial forestry or agriculture, or installing solar photovoltaic modules over large parking lots (Plates 27 and 28 and Figures 11.1 and 11.2).

From an environmental and economic perspective, it is best to 'layer' protection as much as possible. This means that areas protected for scenery should overlap with lands protected for other reasons, such as wildlife habitat, watershed value, or cultural resources. This approach to protecting those areas with the highest concentration of critical values first is more likely to succeed, since it can be expensive and controversial to limit development, especially on privately owned lands.

If we were to accept conceptual approaches such as the European Landscape Convention, we would recognize that all landscapes have value. This attitude tells us that while we need renewable energy, we need to develop it in a way that respects landscape integrity. This likely means that we would avoid constructing extensive wind projects with very large turbines in most landscapes, limiting them to where they will rarely be seen for the lifetime of the project. This is an admittedly high, and probably overly limiting, standard.

Federal land-management agencies in the western US have used integrated, programmatic environmental impact statements as a screening process to identify lands that have renewable energy potential, and then study these for the presence of natural resource conflicts, including scenery. Lands with

FIGURE 11.1 Solar sharing in Germany. Panels make electricity while allowing animal grazing. Near Saarbrucken.

(Photo: M. Pasqualetti.)

FIGURE 11.2 Wind sharing in Illinois. Wind turbines and farming coexist near Dillon.

(Photo: M. Pasqualetti.)

multiple resource conflicts tend to get the highest levels of protection from energy development. Lands with the fewest resource conflicts are places where permitting of renewable energy can be expedited.

This approach has been used across regions with a wide range of land ownership and jurisdictions. For example, the California Desert Renewable Energy Plan is investigating millions of acres to identify areas where development should be off limits, and where it should be potentially permitted. Once completed, this will allow concentration of development on the least valuable lands, regardless of ownership or jurisdiction, and will provide developers with a clear map of where they can expect few permitting obstacles. But the sheer scale of this effort, plus disagreement over values and how to assess them, has resulted in numerous delays in its completion. Two federal agencies, two state agencies, and seven county governments have yet to agree on a final plan even after seven years of work (Plate 29).

11.6 Best practices framework

With or without regional scale zoning or integrated planning, renewable energy developers, consultants, and any jurisdictions that oversee them should commit to using a 'best practices' approach on renewable energy projects; that is, practices that have been recognized as minimizing impact, which can include siting, project design, and visual impact assessment. The details offered throughout this book will help guide this approach, especially in Canada, Australia, Europe, and the United States. All have systems that can be built upon or adapted.

Jurisdictions and regulators with little experience in scenic resource management often make the mistake of assuming that aesthetic value is 'in the eye of the beholder,' as if everyone's sense of beauty varies infinitely from person to person, and thus fails to represent objective patterns. However, years of research have shown that there is often wide agreement on the most and least scenic landscape, and what makes them that way. We believe there is enough agreement, especially within culturally similar groups, to make such a step relatively easy. In other words, the exercise of common sense is the foundation of a best practices framework.

Throughout this book we have offered several summaries of critical elements of useful best practices. For example, Chapter 2 covered the development of the field of visual resource assessment, illustrating many good examples of scenic conservation programs around the world that can and should be adapted to areas or jurisdictions that, to date, lack scenic resource conservation measures.

Chapter 3 presents regulatory oversight and incentives for renewable energy development in the US, Canada, and Australia. It presents multiple jurisdictional challenges for large-scale renewable energy (wind and solar) development particularly for the US.

Chapter 4 presents regulatory oversight for commercial scale renewable energy development in Europe with a major focus on onshore and offshore wind development in the UK.

Chapter 5 proposes a range of factors associated with public acceptance of renewable energy projects. These factors include:

- Physical attributes of the project as well as the project's landscape context;
- Social considerations of who benefits or suffers as a result of the project;
- Whether the process of making the decision is considered to be essentially fair; and
- How, and whether, opinions change over time, especially as development criteria change and more impact research becomes available.

Chapter 6 describes the importance of size, color, shape, lighting, and movement; all are important and inherent features of renewable energy facilities. These objective attributes create contrasts with natural and cultural landscapes, so reducing these contrasts is a key consideration in minimizing objections, at least within the limits of the authority of various agencies such as the Federal Aeronautics Administration (FAA) for warning lighting on structures over a certain height.

Chapter 7 presents various siting, design, and mitigation options available to developers and regulators to reduce visual impacts. Such options include pre-visual assessment mitigation through design as well as post-visual impact assessment (VIA) mitigation.

The BLM has recently published *Best Management Practices for Reducing Visual Impacts of Renewable Energy Facilities on BLM Administered Lands* (USDI BLM, 2013). It provides detailed guidance on best management practices for siting and permitting renewable energy projects and includes more than a hundred mitigation measures for onshore wind, solar, and geothermal facilities, as well as transmission lines, roads, and other infrastructure common to renewable energy facilities. While we recognize that not all of these measures can be used everywhere, many of them can be used almost anywhere.

Chapter 8 demonstrates how the key factors of visual sensitivity and contrast can be used to measure impacts objectively. It includes a step-by step review of best practices for determining landscape sensitivity, magnitude of visual impacts, and assessing cumulative impacts. Recognizing this, the National Park Service has recently published a *Guide to Evaluating Visual Impact Assessments for Renewable Energy Projects* (Sullivan and Meyer 2014). It provides detailed guidance on what should be included in a VIA, and best practices for producing and evaluating visual impact simulations.

The US Department of the Interior has also developed its *Smart from Start Program* (USDI 2013), developed to manage environmental impacts to views from National Park Service and BLM lands. Not surprisingly, these documents include many of the suggestions we mention throughout the book, including avoidance through the use of appropriate siting, minimizing impacts through use of design features, offsetting impacts through mitigation, and developing increased understanding through constant monitoring.

Engineering layouts should be adjusted to create a better fit of renewable energy projects to the landscape and the people for whom this landscape creates a sense of place. Such advice is especially applicable for wind turbines, for obvious reasons. Clustering, breaking up long strings, avoiding interference with prominent features, and reducing night lighting are techniques that are immediately available and should be required by regulatory authorities where landscape aesthetics are especially sensitive but lack full legal protection.

In other words, while design should never be prescriptive, guidelines for reducing impacts already exist.

An important next step in mitigating landscape impacts is to increase coordination between federal and state agencies to both bridge the policy and regulatory gaps, and minimize the jurisdictional turf battles that detract from the opportunities of high-quality projects. In the absence of resolution of these conflicts, local jurisdictions can delay large renewable energy projects based on questionable siting criteria.

Conversely, locally important landscapes and scenic resources (not federal or state recognized) generally have no *conservation standing* in many states that have public utility commissions, Therefore, they are not given adequate consideration in siting decisions. Monitoring of the efficiency and comprehensiveness of state, local, and mixed siting control could help determine which approach works best. Processes and procedures for documenting locally significant landscapes need to be developed or improved. The Cape Cod Commission study, mentioned earlier, provides a good example that could be emulated for other offshore wind development.

Regulatory and financial incentives may not produce desired results, and can be confusing because they vary so much from place to place. Our research and experience has uncovered a crazy quilt of energy incentives (see Chapters 3 and 4). Renewable Energy Portfolio targets and tax benefits should be designed to better reflect regional renewable potential and likely conflicts with valued scenic (and other) resources.

11.7 Tightening visual and scenic analysis methods

While commonalities among visual analysis methods exist, there are also important differences, and none have been developed specifically to deal with the unique problems renewable energy projects present. We need a more consistent methodological framework. Good examples include the six to eight steps and consistent vocabulary presented in Smardon et al. (1986) and the new National Park Service *Guide to Evaluating Visual Impact Assessments for Renewable Energy Projects* (Sullivan and Meyer 2014). In Europe, the UK Department of Trade and Industry (2005) approach for coastal and seascapes, and the UK Landscape Institute's (LI/IEMA 2002) *Guidelines for Landscape and Visual Assessment* provide overall framework for landscape characterization and visual impact assessment.

Another helpful approach, proposed by the National Research Council (2007), recommends six basic steps for any wind project:

1. Project description and appearance,
2. Project visibility and landscape context,
3. Scenic resource values and sensitivity levels,
4. Assessment of aesthetic impacts,
5. Mitigation techniques, and
6. Determination of acceptability or undue aesthetic impacts.

These steps incorporate several techniques, among them the use of visual simulations. As Chapter 9 points out, this is a field of rapidly improving technology that has provided new tools but also new challenges. The quality and appropriate use of visual simulations is a crucial aspect of gaining a proper understanding of scenic impacts of renewable energy.

Over the course of many years and many proposed projects, we have seen multiple instances of poor quality, or incorrectly scaled simulations used by lay authorities to determine the final word on impacts. For these reasons, regulators need to understand that when they are looking at images presented by a proponent of an energy project, they need to view these carefully and ask the right questions. Were the viewpoints appropriate? Do the simulations provide a comprehensive enough picture of the project? Are they accurate with respect to scale, color, placement of facilities and various project features? Do they consider motion, glare, and night lighting? We encourage practitioners and regulators to consult Chapter 9, and its list of references, and to consult independent experts to review simulations before over relying on them.

Despite some shortcomings, there continues to be incremental advances to visual impact assessment methods, particularly in Europe. The Landscape Institute and Institute for Environmental Assessment cooperated about 20 years ago to produce the first edition of *Guidelines for Landscape and Visual Impact Assessment* (LI/IEMA 1995), followed by substantial revisions in 2002 and 2013. This work has served as the launching point for proposing methods specific to offshore wind assessment (DTI 2005; Hill et al. 2001).

At about the same time in the US, the Forest Service's (1995) Scenery Management System began to explicitly incorporate landscape ecology and the sustainability of scenery management decision (USDA FS 2006a). Recent research by the Argonne National Laboratory for the BLM, National Park Service,

Bureau of Ocean and Energy Management, and Department of Energy has begun looking specifically at renewable energy impacts to scenic landscapes (Sullivan 2011; Sullivan and Abplanalp 2013, 2015; Sullivan et al. 2012a, 2012b, 2012c, 2013a, 2013b, 2014).

In response to a private renewable energy proposal, the Cape Cod Commission realized the need to create a new scenery impact assessment procedure for offshore wind development that improved on any of the established methods. The US Corps of Engineers *Visual Resource Assessment Procedure* (VRAP) method (Smardon et al. 1988), which was designed for large-scale water development projects, does not address shorelines and seascape contexts, sensitivity issues, or impact levels. While the DTI (2005) does a better job of addressing those concerns, new methodologies more specific to renewable energy development can and should build on established methods to be relevant to assessing impacts from these new technologies.

The State of Maine has taken a fresh, even radical, look at how to balance the visual impacts of utility-scale wind energy development with the need for clean energy when the legislature passed the *Wind Energy Act*.[1] The Act acknowledged that wind development had "at times proven controversial" and that this was exacerbated by a "lack of clarity and direction in state policy." In response, the Act pioneers a new approach to the visual assessment of utility-scale wind projects. The following are the steps utilized within the State of Maine review process:

1. Identify an area for expedited permitting that excludes the very highest quality scenic areas with a protecting buffer.
2. Accept that explicitly recognizing that wind turbines are highly visible features in the landscape is not a solely sufficient basis for determination that an expedited wind energy project has an unreasonable adverse effect.
3. Consider visual impacts only for designated *State or Nationally Significant Scenic Resources* (SRSNS), and address these resources in an unambiguous way.
4. Include specific evaluation criteria:

   (a) The significance of the potentially affected SRSNS;
   (b) The existing character of the surrounding area;
   (c) The expectations of the typical viewer;
   (d) The expedited wind energy development's purpose and the context of the proposed activity;
   (e) The extent, nature and duration of potentially affected public uses of the scenic resource of state or national significance;
   (f) The potential effect of the generating facilities' presence on the public's continued use and enjoyment of the scenic resource of state or national significance; and
   (g) The scope and scale of the potential effect of views of the generating facilities on the scenic resource of state or national significance, including but not limited to issues.

5. Evaluate the standard based on whether the impact to SRSNS has an unreasonable adverse effect on the scenic character or existing uses related to scenic character of SRSNS.

In practice, several measurable indicators are identified for each of the criteria, and then thresholds are proposed to distinguish among impacts that are not adverse or unreasonably adverse (Palmer 2014).

The Maine example is a more radical departure than those found in other locations. It recognizes that turbines are highly visible, but that, by themselves, they are not unreasonable landscape additions. Impacts are only to SRSNS, all of which are listed. The criteria are different from the contrast/scale/

dominance judgments of professionals. In practice, the criteria are measured by two or three indicators with thresholds to determine whether the impact is unreasonable (see Chapter 8).

We might also wish to examine procedures developed from outside the world of renewable energy. For example, Churchward et al. (2013, p. 6), in a study of US transportation project visual impact assessment practice, offer a sensible first step in building a useful framework in the 10 criteria they propose as the elements of an initial VIA procedure:

1. *Objective* – the procedure is designed to eliminate individual bias.
2. *Valid* – the procedure can be defended as measuring what it intends to measure.
3. *Reliable* – adequately trained professionals following the procedure reach the same conclusion.
4. *Precise* – the data required by the procedure are measured at a grain or scale sufficiently fine to validly measure or describe characteristics of substantive interest, and sufficiently coarse to be pragmatically implemented.
5. *Versatile* – the procedure supports valid assessment of different types of proposed changes from the perspectives of different viewer groups interacting with different landscape settings.
6. *Pragmatic* – the procedure can be easily and efficiently implemented, by a trained professional.
7. *Understood easily* – the procedure and resultant assessments are accessible by the public and decision makers.
8. *Useful* – the procedure and resultant assessments affect location, design, or mitigation decisions.
9. *Implemented consistently* – the procedure can be applied consistently among different projects, and individual assessments are consistent with the chosen procedure.
10. *Legitimate* – the procedure is supported by laws, regulations or other legal mechanisms, uses socially/culturally-accepted standards, and uses scientifically accepted standards.

11.8 Determination of acceptability or undue aesthetic impacts

The phrase 'undue aesthetic impacts' stems from the State of Vermont and its new legislation regulating wind development that states in Act 250's Criterion 8 that it "will not have an undue adverse effect on the scenic or natural beauty of the area". It is further defined by the Vermont Environmental Board's *Quechee Lakes* case, which has stood against challenges in the Vermont Supreme Court (Vermont Superior Court 1990). Wind and solar projects are permitted by the Public Service Board, which uses the principles of the *Quechee* analysis.

Research for this law included reviews of over 200 projects, and is a rare example of a state getting ahead of the problem. The Massachusetts Cape Cod Commission criteria for assessing aesthetic impacts of offshore development, and the State of Maine Wind Energy Regulations (Palmer 2012) addressed cumulative impact of renewable energy development. However, our review of siting procedures (see Chapters 3, 5, and 8) for renewable energy development revealed very few aesthetic impact criteria. We believe there is a strong need for these criteria, especially where landscape regions are important for tourism and recreation.

Updates to existing assessment methods, or development of a new method should be more relevant to new energy and the types of landscapes being most impacted. For example, offshore impacts are new and unique. It may be that 'new and unique' renewable energy technologies that are huge, with lots of motion, blinking lights, or expansive glare cannot be adequately evaluated by revisions to existing methods, and that a whole new approach needs to be developed. That was what happened in Maine.

Although wind power is attracting the greatest amount of attention, solar technologies of various types are introducing several new issues, including glare and even threats to wildlife. In addition, rural, cultural landscapes have their own set of values, distinct from wild landscapes. Present assessment methods could be improved to anticipate the vertical and horizontal scale of such solar installation, the visibility distance, and the design inflexibility of renewable energy systems.

11.9 Potential assessment and mitigation needs

As interest in renewable energy development continues to grow, so too will the specificity and extent of potential assessment and mitigation needs. Several questions still to be answered include:

1. *Should we do the research to improve VIA method validity or reliability or fight it out in the courts and quasi-judicial hearings, or both?* Public perceptions may change over time, although there is little evidence that they do with regard to a specific cohort reacting to a specific project. What may change in the evolution of opinions as new generation matures? We need landscape perception studies (see Palmer 2000; Palmer and Hoffman 2001) to validate both landscape value as well as 'ground' the assessment process.
2. *Should we develop VIA databases by project, including mitigation effectiveness and landscape types?* This would seem to be a good idea, although only if there are actual mitigation (other than resisting) measures that work. The relevant example here is the new BLM publication (USDI BLM 2013) on mitigation tools for wind solar and geothermal energy development out of the Wyoming BLM office. The US Federal agencies, such as the BLM, Bureau of Ocean and Energy Management, US Forest Service, US Department of Transportation, US National Park Service and other equivalent national/regional agencies could have 'before-and-after' databases for renewable energy projects as well as successful or unsuccessful mitigation measures.
3. *Should we develop better public participatory procedures and methods that communicate landscape change to affected publics such as those covered in Chapter 10?* We think the answer is a resounding yes, particularly for countries that have signed the European Landscape Convention, but also in other countries (Sheppard 2012).
4. *Does all of the above argue for a renewable energy impact assessment guidance development project that expands on that of the National Research Council, one that broadens the scope to address scenic and other aesthetic impacts from all types of renewable energy development?* To us the obvious answer is yes but we need to frame this in the most powerful way. An example is to conjoin the work of the National Research Council and US Environmental Protection Agency to quantify ecological-cultural services for accounting purposes, plus aesthetic compensation and mitigation.

This brings us to the final point: *How we should address landscape impacts of climate change over time?* The US Forest Service has been addressing this question with a new Appendix J (Spencer et al. 2008; USDA FS 2006a) to their Scenery Management Systems. It incorporates landscape ecology principles such as fire and insect infestations plus climate change on forested landscapes.

The original *Economic and Environmental Principles and Guidelines for Water and Land Related Resources Implementation Studies* (Water Resources Council 1983) stipulated that impact studies should compare the future with and without the proposed project or other alternatives.

Therefore, this ties climate change with the mitigation effects of large renewable energy projects. Sheppard's 2012 book *Visualizing Climate Change* provides graphic tools for presenting alternate climate futures. On a smaller scale, the State of California has mandated (under the *California Environmental Quality Act*) that all environmental assessments, regional land use plans and transportation plans must address climate change impacts. Another applicable study is that of Palmer (2012), which assessed the cumulative impact of power lines and energy development in the unincorporated areas of upper Maine.

11.10 Research needs

Modern development of renewable energy is still in its early stages, both on land and sea, and much remains to be learned about the visual impacts it will have on the landscape. Below we identify some areas ripe for research.

Visibility and context

The Argonne National Laboratory study (Sullivan et al. 2013b) and work in Australia by Bishop (Bishop 2002, Bishop and Miller 2007) examined offshore wind, and logically concluded that you cannot have visibility without contrast and you cannot have contrast without visibility. Moreover, you cannot have any impact at all without visibility or contrast. In addition, the work by the Cape Cod Commission (Rooney et al. 2012) indicated that *landscape context* continues to be important to some groups. Because we can see both large-scale wind and solar installations in the western states over great distances, appropriate visibility zones or thresholds for regulation need additional research. Weather and atmospheric conditions need to be factored in for offshore large-scale wind facilities, yet they are particularly vexing issues. For example, a developer may argue that a project will only be clearly visible 20% of the time because of weather. An opponent will say that those clear beautiful days are the only ones that matter, given that beach use is typically low on rainy days.

Scenic-resource values and sensitivity levels

This is perhaps one of the most problematic areas needing additional research. In Thayer's early study of windfarms in California (Thayer 1995) people liked the *novelty* of wind turbines; however, in many countries the novelty may be presumed to be wearing off. In Europe, early installations in many places such as the Netherlands and the UK did not attract much public resistance, but this tendency has diminished and in some places, reversed, in part because of cumulative impacts.

Among the significant research questions that remain:

- What is the acceptance and, more importantly, the criteria for acceptance of renewable energy development outside of protected landscapes?
- What are the thresholds of development, past which renewable energy facilities become an aesthetic problem?
- Where and under what conditions do people find renewable energy most appropriate?
- How might renewable energy affect economic development, such as tourism and land values?
- Are people more apt to accept energy development in already developed settings, or do they prefer these developments be sited in places they do not frequent?
- Are there visual impact thresholds where cumulative impact of multiple renewable energy facilities or expansion of facilities occurs?
- How effective are different visual simulation technologies in representing the experience of different types of renewable energy projects, in different locations, and within different cultures?
- How stable are public judgments concerning a proposed project compared to judgments after the project is built?

11.11 Final recommendations

We recommend national and regional efforts to develop and implement best practices and standards for visual/scenic resource assessment for public landscape management. We also recommend the use of visual impact assessment, under the *National Environmental Policy Act* (NEPA) and parallel state mini-NEPAs or equivalents in other countries for large-scale renewable energy projects.

There is urgency for better and more consistent evaluation of visual impacts on landscapes. Indeed many countries, including the US, are facing major resource conflicts in developing large-scale renewable energy. In the US this includes resource development throughout the middle to western part of the country, non-renewable energy development of oil and natural gas in the eastern and middle parts of the country, plus energy transmission power lines and pipelines across much of the country. In Europe, and particularly in the UK, offshore wind energy is creating challenges. Failure to properly consider scenic resource impacts creates siting conflicts that affect the sustainability and security of energy development. In the US there are also multi-jurisdictional issues from such large projects involving multi-agency jurisdictions and conflicting VIA methodology. Similar problems may be experienced in other countries.

We propose the following recommendations on scenic/visual resource assessment principles and standards:

- The need for *common language* and *methodological concepts* – especially among federal and state land management agencies;
- The need for *updating* VIA methodologies utilized by federal, state, and regional agencies as well as practitioners to improve VIA reliability and validity;
- The need for a VIA database containing project *BMP examples, procedures*, and *mitigation measures*;
- The need for appropriate *public involvement process methods*, procedures and resources;
- The need for *integration of landscape context* (urban, suburban, urban fringe, rural and wildland) for local to regional Smart Growth planning as well as VIA assessment;
- The need for *appropriate training* and *practice standards* for professionals doing VIA-related work.

An important goal of this book has been to identify the landscape impacts that are arising with the increased use of renewable energy resources, and to identify ways to assess and avoid or mitigate these impacts to the best of our ability in an effort to smooth the way forward. Doing so is a critical action to allow more widespread utilization of renewable energy, and should help facilitate our transition to a more sustainable energy future, while respecting the treasured gifts of natural beauty, and historic and cultural heritage that we value. Although there remains much to do, we are well on our way to fulfilling this ambition.

Note

1 35-A MRSA §3451-2.

Bibliography

Bishop, I.D. 2002. Determination of thresholds of visual impact; the case of wind turbines. *Environment and Planning B: Planning and Design 29*(5): 707–718.

Bishop, I.D. and D.R. Miller. 2007. Visual assessment of off shore wind turbines: The influence of distance, contrast, movement and social variables. *Renewable Energy 32*, 814–831.

Cape Wind Associates LLC and Environmental Design and Research. 2006. Seascape and Shoreline Visibility Assessment, Cape Wind Energy Project. Environmental Design and Research, Syracuse, NY.

Churchward, C., J.F. Palmer, J.I. Nassauer, and C.A. Swanwick. 2013. Evaluation of Methodologies for Visual Impact Assessments, NCHRP Report 741. National Cooperative Highway Research Program and Transportation Research Board, Washington, DC.

Council of Europe. 2000. European Landscape Convention, ETS No. 176, Florence, Italy.

Department of Trade and Industry (DTI). 2005. *Guidance on the Assessment of the Impact of Offshore Wind Farms: Seascape and Visual Impact.* Report of the Dept. of Trade and Industry, Publication 806666, London.

Environmental Law Institute (ELI). 2011. State Enabling Legislation for Commercial Scale Wind Power Siting and the Local Government Role. Environmental Law Institute, Washington, DC.

Hill, M., J. Briggs, P. Minto, D. Bagnall, K. Foley, and A. Williams. 2001. *Guide to Best Practice in Seascape Assessment.* Countryside Commission of Wales, Brady Shipman Marsh and University College Dublin.

Horner & Maclennan and Envision. 2006. *Visual Representation of Wind Farms: Good Practice Guidance.* Report for Scottish Natural Heritage, The South Scottish Forum and Scottish Society for Direction and Planning.

Landscape Institute/Institute of Environmental Management and Assessment (LI/IEMA). 1995, 2002, 2013. *Guidelines for Landscape and Visual Impact Assessment*, SPON Press, London.

Maguire M. and L. Gaillard (eds.). 2004. *Symposium on Scenic Conservation: Public Agency Practice.* Scenic America, June 9–10, River Farm, Alexandria, VA.

National Research Council. 2007. *Environmental Impacts of Wind Energy Projects.* National Academy Press, Washington, DC.

National Wind Coordinating Council (NWCC). 2006. *State Siting and Permitting of Wind Energy Facilities.* National Wind Coordinating Council and National Conference of State Legislatures, Washington, DC.

Palmer, J.F. 2000. Reliability on rating visible landscape qualities. *Landscape Journal 12*(1/2): 166–178.

Palmer, J. F. 2012. Maine's experience evaluating when scenic impacts from wind energy development are reasonably adverse. In *Science, Politics and Policy: Environmental Nexus.* Proceedings of the 37th Annual National Association of Environmental Professionals Conference, May 21–24, 2012, Portland, OR, pp. 602–620.

Palmer, J.F. 2014. A criteria-indicator-threshold framework for visual impact assessment. Presented at the 39th Annual National Association of Environmental Professionals Conference, *Changing Tides and Shifting Sands Conference Proceedings*, April 7–10, 2014, St. Petersburg, FL.

Palmer, J.F. and R.E. Hoffman. 2001. Rating reliability and representation validity in scenic landscape assessments. *Landscape and Urban Planning 54*(2001): 149–161.

Palmer, J.F., C. Churchward, J.I. Nassauer, and C.A. Swanwick. 2012. An evaluation of the visual impact assessment process as used by State Departments of Transportation in the USA. In *Science, Politics and Policy: Environmental Nexus.* Proceedings of the 37th Annual National Association of Environmental Professionals Conference, May 21–24, 2012, Portland, OR, pp. 735–750.

Rooney, S., P. Dascomb, and S. Korjeff, with R. Smardon. 2012. *Visual Impact Assessment Methodology for Offshore Wind Energy Conversion Facilities. Technical Bulletin 12-001.* Cape Cod Commission, Barnstable, MA. Available at www.esf.edu/via

Sheppard, S.R.J. 2012. *Visualizing Climate Change: A Guide to Visual Communication of Climate Change and Developing Local Solutions.* Routledge, London.

Smardon, R.C. 2009. The Wadden Sea wetlands: A multi-jurisdictional challenge. In *Sustaining the World's Wetlands: Setting Policy and Resolving Conflicts.* Springer, New York, pp. 21–56.

Smardon, R.C., J.F. Palmer, and J.F. Felleman. 1986. *Foundations for Visual Project Analysis.* Wiley, NY. Available at www.esf.edu/via

Smardon, R.C., J.F. Palmer, J. Knopf, K. Grinde, J.E. Henderson and L.D. Peyman-Dove. 1988. Visual Resources Assessment Procedure for US Army Corps of Engineers. Instruction Report EL-88-1, USACOE, Waterways Experiment Station, Vicksburg, MS. Available at www.esf.edu/via

Spencer, J., J. Mosier, and N. Laughlin. 2008. Scenery Management System, Appendix J: Project Level Scenery Analysis Template. US Forest Service.

Sullivan, R.G. 2011. Visual impacts of concentrating solar power facilities on desert landscapes of the American Southwest. In *Proceedings of Solar Paces 2011*, Sept. 20–23, Granada, Spain.

Sullivan, R.G. and J.M. Abplanalp. 2013. Utility-scale Solar Energy Facility Visual Impact Characterization and Mitigation: Project Report. Argonne National Laboratory, Argonne, IL.

Sullivan, R.G. and J.M. Abplanalp. 2015. Visibility and Visual Characteristics of the Ivanpah Solar Electric Generating System Power Tower Facility: Project Final Report. Argonne National Laboratory, Argonne, IL.
Sullivan, R.G. and M. Meyer. 2014. Guide to Evaluating Visual Impact Assessments for Renewable Energy Projects. Natural Resource Report NPS/ARD/NRR-2014/836, USDI National Park Service, Fort Collins, CO.
Sullivan, R., J. M, Abplanalp, B. Cantwell, and K. Beckman. 2013a. Potential Visual Impacts of Utility-scale Solar Energy Development within Solar Energy Zones on Selected Viewpoints in Death Valley and Joshua Tree National Parks, and El Camino Real De Tierra Adentro National Historic Trail. ANL/EVS/R-12/6, June.
Sullivan, R., J.M. Abplanalp, S. Lahti, K. Beckman, B. Cantwell, and P. Richmond. 2014. Electric transmission visibility and visual contrast threshold distances in western landscapes. Presented at the 39th Annual National Association of Environmental Professionals Conference, *Changing Tides and Shifting Sands Conference Proceedings*, April 7–10, 2014, St. Petersburg, FL.
Sullivan, R.G., J. Cothren, M. Williamson, P. Smith, J. McCarty and L.B. Kirchler. 2102a. Visual impact risk assessment and mitigation mapping system for utility-scale wind energy facilities. In *Science, Politics and Policy: Environmental Nexus*. Proceedings of the 37th Annual National Association of Environmental Professionals Conference, May 21–24, 2012, Portland, OR, pp. 667–692.
Sullivan, R.G., L.B. Kirchler, J. Cothren, and S.L. Winters. 2013b. Offshore wind turbine visibility and visual impact threshold distances. *Environment Practice 15*(1), 33–49.
Sullivan, R.G., L.B. Kirchler, T. Lahti, S. Roche, K. Beckman, B. Cantwell, and P. Richmond. 2012b. Wind turbine visibility and visual impact distances in western landscapes. In *Science, Politics and Policy: Environmental Nexus*. Proceedings of the 37th Annual National Association of Environmental Professionals Conference, May 21–24, 2012, Portland, OR, pp. 897–942.
Sullivan, R.G., L.B. Kirchler, C. McCoy, J. McCarty, K. Beckman, and P.D. Richmond. 2012c. Visual impacts of utility-scale solar energy facilities on southwestern desert landscapes. Paper presented at the National Association of Environmental Professionals, 37th Annual Conference. May 21–24, 2012, Portland, OR.
Thayer, R. 1995. *Gray World, Green Heart*. Wiley, NY.
Tzoulas, K., K. Korpela, S. Venn, V. Yli-Pelkonen, A. Kazmierczak, J. Niemela, and P. James. 2007. Promoting ecosystem and human health in urban areas using green infrastructure: A literature review. *Landscape and Urban Planning 81*(3): 167–178.
Ulrich, Roger. 1984. View through a window may influence recovery from surgery. *Science 224*: 420–421.
USDA Forest Service (USDA FS). 1995. *Landscape Aesthetics: A Handbook for Scenery Management*. Rev. ed., Agriculture Handbook 701. US Department of Agriculture, Washington, DC.
USDA Forest Service (USDA FS). 2006a. Scenery Management System, Appendix I: Scenery Management System. Available at www.fs.usda.gov/Internet/FSE_DOCUMENTS/fsm9_005639.pdf
USDA Forest Service (USDA FS). 2006b. Appendix J: Recommended SMS Refinements. (Final Draft) US Forest Service.
USDI. 2013. The Secretary of the Interior Order No. 3330: Improving Mitigation Policies and Practices of the Department of the Interior. US Department of the Interior, Washington, DC.
USDI, Bureau of Land Management (USDI BLM). 1980. BLM Manual sections 6300, 6310 and 6320, Visual Resource Management. US Department of the Interior, Bureau of Land Management, Washington, DC. Available at www.esf.edu/via
USDI, Bureau of Land Management (USDI BLM). 2005. Final Programmatic Environmental Impact Statement on Wind Energy Development on BLM-Administered Lands in the Western United States, US Department of the Interior, Bureau of Land Management, Washington, DC.
USDI, Bureau of Land Management (USDI BLM). 2012. Final Programmatic Environmental Impact Statement (PEIS) for Solar Energy Development in Six Southwestern States. FES 12-24 DOE/EIS-0403. US Department of the Interior, Bureau of Land Management and US Department of Energy, Washington, DC
USDI, Bureau of Land Management (USDI BLM). 2013. Best Management Practices for Reducing Visual Impacts of Renewable Energy Facilities on BLM Administered Lands. US Department of the Interior, Bureau of Land Management, Cheyenne, WY.
USEPA. undated. *Handbook on Siting Renewable Energy Project while Addressing Environmental Issues*. USEPA Office of Solid Waste and Emergency Response's Center for Program Analysis, Washington, DC.

Vermont Department of Natural Resources. 2006. Review and Evaluation of Potential Natural Resources Impacts from Utility-scale Wind Facilities in Vermont. Vermont Department of Natural Resources, Montpelier, VT. Available at www.anr.state.vt.us/site/html/plan/DraftWindGuidelines.pdf

Vermont Superior Court. 1990. Quechee Lakes Corporation, 154 Vt. 543, 550.

Water Resources Council. 1983. Economic and Environmental Principles and Guidelines for Water and Land Related Resources Implementation Studies. Water Resources Council, Washington, DC.

Williams, W. and R. Whitcomb. 2007. *Cape Wind: Money, Celebrity, Class Politics and the Battle for Our Energy Future.* Public Affairs, NY.

EDITORS AND CONTRIBUTORS

Contributing editors

Dean Apostol is a landscape architect and project manager for the Federal government, based in the Pacific Northwest. He also does work with MIG Inc, a design, planning and public engagement firm with offices in California, Oregon, Washington, and Colorado. Mr Apostol researches, consults, and provides analysis and expert witness testimony on the visual and scenic impacts of energy projects. He also consults on natural resource management, restoration ecology, and nature-based recreation. Mr Apostol has taught at the University of Oregon, Oregon State, and Portland State Universities, and has published several books, including *Forest Landscape Analysis and Design*, *Restoring the Pacific Northwest*, and *Designing Sustainable Forest Landscapes*. At the time of this writing, he splits time between a tiny houseboat in Seattle and a farm in Damascus, Oregon.

James F. Palmer has had a distinguished professional career in landscape architecture spanning thirty-five years, focusing on the assessment of landscape character and aesthetic quality. He is among a handful of research landscape architects whose expertise has led the way in informing our scientific understanding of visual impacts, community landscape values, and other landscape aesthetics issues. Dr Palmer holds an MLA and a PhD in forestry/natural resource planning from the University of Massachusetts, Amherst. From 1980 through 2005, he served on the Faculty of Landscape Architecture at the SUNY College of Environmental Science and Forestry in Syracuse, New York, reaching the rank of Professor. Through publications, peer reviews, court expert testimony, and teaching he has raised the standards in the field by investigating the validity and reliability of landscape architectural research and professional practices. Among his many publications on scenic impact analysis is *Foundations for Visual Project Analysis*. He currently serves as Professor Emeritus at SUNY, and is practicing landscape architecture in Vermont with Scenic Quality Consultants and T.J. Boyle Associates. www.tjboyle.com

Martin J. Pasqualetti is Professor in the School of Geographical Sciences and Urban Planning at Arizona State University, where he is also a Senior Sustainability Scientist in the Julie Ann Wrigley Global Institute of Sustainability. For more than forty years, he has contributed innovative yet practical research that has drawn our attention to the creation and adjustments to energy landscapes. On this topic, he has been a frequent guest speaker, both in the United States and abroad, including recent presentations in

England, Scotland, Germany, the Czech Republic, Mexico, Canada, China and Singapore. In recognition of his contributions to our understanding of energy landscapes, the American Geographical Society awarded him the Alexander and Ilse Melamid gold medal in 2015. While his record of publication includes examination of all the principal energy resources, he has given concentrated attention in recent years to wind energy including a prescient book, *Wind Power in View: Energy Landscapes in a Crowded World*, with Paul Gipe and Robert Righter. He was a founding member of the Arizona Solar Center and twice served the Governor as Chairman of the Arizona Solar Energy Advisory Council. He holds a BA from the University of California (Berkeley), an MA from Louisiana State University, and a PhD from the University of California (Riverside): https://geoplan.asu.edu/people/martin-j-pasqualetti

Richard C. Smardon is a SUNY Distinguished Service Professor Emeritus at the State University of New York's College of Environmental Science and Forestry where he was, until recently, Chair of the Department of Environmental Studies. In 1989, the Governor of New York appointed Dr Smardon to the Great Lakes Basin Advisory Council and he is now serving as its chair. He has served on several editorial boards including *Landscape and Urban Planning, Water, and International Journal of Environmental Sciences*. His major areas of expertise include landscape assessment and management, wetland assessment and mitigation, environmental management/citizen participation, law and aesthetics, ecotourism and heritage resource management, and energy sustainability planning implementation. His books include *The Future of Wetlands: Assessing Visual-Cultural Values*; *Foundations for Visual Project Analysis*; *The Legal Landscape: Guidelines for Environmental and Aesthetic Protection;* and *Sustaining the World's Wetlands*, which focuses on the role of NGOs in international wetland management. He holds a PhD in Environmental Planning from the University of California, Berkeley and an MLA and BS in Environmental Design from the University of Massachusetts, Amherst. He created and maintains a website for visual and landscape assessment references at www.esf.edu/via

Robert Sullivan is an environmental scientist in Argonne National Laboratory's Environmental Science Division, where he conducts research, develops programs and guidance, and conducts analyses on the visual impacts of utility-scale energy projects, primarily renewable energy facilities and electric transmission. He has conducted visual resource research for several federal agencies, including the Bureau of Land Management, the National Park Service, the Bureau of Ocean Energy Management, and the Department of Energy. He is co-designer of the National Park Service visual resource inventory system. He is the lead author for major publications on visual impact mitigation for renewable energy and electric transmission facilities, and critical evaluation of visual impact assessments for energy projects. He has conducted programmatic visual impact assessments for onshore and offshore wind energy, wave energy, solar energy, oil shale and tar sands mining, uranium mining, hydropower, and electric and pipeline transmission. He is also a software and web designer specializing in geospatial data management and communication. He has served as Chair of the Climate Change Task Force of the Chicago Wilderness Coalition, and served on two working groups associated with wind energy development on the Great Lakes. Currently, Mr Sullivan is conducting studies that focus on incorporating human factors in public land planning and management, energy facility siting, regional compensatory mitigation, and long-term monitoring. He holds an MLA degree from the University of Illinois at Urbana.

Contributors

Simon Bell is a forester and landscape architect specializing in large-scale landscape planning and design. His focus is on forest landscape planning and design, where the emphasis is on the visual and aesthetic aspects of forestry, outdoor recreation planning and design, and landscape and visual impact assessment of large-scale elements such as wind energy developments. He has published several books and papers on these subjects and has a wide range of project experience gained in the UK and elsewhere. Dr Bell is currently Professor and Head of the Department of Landscape Architecture in the Institute of Agricultural and Environmental Sciences at the Estonian University of Life Sciences (www.emu.ee/) in Tartu, Estonia. He is also Associate Director of OPENspace, the Research Centre for Inclusive Access to Outdoor Environments based in the Edinburgh School of Architecture and Landscape Architecture at Edinburgh College of Art (now part of the University of Edinburgh) (www.openspace.ed.ac.uk/). He holds a PhD in landscape architecture, an MPhil in landscape architecture and a BSc in forestry.

Ian Bishop is an honorary Professorial Fellow in the Department of Infrastructure Engineering, University of Melbourne. He was previously a Professor in Geomatics and prior to that Director of the Centre for GIS and Modelling – a joint initiative of the Melbourne School of Engineering and the Faculty of Architecture, Building and Planning. He has taught GIS, spatial analysis and visualization mainly in geomatics and landscape architecture programs. He is widely published in the areas of landscape assessment, landscape planning, spatial analysis, environmental visualization, modelling and decision support. A particular focus has been the visual impact of energy infrastructure – transmission lines and wind turbines. He is currently working on contract for the Australian Urban Research Infrastructure Network (AURIN) where he advises on a range of issues relating the functionality of the Australia-wide data portal and workbench analysis tools. From the University of Melbourne, he holds a BSc (Hons), MSc, and PhD.

Louise Kling is an environmental planner with AECOM in Portland, Oregon. She began her career with the US Forest Service studying disturbance ecology on the Gifford Pinchot National Forest and Mt St Helens National Volcanic Monument. As a continuation of her interest in large-scale land use and disturbance ecology, she shifted her focus to human health outcomes through her work with the Oregon Health and Science University's (OHSU) Center for Research on Occupational and Environmental Toxicology. In this role, she collaborated with the Columbia Basin Intertribal Fish Commission to understand public health risks to Columbia Basin tribes due to regular consumption of contaminated fish. As an environmental planner at AECOM, she applies her broad biological and social resource experience to prepare environmental documentation for energy and infrastructure development and public land use planning. As an avid supporter of responsible public land management, she has spent the last decade expanding her technical practice to focus on the interplay of land use, ecology, recreation and scenery management, through which she strives to inform sound project siting and design on public lands. She completed her graduate work at Utah State University in landscape ecology with a focus on the Northwest Forest Plan's Survey and Management program, and is currently President of the Northwest Chapter of the River Management Society.

John McCarty works in the federal sector as a national level landscape architect. Central to Mr McCarty's responsibility is the development and oversight of national policy that directs the management of visual/ scenic resources, designation of backcountry byways, and implementation of sustainable landscape architectural design principles for the built environment. He imparts landscape architecture leadership in the national discussion on recreation planning, renewable energy development, electrical transmission routing, fluid minerals, transportation and land use planning. Mr McCarty coordinates and instructs visual/scenic resource management training curricula. He is also engaged with advancing research on new technologies to resolve the visual challenges of large-scale energy development. Prior to federal service, Mr McCarty worked in the private sector where he consulted both the federal government and energy industry in scenic/visual resource design strategies and mitigation planning to meet federal land use planning requirements. He holds a Bachelor's of Science in Landscape Architecture from Colorado State University.

Robert Ribe is a professor of landscape architecture, community and regional planning, and environmental studies at the University of Oregon where he teaches landscape analysis, planning and design. His research focuses on public aesthetic and acceptability perceptions of forest plans and harvest methods aimed at reconciling the spotted owl controversy in the Pacific Northwest, and similar conflicts in Tasmania and British Columbia. Dr Ribe has contributed to enhancing the visual quality of landscapes produced by regional urban growth plans in Portland, Oregon and Vancouver, British Columbia, and along scenic highways in New Mexico, Oregon, Washington and California. Prior to his academic career, he helped plan a variety of energy and related projects in the western United States and Saudi Arabia in the 1980s while with Wirth Associates, the US Army Corps of Engineers, and Washington State University. He continues to serve as an expert witness for permitting of similar projects. He holds a PhD in Land Resources, an MA in Applied Economics and an MLA, all from the University of Wisconsin at Madison.

INDEX

Page numbers in italics denote figures. Page numbers in bold denote tables. Agency and organization acronyms are used where applicable to identify document and visual resource management system associations.

For Product Safety Concerns and Information please contact our EU
representative GPSR@taylorandfrancis.com
Taylor & Francis Verlag GmbH, Kaufingerstraße 24, 80331 München, Germany

www.ingramcontent.com/pod-product-compliance
Ingram Content Group UK Ltd.
Pitfield, Milton Keynes, MK11 3LW, UK
UKHW052224190926
475702UK00007B/1183
* 9 7 8 1 1 3 8 8 0 8 9 8 0 *